D0782232

Applied Mathematical Sciences
Volume 92

Applied Mathematical Sciences

1. *John:* Partial Differential Equations, 4th ed.
2. *Sirovich:* Techniques of Asymptotic Analysis.
3. *Hale:* Theory of Functional Differential Equations, 2nd ed.
4. *Percus:* Combinatorial Methods.
5. *von Mises/Friedrichs:* Fluid Dynamics.
6. *Freiberger/Grenander:* A Short Course in Computational Probability and Statistics.
7. *Pipkin:* Lectures on Viscoelasticity Theory.
8. *Giacoglia:* Perturbation Methods in Non-linear Systems.
9. *Friedrichs:* Spectral Theory of Operators in Hilbert Space.
10. *Stroud:* Numerical Quadrature and Solution of Ordinary Differential Equations.
11. *Wolovich:* Linear Multivariable Systems.
12. *Berkovitz:* Optimal Control Theory.
13. *Bluman/Cole:* Similarity Methods for Differential Equations.
14. *Yoshizawa:* Stability Theory and the Existence of Periodic Solution and Almost Periodic Solutions.
15. *Braun:* Differential Equations and Their Applications, 3rd ed.
16. *Lefschetz:* Applications of Algebraic Topology.
17. *Collatz/Wetterling:* Optimization Problems.
18. *Grenander:* Pattern Synthesis: Lectures in Pattern Theory, Vol. I.
19. *Marsden/McCracken:* Hopf Bifurcation and Its Applications.
20. *Driver:* Ordinary and Delay Differential Equations.
21. *Courant/Friedrichs:* Supersonic Flow and Shock Waves.
22. *Rouche/Habets/Laloy:* Stability Theory by Liapunov's Direct Method.
23. *Lamperti:* Stochastic Processes: A Survey of the Mathematical Theory.
24. *Grenander:* Pattern Analysis: Lectures in Pattern Theory, Vol. II.
25. *Davies:* Integral Transforms and Their Applications, 2nd ed.
26. *Kushner/Clark:* Stochastic Approximation Methods for Constrained and Unconstrained Systems.
27. *de Boor:* A Practical Guide to Splines.
28. *Keilson:* Markov Chain Models—Rarity and Exponentiality.
29. *de Veubeke:* A Course in Elasticity.
30. *Shiatycki:* Geometric Quantization and Quantum Mechanics.
31. *Reid:* Sturmian Theory for Ordinary Differential Equations.
32. *Meis/Markowitz:* Numerical Solution of Partial Differential Equations.
33. *Grenander:* Regular Structures: Lectures in Pattern Theory, Vol. III.
34. *Kevorkian/Cole:* Perturbation Methods in Applied Mathematics.
35. *Carr:* Applications of Centre Manifold Theory.
36. *Bengtsson/Ghil/Källén:* Dynamic Meteorology: Data Assimilation Methods.
37. *Saperstone:* Semidynamical Systems in Infinite Dimensional Spaces.
38. *Lichtenberg/Lieberman:* Regular and Chaotic Dynamics, 2nd ed.
39. *Piccini/Stampacchia/Vidossich:* Ordinary Differential Equations in $\mathbf{R}^n$.
40. *Naylor/Sell:* Linear Operator Theory in Engineering and Science.
41. *Sparrow:* The Lorenz Equations: Bifurcations, Chaos, and Strange Attractors.
42. *Guckenheimer/Holmes:* Nonlinear Oscillations, Dynamical Systems and Bifurcations of Vector Fields.
43. *Ockendon/Taylor:* Inviscid Fluid Flows.
44. *Pazy:* Semigroups of Linear Operators and Applications to Partial Differential Equations.
45. *Glashoff/Gustafson:* Linear Operations and Approximation: An Introduction to the Theoretical Analysis and Numerical Treatment of Semi-Infinite Programs.
46. *Wilcox:* Scattering Theory for Diffraction Gratings.
47. *Hale et al:* An Introduction to Infinite Dimensional Dynamical Systems—Geometric Theory.
48. *Murray:* Asymptotic Analysis.
49. *Ladyzhenskaya:* The Boundary—Value Problems of Mathematical Physics.
50. *Wilcox:* Sound Propagation in Stratified Fluids.
51. *Golubitsky/Schaeffer:* Bifurcation and Groups in Bifurcation Theory, Vol. I.

(continued following index)

Gregory L. Naber

The Geometry of Minkowski Spacetime

An Introduction to the Mathematics of the Special Theory of Relativity

With 43 Illustrations

Springer-Verlag
New York Berlin Heidelberg London Paris
Tokyo Hong Kong Barcelona Budapest

Gregory L. Naber
Department of Mathematics
California State University
Chico, CA 95929
USA

Editors

Mathematics Subject Classifications (1991): 83A05, 83C60, 83-02

Library of Congress Cataloging-in-Publication Data
Naber, Gregory L., 1948–
The geometry of Minkowski spacetime : an introduction to the mathematics of the special theory of relativity / Gregory L. Naber.
p. cm. – (Applied mathematical sciences ; 92)
Includes bibliographical references and index.
ISBN 0-387-97848-8 (New York). – ISBN 3-540-97848-8 (Berlin)
1. Special relativity (Physics)—Mathematics. 2. Spaces, Generalized. I. Title. II. Series.
QA1.A647 vol. 92
[QC173.6]
510 s—dc20
[530.1'1] 92-12652

Printed on acid-free paper.

Production managed by Hal Henglein; manufacturing supervised by Jacqui Ashri.
Photocomposed copy prepared from the author's LaTeX files.
Printed and bound by R.R. Donnelley & Sons, Harrisonburg, VA.
Printed in the United States of America.

9 8 7 6 5 4 3 2 1

ISBN 0-387-97848-8 Springer-Verlag New York Berlin Heidelberg
ISBN 3-540-97848-8 Springer-Verlag Berlin Heidelberg New York

For Debora

Preface

It is the intention of this monograph to provide an introduction to the special theory of relativity that is mathematically rigorous and yet spells out in considerable detail the physical significance of the mathematics. Particular care has been exercised in keeping clear the distinction between a physical phenomenon and the mathematical model which purports to describe that phenomenon so that, at any given point, it should be clear whether we are doing mathematics or appealing to physical arguments to interpret the mathematics.

The Introduction is an attempt to motivate, by way of a beautiful theorem of Zeeman [$\mathbf{Z}_1$], our underlying model of the "event world." This model consists of a 4-dimensional real vector space on which is defined a nondegenerate, symmetric, bilinear form of index one (Minkowski spacetime) and its associated group of orthogonal transformations (the Lorentz group).

The first five sections of Chapter 1 contain the basic geometrical information about this model including preliminary material on indefinite inner product spaces in general, elementary properties of spacelike, timelike and null vectors, time orientation, proper time parametrization of timelike curves, the Reversed Schwartz and Triangle Inequalities, Robb's Theorem on measuring proper spatial separation with clocks and the decomposition of a general Lorentz transformation into a product of two rotations and a special Lorentz transformation. In these sections one will also find the usual kinematic discussions of time dilation, the relativity of simultaneity, length contraction, the addition of velocities formula and hyperbolic motion as well as the construction of 2-dimensional Minkowski diagrams and, somewhat reluctantly, an assortment of the obligatory "paradoxes."

Section 6 of Chapter 1 contains the definitions of the causal and chronological precedence relations and a detailed proof of Zeeman's extraordinary theorem characterizing causal automorphisms as compositions $T \circ K \circ L$, where T is a translation, K is a dilation, and L is an orthochronous orthogonal transformation. The proof is somewhat involved, but the result itself is used only in the Introduction (for purposes of motivation) and in Appendix A to construct the homeomorphism group of the path topology.

Section 1.7 is built upon the one-to-one correspondence between vectors in Minkowski spacetime and 2×2 complex Hermitian matrices and contains a detailed construction of the spinor map [the two-to-one homomorphism of $SL(2, \mathbb{C})$ onto the Lorentz group]. We show that the fractional linear transformation of the "celestial sphere" determined by an element A of $SL(2, \mathbb{C})$ has the same effect on past null directions as the Lorentz transformation corresponding to A under the spinor map. Immediate consequences include Penrose's Theorem [**Pen**] on the apparent shape of a relativistically moving sphere, the existence of invariant null directions for an arbitrary Lorentz transformation, and the fact that a general Lorentz transformation is completely determined by its effect on any three distinct past null directions. The material in this section is required only in Chapter 3 and Appendix B.

In Section 1.8 (which is independent of Sections 1.6 and 1.7) we introduce into our model the additional element of world momentum for material particles and photons and its conservation in what are called contact interactions. With this one can derive most of the well-known results of relativistic particle mechanics and we include a sampler (the Doppler effect, the aberration formula, the nonconservation of proper mass in a decay reaction, the Compton effect and the formulas relevant to inelastic collisions).

Chapter 2 introduces charged particles and uses the classical Lorentz World Force Law ($FU = \frac{m}{e}\frac{dU}{d\tau}$) as motivation for describing an electromagnetic field at a point in Minkowski spacetime as a linear transformation F whose job it is to tell a charged particle with world velocity U passing through that point what change in world momentum it should expect to experience due to the presence of the field. Such a linear transformation is necessarily skew-symmetric with respect to the Lorentz inner product and Sections 2.2, 2.3 and 2.4 analyze the algebraic structure of these in some detail. The essential distinction between regular and null skew-symmetric linear transformations is described first in terms of the physical invariants $\vec{E} \cdot \vec{B}$ and $|\vec{B}|^2 - |\vec{E}|^2$ of the electromagnetic field (which arise as coefficients in the characteristic equation of F) and then in terms of the existence of invariant subspaces. This material culminates in the existence of canonical forms for both regular and null fields that are particularly useful for calculations, e.g., of eigenvalues and principal null directions.

Section 2.5 introduces the energy-momentum transformation for an arbitrary skew-symmetric linear transformation and calculates its matrix entries in terms of the classical energy density, Poynting 3-vector and Maxwell stress tensor. Its principal null directions are determined and the Dominant Energy Condition is proved.

In Section 2.6, the Lorentz World Force equation is solved for charged particles moving in constant electromagnetic fields, while variable fields are introduced in Section 2.7. Here we describe the skew-symmetric bilinear form (bivector) associated with the linear transformation representing the field and use it and its dual to write down Maxwell's (source-free) equations. As sample solutions to Maxwell's equations we consider the Coulomb field, the field of a uniformly moving charge, and a rather complete discussion of simple, plane electromagnetic waves.

Chapter 3 is an elementary introduction to the algebraic theory of spinors in Minkowski spacetime. The rather lengthy motivational Section 3.1 traces the emergence of the spinor concept from the general notion of a (finite dimensional) group representation. Section 3.2 contains the abstract definition of spin space and introduces spinors as complex-valued multilinear functionals on spin space. The Levi-Civita spinor ϵ and the elementary operations of spinor algebra [type changing, sums, components, outer products, (skew-) symmetrization, etc.] are treated in Section 3.3.

In Section 3.4 we introduce the Infeld-van der Waerden symbols (essentially, normalized Pauli spin matrices) and use them, together with the spinor map from Section 1.7, to define natural spinor equivalents for vectors and covectors in Minkowski spacetime. The spinor equivalent of a future-directed null vector is shown to be expressible as the outer product of a spin vector and its conjugate. Reversing the procedure leads to the existence of a future-directed null "flagpole" for an arbitrary nonzero spin vector.

Spinor equivalents for bilinear forms are constructed in Section 3.5 with the skew-symmetric forms (bivectors) playing a particularly prominant role. With these we can give a detailed construction of the geometrical representation "up to sign" of a nonzero spin vector as a null flag (due to Penrose). The sign ambiguity in this representation intimates the "essential 2-valuedness" of spinors which we discuss in some detail in Appendix B.

Chapter 3 culminates with a return to the electromagnetic field. We introduce the electromagnetic spinor ϕ_{AB} associated with a skew-symmetric linear transformation F and find that it can be decomposed into a symmetrized outer product of spin vectors α and β. The flagpoles of these spin vectors are eigenvectors for the electromagnetic field transformation, i.e., they determine its principal null directions. The solution to the eigenvalue problem for ϕ_{AB} yields two elegant spinor versions of the "Petrov type" classification theorems of Chapter 2. Specifically, we prove that a skew-symmetric linear transformation F on M is null if and only if $\lambda = 0$ is the only eigenvalue of the associated electromagnetic spinor ϕ_{AB} and that this, in turn, is the case if and only if the associated spin vectors α and β are linearly dependent. Next we find that the energy-momentum transformation has a beautifully simple spinor equivalent and use it to give another proof of the Dominant Energy Condition. Finally, we derive the elegant spinor form of Maxwell's equations and briefly discuss its generalizations to massless free field equations for arbitrary spin $\frac{1}{2}n$ particles.

The background required for an effective reading of the body of the text is a solid course in linear algebra and the usual supply of "mathematical maturity." For the two appendices we must increment our demands upon the reader and assume some familiarity with elementary point-set topology. Appendix A describes, in the special case of Minkowski spacetime, a remarkable topology devised by Hawking, King and McCarthy [**HKM**] and based on ideas of Zeeman [$\mathbf{Z}_2$] whose homeomorphisms are just compositions of translations, dilations and Lorentz transformations. Only quite routine point-set topology is required, but the construction of the homeomorphism group depends on Zeeman's Theorem from Section 1.6.

In Appendix B we elaborate upon the "essential 2-valuedness" of spinors and its significance in physics for describing, for example, the quantum mechanical state of an electron with spin. Paul Dirac's ingenious "Scissors Problem" is used, as Dirac himself used it, to suggest, in a more familiar context, the possibility that certain aspects of a physical system's state may be invariant under a rotation of the system through $720°$, but *not* under a $360°$ rotation. To fully appreciate such a phenomenon one must see its reflection in the mathematics of the rotation group (the "configuration space" of the scissors). For this we briefly review the notion of homotopy for paths and the construction of the fundamental group. Noting that the 3-sphere S^3 is the universal cover for real projective 3-space $\mathbb{R}P^3$ and that $\mathbb{R}P^3$ is homeomorphic to the rotation group $SO(3)$ we show that $\pi_1(SO(3)) \cong \mathbb{Z}_2$. One then sees Dirac's demonstration as a sort of physical model of the two distinct homotopy classes of loops in $SO(3)$. But there is a great deal more to be learned here. By regarding the elements of SU_2 (Section 1.7) as unit quaternions we find that, topologically, it is S^3 and then recognize SU_2 and the restriction of the spinor map to it as a concrete realization of the covering space for $SO(3)$ that we just used to calculate $\pi_1(SO(3))$. One is then led naturally to SU_2 as a model for the "state space" (as distinguished from the "configuration space") of the system described in Dirac's demonstration. Recalling our discussion of group representations in Section 3.1 we find that it is the representations of SU_2, i.e., the spinor representations of $SO(3)$, that contain the physically significant information about the system. So it is with the quantum mechanical state of an electron, but in this case one requires a relativistically invariant theory and so one looks, not to SU_2 and the restriction of the spinor map to it, but to the full spinor map which carries $SL(2, \mathbb{C})$ onto the Lorentz group.

Lemmas, Propositions, Theorems and Corollaries are numbered sequentially within each section so that "p.q.r" will refer to result #r in Section #q of Chapter #p. Exercises and equations are numbered in the same way, but with equation numbers enclosed in parentheses. There are 191 exercises scattered throughout the text and no asterisks appear to designate those that are used in the sequel; they are all used and must be worked conscientiously. Finally, we shall make extensive use of the *Einstein summation convention* according to which a repeated index, one subscript and one superscript, indicates a sum over the range of values that the index can assume. For example, if a and b are indices that range over $1, 2, 3, 4$, then

$$\begin{aligned} x^a e_a &= \sum_{a=1}^{4} x^a e_a = x^1 e_1 + x^2 e_2 + x^3 e_3 + x^4 e_4, \\ \Lambda^a{}_b x^b &= \sum_{b=1}^{4} \Lambda^a{}_b x^b = \Lambda^a{}_1 x^1 + \Lambda^a{}_2 x^2 + \Lambda^a{}_3 x^3 + \Lambda^a{}_4 x^4, \\ \eta_{ab} v^a w^b &= \eta_{11} v^1 w^1 + \eta_{12} v^1 w^2 + \eta_{13} v^1 w^3 + \eta_{14} v^1 w^4 \\ &\quad + \eta_{21} v^2 w^1 + \cdots + \eta_{44} v^4 w^4, \end{aligned}$$

and so on.

Gregory L. Naber

Acknowledgments

Some debts are easy to describe and acknowledge with gratitude. To the California State University at Chico, its Department of Mathematics and, more particularly, to Tom McCready, the Chair of that department, go my sincere thanks for the support, financial and otherwise, that they provided throughout the period during which the manuscript was being written. On the other hand, my indebtedness to my wife, Debora, is not so easily expressed in a few words. She saw to it that I had the time and the peace to think and to write and she bore me patiently when one or the other of these did not go well. She took upon herself what would have been, for me, the onerous task of mastering the software required to produce a beautiful typescript from my handwritten version and, while producing it, held me to standards that I would surely have abandoned just to be done with the thing. Let it be enough to say that the book would not exist were it not for Debora. With love and gratitude, it is dedicated to her.

Contents

Introduction

> All beginnings are obscure. Inasmuch as the mathematician operates with his conceptions along strict and formal lines, he, above all, must be reminded from time to time that the origins of things lie in greater depths than those to which his methods enable him to descend.
>
> Hermann Weyl, *Space, Time, Matter*

Minkowski spacetime is generally regarded as the appropriate arena within which to formulate those laws of physics that do not refer specifically to gravitational phenomena. We would like to spend a moment here at the outset briefly examining some of the circumstances which give rise to this belief.

We shall adopt the point of view that the basic problem of science in general is the description of "events" which occur in the physical universe and the analysis of relationships between these events. We use the term "event," however, in the idealized sense of a "point-event," that is, a physical occurrence which has no spatial extension and no duration in time. One might picture, for example, an instantaneous collision or explosion or an "instant" in the history of some (point) material particle or photon (to be thought of as a "particle of light"). In this way the existence of a material particle or photon can be represented by a continuous sequence of events called its "worldline." We begin then with an abstract set $\mathcal{M}$ whose elements we call "events." We shall provide $\mathcal{M}$ with a mathematical structure which reflects certain simple facts of human experience as well as some rather nontrivial results of experimental physics.

Events are "observed" and we will be particularly interested in a certain class of observers (called "admissible") and the means they employ to describe events. Since it is in the nature of our perceptual apparatus that we identify events by their "location in space and time" we must specify the means by which an observer is to accomplish this in order to be deemed "admissible."

> *Each admissible observer presides over a 3-dimensional, right-handed, Cartesian spatial coordinate system based on an agreed*

unit of length and relative to which photons propagate rectilinearly in any direction.

A few remarks are in order. First, the expression "presides over" is not to be taken too literally. An observer is in no sense ubiquitous. Indeed, we generally picture the observer as just another material particle residing at the origin of his spatial coordinate system; any information regarding events which occur at other locations must be communicated to him by means we will consider shortly. Second, the restriction on the propagation of photons is a real restriction. The term "straight line" has meaning only relative to a given spatial coordinate system and if, in one such system, light does indeed travel along straight lines, then it certainly will not in another system which, say, rotates relative to the first. Notice, however, that this assumption does not preclude the possibility that two admissible coordinate systems are in relative motion. We shall denote the spatial coordinate systems of observers $\mathcal{O}$, $\hat{\mathcal{O}}$,... by $\sum(x^1, x^2, x^3)$, $\hat{\sum}(\hat{x}^1, \hat{x}^2, \hat{x}^3)$,... .

We take it as a fact of human experience that each observer has an innate, intuitive sense of temporal order which applies to events which he experiences directly, i.e., to events on his worldline. This sense, however, is not quantitative; there is no precise, reliable sense of "equality" for "time intervals." We remedy this situation by giving him a watch.

Each admissible observer is provided with an ideal standard clock based on an agreed unit of time with which to provide a quantitative temporal order to the events on his worldline.

Notice that thus far we have assumed only that an observer can assign a time to each event *on his worldline.* In order for an observer to be able to assign times to arbitrary events we must specify a procedure for the placement and synchronization of clocks throughout his spatial coordinate system. One possibility is simply to mass-produce clocks at the origin, synchronize them and then move them to various other points throughout the coordinate system. However, it has been found that moving clocks about has a most undesirable effect upon them. Two identical and very accurate atomic clocks are manufactured in New York and synchronized. One is placed aboard a passenger jet and flown around the world. Upon returning to New York it is found that the two clocks, although they still "tick" at the same rate, are no longer synchronized. The travelling clock lags behind its stay-at-home twin. Strange, indeed, but it is a fact and we shall come to understand the reason for it shortly.

To avoid this difficulty we shall ask our admissible observers to build their clocks at the origins of their coordinate systems, transport them to the desired locations, set them down and return to the master clock at the origin. We assume that each observer has stationed an assistant at the location of each transported clock. Now our observer must "communicate" with each assistant, telling him the time at which his clock should be set in order that it be sychronized with the clock at the origin. As a means of communication we select a signal which seems, among all the possible choices, to be least susceptible to annoying fluctuations in reliability, i.e., light signals. To persuade the reader that this is an appropriate choice we shall record some of the experimentally documented properties of light signals, but first, a little experiment. From his location at the origin O an observer $\mathcal{O}$ emits a light signal at the instant his clock reads t_0. The signal is reflected back to him at a point P and arrives again at O at the instant t_1. Assuming there is no delay at P when the signal is bounced back, $\mathcal{O}$ will calculate the speed of the signal to be *distance* $(O,P)/\frac{1}{2}(t_1 - t_0)$. This technique for measuring the speed of light we call the *Fizeau procedure* in honor of the gentleman who first carried it out with care (notice that we *must* bounce the signal back to O since we do not yet have a clock at P that is synchronized with that at O).

> *For each admissible observer the speed of light in vacuo as determined by the Fizeau procedure is independent of when the experiment is performed, the arrangement of the apparatus (i.e., the choice of P), the frequency (energy) of the signal and, moreover, has the same numerical value c (approximately* 3.0×10^8 *meters per second) for all such observers.*

Here we have the conclusions of numerous experiments performed over the years, most notably those first performed by Michelson-Morley and Kennedy-Thorndike (see Ex. 33 and Ex. 34 of [**TW**] for a discussion of these experiments). The results may seem odd. Why is a photon so unlike an electron whose speed certainly will not have the same numerical value for two observers in relative motion? Nevertheless, they are incontestable facts of nature and we must deal with them. We shall exploit these rather remarkable properties of light signals immediately by asking all of our observers to multiply each of their time readings by the constant c and thereby *measure time in units of distance* (light travel time, e.g., "one meter of time" is the amount of time required by a light signal to travel one meter *in vacuo*). With these units all speeds are dimensionless and $c = 1$. Such time readings for observers $\mathcal{O}, \hat{\mathcal{O}}, \ldots$ will be designated $x^4\,(=ct)$, $\hat{x}^4\,(=c\hat{t}\,), \ldots$.

Now we provide each of our observers with a system of synchronized clocks in the following way: At each point P of his spatial coordinate system place a clock identical to that at the origin. At some time x^4 at O emit a spherical electromagnetic wave (photons in all directions). As the wavefront encounters P set the clock placed there at time $x^4 + \text{distance}\,(O,P)$ and set it ticking, thus synchronized with the clock at the origin.

At this point each of our observers $\mathcal{O}, \hat{\mathcal{O}}, \ldots$ has established a *frame of reference* $\mathcal{S}(x^1, x^2, x^3, x^4)$, $\hat{\mathcal{S}}(\hat{x}^1, \hat{x}^2, \hat{x}^3, \hat{x}^4), \ldots$. A useful intuitive visualization of such a reference frame is as a latticework of spatial coordinate lines with, at each lattice point, a clock and an assistant whose task it is to record locations and times for events occurring in his immediate vicinity; the data can later be collected for analysis by the observer.

How are the $\hat{\mathcal{S}}$-coordinates of an event related to its $\mathcal{S}$-coordinates? That is, what can be said about the mapping $\mathcal{F}: \mathbb{R}^4 \to \mathbb{R}^4$ defined by $\mathcal{F}(x^1, x^2, x^3, x^4) = (\hat{x}^1, \hat{x}^2, \hat{x}^3, \hat{x}^4)$? Certainly, it must be one-to-one and onto. Indeed, $\mathcal{F}^{-1}: \mathbb{R}^4 \to \mathbb{R}^4$ must be the coordinate transformation from hatted to unhatted coordinates. To say more we require a seemingly innocuous "causality assumption."

> *Any two admissible observers agree on the temporal order of any two events on the worldline of a photon, i.e., if two such events have coordinates (x^1, x^2, x^3, x^4) and $(x_0^1, x_0^2, x_0^3, x_0^4)$ in $\mathcal{S}$ and $(\hat{x}^1, \hat{x}^2, \hat{x}^3, \hat{x}^4)$ and $(\hat{x}_0^1, \hat{x}_0^2, \hat{x}_0^3, \hat{x}_0^4)$ in $\hat{\mathcal{S}}$, then $x^4 - x_0^4$ and $\hat{x}^4 - \hat{x}_0^4$ have the same sign.*

Notice that we do not prejudge the issue by assuming that Δx^4 and $\Delta \hat{x}^4$ are equal, but only that they have the same sign, i.e., that $\mathcal{O}$ and $\hat{\mathcal{O}}$ agree as to which of the two events occurred first. Thus, $\mathcal{F}$ preserves order in the fourth coordinate, at least for events which lie on the worldline of some photon. How are two such events related? Since photons propagate rectilinearly with speed 1, two events on the worldline of a photon have coordinates in $\mathcal{S}$ which satisfy

$$x^i - x_0^i = v^i(x^4 - x_0^4), \quad i = 1, 2, 3,$$

for some constants v^1, v^2 and v^3 with $(v^1)^2 + (v^2)^2 + (v^3)^2 = 1$ and consequently

$$(x^1 - x_0^1)^2 + (x^2 - x_0^2)^2 + (x^3 - x_0^3)^2 - (x^4 - x_0^4)^2 = 0. \qquad (0.1)$$

Geometrically, we think of (0.1) as the equation of a "cone" in $\mathbb{R}^4$ with vertex at $(x_0^1, x_0^2, x_0^3, x_0^4)$ [compare $(z - z_0)^2 = (x - x_0)^2 + (y - y_0)^2$ in $\mathbb{R}^3$]. But all of this must be true in *any* admissible frame of reference so $\mathcal{F}$ must preserve the cone (0.1). We summarize:

The coordinate transformation map $\mathcal{F}: \mathbb{R}^4 \to \mathbb{R}^4$ carries the cone (0.1) onto the cone

$$(\hat{x}^1 - \hat{x}_0^1)^2 + (\hat{x}^2 - \hat{x}_0^2)^2 + (\hat{x}^3 - \hat{x}_0^3)^2 - (\hat{x}^4 - \hat{x}_0^4)^2 = 0 \qquad (0.2)$$

and satisfies $\hat{x}^4 > \hat{x}_0^4$ whenever $x^4 > x_0^4$ and (0.1) is satisfied.

Being simply the coordinate transformation from hatted to unhatted coordinates, $\mathcal{F}^{-1}: \mathbb{R}^4 \to \mathbb{R}^4$ has the obvious analogous properties. In 1964, Zeeman [**Z₁**] called such a mapping $\mathcal{F}$ a "causal automorphism" and proved the remarkable fact that any causal automorphism is a composition of the following three basic types:

1. Translations: $\hat{x}^a = x^a + \Lambda^a,\ a = 1, 2, 3, 4,$ for some constants Λ^a.

2. Positive scalar multiples: $\hat{x}^a = kx^a,\ a = 1, 2, 3, 4,$ for some positive constant k.

3. Linear transformations

$$\hat{x}^a = \Lambda^a{}_b x^b,\ a = 1, 2, 3, 4, \tag{0.3}$$

where the matrix $\Lambda = [\Lambda^a{}_b]_{a,b=1,2,3,4}$ satisfies the two conditions

$$\Lambda^T \eta \Lambda = \eta, \tag{0.4}$$

where T means "transpose" and

$$\eta = \begin{bmatrix} 1 & 0 & 0 & 0 \\ 0 & 1 & 0 & 0 \\ 0 & 0 & 1 & 0 \\ 0 & 0 & 0 & -1 \end{bmatrix},$$

and

$$\Lambda^4{}_4 \geq 1. \tag{0.5}$$

This result is particularly remarkable in that it is not even assumed at the outset that $\mathcal{F}$ is continuous (much less, linear). We provide a proof in Section 1.6.

Since two frames of reference related by a mapping of type 2 differ only by a trivial and unnecessary change of scale we shall banish them from further consideration. Moreover, since the constants Λ^a in maps of type 1 can be regarded as the $\hat{\mathcal{S}}$-coordinates of $\mathcal{S}$'s spacetime origin we may request that all of our observers cooperate to the extent that they select a common event to act as origin and thereby take $\Lambda^a = 0$ for $a = 1, 2, 3, 4$. All that remain for consideration then are the admissible frames of reference related by transformations of the form (0.3) subject to (0.4) and (0.5). These are the so-called "orthochronous Lorentz transformations" and, as we shall prove in Chapter 1, are precisely the maps which leave invariant the quadratic form $(x^1)^2 + (x^2)^2 + (x^3)^2 - (x^4)^2$ (analogous to orthogonal transformations of $\mathbb{R}^3$ which leave invariant the usual squared length $x^2 + y^2 + z^2$) and which preserve "time orientation" in the sense described immediately after (0.2). It is the geometry of this quadratic form, the structure of the group of Lorentz transformations and their various physical interpretations that will be our concern in the text.

With this we conclude our attempt at motivation for the definitions that confront the reader in Chapter 1. There is, however, one more item on the agenda of our introductory remarks. It is the cornerstone upon which the special theory of relativity is built.

The Relativity Principle: All admissible frames of reference are completely equivalent for the formulation of the laws of physics.

The Relativity Principle is a powerful tool for building the physics of special relativity. Since our concern is primarily with the mathematical structure of the theory we shall have few occasions to call explicitly upon the Principle except for the physical interpretation of the mathematics and here it is vital. We regard the Relativity Principle primarily as an heuristic principle asserting that there are no "distinguished" admissible observers, i.e., that none can claim to have a privileged view of the universe. In particular, no such observer can claim to be "at rest" while the others are moving; they are all simply in relative motion. We shall see that admissible observers can disagree about some rather startling things (e.g., whether or not two given events are "simultaneous") and the Relativity Principle will prohibit us from preferring the judgment of one to any of the others. Although we will not dwell on the experimental evidence in favor of the Relativity Principle it should be observed that its roots lie in such commonplace observations as the fact that a passenger in a (smooth, quiet) airplane travelling at constant groundspeed in a straight line cannot "feel" his motion relative to the earth, i.e., that no physical effects are apparent in the plane which would serve to distinguish it from the (quasi-) admissible frame rigidly attached to the earth.

Our task then is to conduct a serious study of these "admissible frames of reference". Before embarking on such a study, however, it is only fair to concede that, in fact, no such thing exists. As is the case with any intellectual construct with which we attempt to model the physical universe, the notion of an admissible frame of reference is an idealization, a rather fanciful generalization of circumstances which, to some degree of accuracy, are encountered in the world. In particular, it has been found that the existence of gravitational fields imposes severe restrictions on the "extent" (both in space and in time) of an admissible frame (see [**TW**] for more on this). Knowing this we intentionally avoid the difficulty by restricting our attention to situations in which the effects of gravity are "negligible".

1

Geometrical Structure of $\mathcal{M}$

1.1 Preliminaries

We denote by $\mathcal{V}$ an arbitrary vector space of dimension $n \geq 1$ over the real numbers. A *bilinear form* on $\mathcal{V}$ is a map $g : \mathcal{V} \times \mathcal{V} \to \mathbb{R}$ that is linear in each variable, i.e., such that $g(a_1 v_1 + a_2 v_2, w) = a_1 g(v_1, w) + a_2 g(v_2, w)$ and $g(v, a_1 w_1 + a_2 w_2) = a_1 g(v, w_1) + a_2 g(v, w_2)$ whenever the a's are real numbers and the v's and w's are elements of $\mathcal{V}$. g is *symmetric* if $g(w, v) = g(v, w)$ for all v and w and *nondegenerate* if $g(v, w) = 0$ for all w in $\mathcal{V}$ implies $v = 0$. A nondegenerate, symmetric, bilinear form g is generally called an *inner product* and the image of (v, w) under g is often written $v \cdot w$ rather than $g(v, w)$. The standard example is the usual inner product on $\mathbb{R}^n$: if $v = (v^1, \ldots, v^n)$ and $w = (w^1, \ldots, w^n)$, then $g(v, w) = v \cdot w = v^1 w^1 + \cdots + v^n w^n$. This particular inner product is *positive definite*, i.e., has the property that if $v \neq 0$, then $g(v, v) > 0$. Not all inner products share this property, however.

Exercise 1.1.1. Define a map $g_1 : \mathbb{R}^n \times \mathbb{R}^n \to \mathbb{R}$ by $g_1(v, w) = v^1 w^1 + v^2 w^2 + \cdots + v^{n-1} w^{n-1} - v^n w^n$. Show that g_1 is an inner product and exhibit nonzero vectors v and w such that $g_1(v, v) = 0$ and $g_1(w, w) < 0$.

An inner product g for which $v \neq 0$ implies $g(v, v) < 0$ is said to be *negative definite*, whereas if g is neither positive definite nor negative definite it is said to be *indefinite*.

If g is an inner product on $\mathcal{V}$, then two vectors v and w for which $g(v, w) = 0$ are said to be *g-orthogonal*, or simply *orthogonal* if there is no ambiguity as to which inner product is intended. If $\mathcal{W}$ is a subspace of $\mathcal{V}$, then the *orthogonal complement* $\mathcal{W}^\perp$ of $\mathcal{W}$ in $\mathcal{V}$ is defined by $\mathcal{W}^\perp = \{v \in \mathcal{V}: \ g(v, w) = 0 \text{ for all } w \in \mathcal{W}\}$.

Exercise 1.1.2. Show that $\mathcal{W}^\perp$ is a subspace of $\mathcal{V}$.

The *quadratic form* associated with the inner product g on $\mathcal{V}$ is the map $\mathcal{Q} : \mathcal{V} \to \mathbb{R}$ defined by $\mathcal{Q}(v) = g(v, v) = v \cdot v$ (often denoted v^2). We ask the reader to show that distinct inner products on $\mathcal{V}$ cannot give rise to the same quadratic form.

Exercise 1.1.3. Show that if g_1 and g_2 are two inner products on $\mathcal{V}$ which satisfy $g_1(v,v) = g_2(v,v)$ for all v in $\mathcal{V}$, then $g_1(v,w) = g_2(v,w)$ for all v and w in $\mathcal{V}$. *Hint:* The map $g_1 - g_2 : \mathcal{V} \times \mathcal{V} \to \mathbb{R}$ defined by $(g_1 - g_2)(v,w) = g_1(v,w) - g_2(v,w)$ is bilinear and symmetric. Evaluate $(g_1 - g_2)(v+w, v+w)$.

A vector v for which $\mathcal{Q}(v)$ is either 1 or -1 is called a *unit vector*. A basis $\{e_1, \ldots, e_n\}$ for $\mathcal{V}$ which consists of mutually orthogonal unit vectors is called an *orthonormal basis* for $\mathcal{V}$ and we shall now prove that such bases always exist.

Theorem 1.1.1. *Let $\mathcal{V}$ be an n-dimensional real vector space on which is defined a nondegenerate, symmetric, bilinear form $g \colon \mathcal{V} \times \mathcal{V} \to \mathbb{R}$. Then there exists a basis $\{e_1, \ldots, e_n\}$ for $\mathcal{V}$ such that $g(e_i, e_j) = 0$ if $i \neq j$ and $\mathcal{Q}(e_i) = \pm 1$ for each $i = 1, \ldots, n$. Moreover, the number of basis vectors e_i for which $\mathcal{Q}(e_i) = -1$ is the same for any such basis.*

Proof: We begin with an observation. Since g is nondegenerate there exists a pair of vectors (v,w) for which $g(v,w) \neq 0$. We claim that, in fact, there must be a single vector u in $\mathcal{V}$ with $\mathcal{Q}(u) \neq 0$. Of course, if one of $\mathcal{Q}(v)$ or $\mathcal{Q}(w)$ is nonzero we are done. On the other hand, if $\mathcal{Q}(v) = \mathcal{Q}(w) = 0$, then $\mathcal{Q}(v+w) = \mathcal{Q}(v) + 2g(v,w) + \mathcal{Q}(w) = 2g(v,w) \neq 0$ so we may take $u = v + w$.

The proof of the theorem is by induction on n. If $n = 1$ we select any u in $\mathcal{V}$ with $\mathcal{Q}(u) \neq 0$ and define $e_1 = (|\, \mathcal{Q}(u) \,|)^{-1/2} u$. Then $\mathcal{Q}(e_1) = \pm 1$ so $\{e_1\}$ is the required basis.

Now we assume that $n > 1$ and that every inner product on a vector space of dimension less than n has a basis of the required type. Let the dimension of $\mathcal{V}$ be n. Again we begin by selecting a u in $\mathcal{V}$ such that $\mathcal{Q}(u) \neq 0$ and letting $e_n = (|\, \mathcal{Q}(u) \,|)^{-1/2} u$ so that $\mathcal{Q}(e_n) = \pm 1$. Now we let $\mathcal{W}$ be the orthogonal complement in $\mathcal{V}$ of the subspace $\text{Span}\{e_n\}$ of $\mathcal{V}$ spanned by $\{e_n\}$. By Exercise 1.1.2, $\mathcal{W}$ is a subspace of $\mathcal{V}$ and since e_n is not in $\mathcal{W}$, $\dim \mathcal{W} < n$. The restriction of g to $\mathcal{W} \times \mathcal{W}$ is an inner product on $\mathcal{W}$ so the induction hypothesis assures us of the existence of a basis $\{e_1, \ldots, e_m\}$, $m = \dim \mathcal{W}$, for $\mathcal{W}$ such that $g(e_i, e_j) = 0$ if $i \neq j$ and $\mathcal{Q}(e_i) = \pm 1$ for $i = 1, \ldots, m$. We claim that $m = n - 1$ and that $\{e_1, \ldots, e_m, e_n\}$ is a basis for $\mathcal{V}$.

Exercise 1.1.4. Show that the vectors $\{e_1, \ldots, e_m, e_n\}$ are linearly independent.

Since the number of elements in the set $\{e_1, \ldots, e_m, e_n\}$ is $m + 1 \leq n$, both of our assertions will follow if we can show that this set spans $\mathcal{V}$. Thus, we let v be an arbitrary element of $\mathcal{V}$ and consider the vector $w = v - (\mathcal{Q}(e_n) g(v, e_n)) e_n$. Then w is in $\mathcal{W}$ since $g(w, e_n) = g(v -$

$(\mathcal{Q}(e_n)g(v,e_n))e_n,\ e_n) = g(v,e_n) - (\mathcal{Q}(e_n))^2\, g(v,e_n) = 0$. Thus, we may write $v = w^1 e_1 + \cdots + w^m e_m + (\mathcal{Q}(e_n)g(v,e_n))e_n$ so $\{e_1, \ldots, e_m, e_n\}$ spans $\mathcal{V}$.

To show that the number r of e_i for which $\mathcal{Q}(e_i) = -1$ is the same for any orthonormal basis we proceed as follows: If $r = 0$ the result is clear since $\mathcal{Q}(v) \geq 0$ for every v in $\mathcal{V}$, i.e., g is positive definite. If $r > 0$, then $\mathcal{V}$ will have subspaces on which g is negative definite and so will have subspaces of maximal dimension on which g is negative definite. We will show that r is the dimension of any such maximal subspace $\mathcal{W}$ and thereby give an invariant (basis-independent) characterization of r. Number the basis elements so that $\{e_1, \ldots, e_r, e_{r+1}, \ldots, e_n\}$, where $\mathcal{Q}(e_i) = -1$ for $i = 1, \ldots, r$ and $\mathcal{Q}(e_i) = 1$ for $i = r+1, \ldots, n$. Let $\mathcal{X} = \mathrm{Span}\{e_1, \ldots, e_r\}$ be the subspace of $\mathcal{V}$ spanned by $\{e_1, \ldots, e_r\}$. Then, since g is negative definite on $\mathcal{X}$ and $\dim \mathcal{X} = r$, we find that $r \leq \dim \mathcal{W}$. To show that $r \geq \dim \mathcal{W}$ as well we define a map $T\colon \mathcal{W} \to \mathcal{X}$ as follows: If $w = \sum_{i=1}^n w^i e_i$ is in $\mathcal{W}$ we let $Tw = \sum_{i=1}^r w^i e_i$. Then T is obviously linear. Suppose w is such that $Tw = 0$. Then for each $i = 1, \ldots, r$, $w^i = 0$. Thus,

$$\mathcal{Q}(w) = g\left(\sum_{i=r+1}^n w^i e_i, \sum_{j=r+1}^n w^j e_j\right) = \sum_{i,j=r+1}^n g(e_i, e_j) w^i w^j = \sum_{i=r+1}^n (w^i)^2$$

which is greater than or equal to zero. But g is negative definite on $\mathcal{W}$ so we must have $w^i = 0$ for $i = r+1, \ldots, n$, i.e., $w = 0$. Thus, the null space of T is $\{0\}$ and T is therefore an isomorphism of $\mathcal{W}$ onto a subspace of $\mathcal{X}$. Consequently, $\dim \mathcal{W} \leq \dim \mathcal{X} = r$ as required. ■

The number r of e_i in any orthonormal basis for g with $\mathcal{Q}(e_i) = -1$ is called the *index* of g. Henceforth we will assume that all orthonormal bases are indexed in such a way that these e_i appear at the end of the list and so are numbered as follows:

$$\{e_1, e_2, \ldots, e_{n-r}, e_{n-r+1}, \ldots, e_n\}$$

where $\mathcal{Q}(e_i) = 1$ for $i = 1, 2, \ldots, n-r$ and $\mathcal{Q}(e_i) = -1$ for $i = n-r+1, \ldots, n$. Relative to such a basis if $v = v^i e_i$ and $w = w^i e_i$, then we have

$$g(v,w) = v^1 w^1 + \cdots + v^{n-r} w^{n-r} - v^{n-r+1} w^{n-r+1} - \cdots - v^n w^n.$$

1.2 Minkowski Spacetime

Minkowski spacetime is a 4-dimensional real vector space $\mathcal{M}$ on which is defined a nondegenerate, symmetric, bilinear form g of index 1. The elements of $\mathcal{M}$ will be called *events* and g is referred to as a *Lorentz inner product* on $\mathcal{M}$. Thus, there exists a basis $\{e_1, e_2, e_3, e_4\}$ for $\mathcal{M}$ with the property that if $v = v^a e_a$ and $w = w^a e_a$, then

$$g(v,w) = v^1w^1 + v^2w^2 + v^3w^3 - v^4w^4.$$

The elements of $\mathcal{M}$ are "events" and, as we suggested in the Introduction, are to be thought of intuitively as actual or physically possible point-events. An orthonormal basis $\{e_1, e_2, e_3, e_4\}$ for $\mathcal{M}$ "coordinatizes" this event world and is to be identified with a "frame of reference". Thus, if $x = x^1e_1 + x^2e_2 + x^3e_3 + x^4e_4$, we regard the coordinates (x^1, x^2, x^3, x^4) of x relative to $\{e_a\}$ as the spatial (x^1, x^2, x^3) and time (x^4) coordinates supplied the event x by the observer who presides over this reference frame. As we proceed with the development we will have occasion to expand upon, refine and add additional elements to this basic physical interpretation, but, for the present, this will suffice.

In the interest of economy we shall introduce a 4×4 matrix η defined by

$$\eta = \begin{bmatrix} 1 & 0 & 0 & 0 \\ 0 & 1 & 0 & 0 \\ 0 & 0 & 1 & 0 \\ 0 & 0 & 0 & -1 \end{bmatrix},$$

whose entries will be denoted either η_{ab} or η^{ab}, the choice in any particular situation being dictated by the requirements of the summation convention. Thus, $\eta_{ab} = \eta^{ab} = 1$ if $a = b = 1, 2, 3$, -1 if $a = b = 4$ and 0 otherwise. As a result we may write $g(e_a, e_b) = \eta_{ab} = \eta^{ab}$ and, with the summation convention, $g(v,w) = \eta_{ab}v^aw^b$.

Since our Lorentz inner product g on $\mathcal{M}$ is not positive definite there exist nonzero vectors v in $\mathcal{M}$ for which $g(v,v) = 0$, e.g., $v = e_1 + e_4$ is one such since $g(v,v) = \mathcal{Q}(e_1) + 2g(e_1, e_4) + \mathcal{Q}(e_4) = 1 + 0 - 1 = 0$. Such vectors are said to be *null* (or *lightlike*, for reasons which will become clear shortly) and $\mathcal{M}$ actually has bases which consist exclusively of this type of vector.

Exercise 1.2.1. Construct a *null basis* for $\mathcal{M}$, i.e., a set of four linearly independent null vectors.

Such a null basis cannot consist of mutually orthogonal vectors, however.

Theorem 1.2.1. *Two nonzero null vectors v and w in $\mathcal{M}$ are orthogonal if and only if they are parallel, i.e., iff there is a t in $\mathbb{R}$ such that $v = tw$.*

Exercise 1.2.2. Prove Theorem 1.2.1. *Hint:* The *Schwartz Inequality* for $\mathbb{R}^3$ asserts that if $x = (x^1, x^2, x^3)$ and $y = (y^1, y^2, y^3)$, then

$$(x^1y^1 + x^2y^2 + x^3y^3)^2 \leq ((x^1)^2 + (x^2)^2 + (x^3)^2)\,((y^1)^2 + (y^2)^2 + (y^3)^2)$$

and that equality holds if and only if x and y are linearly dependent. ■

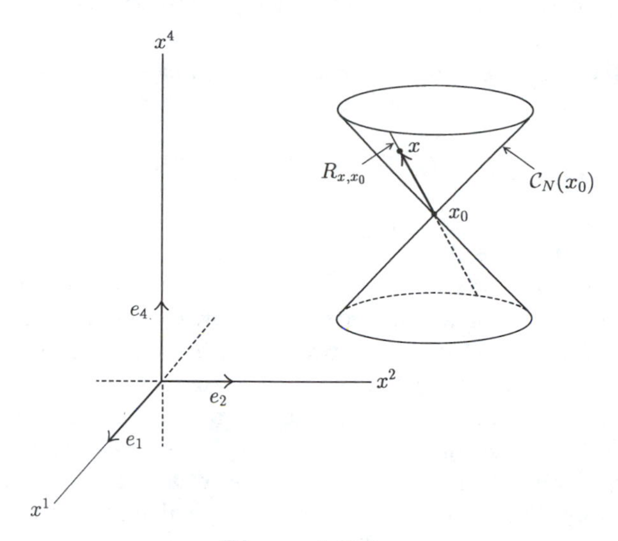

Figure 1.2.1

Next consider two distinct events x_0 and x for which the *displacement vector* $v = x - x_0$ from x_0 to x is null, i.e., $\mathcal{Q}(v) = \mathcal{Q}(x - x_0) = 0$. Relative to any orthonormal basis $\{e_a\}$, if $x = x^a e_a$ and $x_0 = x_0^a e_a$, then

$$(x^1 - x_0^1)^2 + (x^2 - x_0^2)^2 + (x^3 - x_0^3)^2 - (x^4 - x_0^4)^2 = 0. \qquad (1.2.1)$$

But we have seen this before. It is precisely the condition which, in the Introduction, we decided describes the relationship between two events that lie on the worldline of some photon. For this reason, and because of the formal similarity between (1.2.1) and the equation of a right circular cone in $\mathbb{R}^3$, we define the *null cone* (or *light cone*) $\mathcal{C}_N(x_0)$ *at* x_0 in $\mathcal{M}$ by

$$\mathcal{C}_N(x_0) = \{x \in \mathcal{M} : \mathcal{Q}(x - x_0) = 0\}$$

and picture it by suppressing the third spatial dimension x^3 (see Figure 1.2.1). $\mathcal{C}_N(x_0)$ therefore consists of all those events in $\mathcal{M}$ that are "connectible to x_0 by a light ray". For any such event x (other than x_0 itself) we define the *null worldline* (or *light ray*) $R_{x_0,x}$ containing x_0 and x by

$$R_{x_0,x} = \{x_0 + t(x - x_0) : t \in \mathbb{R}\}$$

and think of it as the worldline of that particular photon which experiences both x_0 and x.

Exercise 1.2.3. Show that if $\mathcal{Q}(x - x_0) = 0$, then $R_{x,x_0} = R_{x_0,x}$.

$\mathcal{C}_N(x_0)$ is just the union of all the light rays through x_0. Indeed,

Theorem 1.2.2. *Let x_0 and x be two distinct events with $\mathcal{Q}(x-x_0) = 0$. Then*

$$R_{x_0,x} = \mathcal{C}_N(x_0) \cap \mathcal{C}_N(x). \tag{1.2.2}$$

Proof: First let $z = x_0 + t(x - x_0)$ be an element of $R_{x_0,x}$. Then $z - x_0 = t(x - x_0)$ so $\mathcal{Q}(z - x_0) = t^2\mathcal{Q}(x - x_0) = 0$ so z is in $\mathcal{C}_N(x_0)$. With Exercise 1.2.3 it follows in the same way that z is in $\mathcal{C}_N(x)$ and so $R_{x_0,x} \subseteq \mathcal{C}_N(x_0) \cap \mathcal{C}_N(x)$. To prove the reverse containment we assume that z is in $\mathcal{C}_N(x_0) \cap \mathcal{C}_N(x)$. Then each of the vectors $z - x$, $z - x_0$ and $x_0 - x$ is null. But $z - x_0 = (z - x) - (x_0 - x)$ so $0 = \mathcal{Q}(z - x_0) = \mathcal{Q}(z - x) - 2g(z - x, x_0 - x) + \mathcal{Q}(x_0 - x) = -2g(z - x, x_0 - x)$. Thus, $g(z - x, x_0 - x) = 0$. If $z = x$ we are done. If $z \neq x$, then, since $x \neq x_0$, we may apply Theorem 1.2.1 to the orthogonal null vectors $z - x$ and $x_0 - x$ to obtain a t in $\mathbb{R}$ such that $z - x = t(x_0 - x)$ and it follows that z is in $R_{x_0,x}$ as required. ∎

For reasons which may not be apparent at the moment, but will become clear shortly, a vector v in $\mathcal{M}$ is said to be *timelike* if $\mathcal{Q}(v) < 0$ and *spacelike* if $\mathcal{Q}(v) > 0$.

Exercise 1.2.4. Use an orthonormal basis for $\mathcal{M}$ to construct a few vectors of each type.

If v is the displacement vector $x - x_0$ between two events, then, relative to *any* orthonormal basis for $\mathcal{M}$, $\mathcal{Q}(x - x_0) < 0$ becomes $(\Delta x^1)^2 + (\Delta x^2)^2 + (\Delta x^3)^2 < (\Delta x^4)^2$ ($x - x_0$ is *inside* the null cone at x_0). Thus, the (squared) spatial separation of the two events is less than the (squared) distance light would travel during the time lapse between the events (remember that x^4 is measured in light travel time). If $x - x_0$ is spacelike the inequality is reversed, we picture $x - x_0$ *outside* the null cone at x_0 and the spatial separation of x_0 and x is so great that not even a photon travels quickly enough to experience both events.

If $\{e_1, e_2, e_3, e_4\}$ and $\{\hat{e}_1, \hat{e}_2, \hat{e}_3, \hat{e}_4\}$ are two orthonormal bases for $\mathcal{M}$, then there is a unique linear transformation $L : \mathcal{M} \to \mathcal{M}$ such that $L(e_a) = \hat{e}_a$ for each $a = 1, 2, 3, 4$. As we shall see, such a map "preserves the inner product of $\mathcal{M}$", i.e., is of the following type: A linear transformation $L : \mathcal{M} \to \mathcal{M}$ is said to be an *orthogonal transformation* of $\mathcal{M}$ if $g(Lx, Ly) = g(x, y)$ for all x and y in $\mathcal{M}$.

Exercise 1.2.5. Show that, since the inner product on $\mathcal{M}$ is nondegenerate, an orthogonal transformation is necessarily one-to-one and therefore an isomorphism.

Lemma 1.2.3. *Let $L\colon \mathcal{M} \to \mathcal{M}$ be a linear transformation. Then the following are equivalent:*

(a) L is an orthogonal transformation.

(b) L preserves the quadratic form of $\mathcal{M}$, i.e., $\mathcal{Q}(Lx) = \mathcal{Q}(x)$ for all x in $\mathcal{M}$.

(c) L carries any orthonormal basis for $\mathcal{M}$ onto another orthonormal basis for $\mathcal{M}$.

Exercise 1.2.6. Prove Lemma 1.2.3. *Hint:* To prove that (b) implies (a) compute $L(x+y)\cdot L(x+y) - L(x-y)\cdot L(x-y)$. ■

Now let $L\colon \mathcal{M} \to \mathcal{M}$ be an orthogonal transformation of $\mathcal{M}$ and $\{e_1, e_2, e_3, e_4\}$ an orthonormal basis for $\mathcal{M}$. By Lemma 1.2.3, $\hat{e}_1 = Le_1$, $\hat{e}_2 = Le_2$, $\hat{e}_3 = Le_3$ and $\hat{e}_4 = Le_4$ also form an orthonormal basis for $\mathcal{M}$. In particular, each e_u, $u = 1,2,3,4$, can be expressed as a linear combination of the $\hat{e}_a$:

$$e_u = \Lambda^1{}_u \hat{e}_1 + \Lambda^2{}_u \hat{e}_2 + \Lambda^3{}_u \hat{e}_3 + \Lambda^4{}_u \hat{e}_4 = \Lambda^a{}_u \hat{e}_a, \qquad u = 1,2,3,4, \qquad (1.2.3)$$

where the $\Lambda^a{}_u$ are constants. Now, the orthogonality conditions $g(e_c, e_d) = \eta_{cd}$, $c, d = 1,2,3,4$, can be written

$$\Lambda^1{}_c\Lambda^1{}_d + \Lambda^2{}_c\Lambda^2{}_d + \Lambda^3{}_c\Lambda^3{}_d - \Lambda^4{}_c\Lambda^4{}_d = \eta_{cd} \qquad (1.2.4)$$

or, with the summation convention,

$$\Lambda^a{}_c\Lambda^b{}_d\eta_{ab} = \eta_{cd}, \qquad c, d = 1,2,3,4. \qquad (1.2.5)$$

Exercise 1.2.7. Show that (1.2.5) is equivalent to

$$\Lambda^a{}_c\Lambda^b{}_d\eta^{cd} = \eta^{ab}, \qquad a, b = 1,2,3,4. \qquad (1.2.6)$$

We define the *matrix* $\Lambda = [\Lambda^a{}_b]_{a,b=1,2,3,4}$ *associated with* the orthogonal transformation L and the orthonormal basis $\{e_a\}$ by

$$\Lambda = \begin{bmatrix} \Lambda^1{}_1 & \Lambda^1{}_2 & \Lambda^1{}_3 & \Lambda^1{}_4 \\ \Lambda^2{}_1 & \Lambda^2{}_2 & \Lambda^2{}_3 & \Lambda^2{}_4 \\ \Lambda^3{}_1 & \Lambda^3{}_2 & \Lambda^3{}_3 & \Lambda^3{}_4 \\ \Lambda^4{}_1 & \Lambda^4{}_2 & \Lambda^4{}_3 & \Lambda^4{}_4 \end{bmatrix}.$$

Observe that Λ is actually the matrix of L^{-1} relative to the basis $\{\hat{e}_a\}$. Heuristically, conditions (1.2.5) assert that "the columns of Λ are mutually orthogonal unit vectors", whereas (1.2.6) makes the same statement about the rows.

We regard the matrix Λ associated with L and $\{e_a\}$ as a coordinate transformation matrix in the usual way. Specifically, if the event x in $\mathcal{M}$ has coordinates $x = x^1e_1 + x^2e_2 + x^3e_3 + x^4e_4$ relative to $\{e_a\}$, then its coordinates relative to $\{\hat{e}_a\} = \{Le_a\}$ are $x = \hat{x}^1\hat{e}_1 + \hat{x}^2\hat{e}_2 + \hat{x}^3\hat{e}_3 + \hat{x}^4\hat{e}_4$, where

$$\begin{aligned}
\hat{x}^1 &= \Lambda^1{}_1x^1 + \Lambda^1{}_2x^2 + \Lambda^1{}_3x^3 + \Lambda^1{}_4x^4,\\
\hat{x}^2 &= \Lambda^2{}_1x^1 + \Lambda^2{}_2x^2 + \Lambda^2{}_3x^3 + \Lambda^2{}_4x^4,\\
\hat{x}^3 &= \Lambda^3{}_1x^1 + \Lambda^3{}_2x^2 + \Lambda^3{}_3x^3 + \Lambda^3{}_4x^4,\\
\hat{x}^4 &= \Lambda^4{}_1x^1 + \Lambda^4{}_2x^2 + \Lambda^4{}_3x^3 + \Lambda^4{}_4x^4,
\end{aligned}$$

which we generally write more concisely as

$$\hat{x}^a = \Lambda^a{}_bx^b, \qquad a = 1,2,3,4. \tag{1.2.7}$$

Exercise 1.2.8. By performing the indicated matrix multiplications show that (1.2.5) [and therefore (1.2.6)] is equivalent to

$$\Lambda^T\eta\Lambda = \eta, \tag{1.2.8}$$

where T means "transpose".

Notice that we have seen (1.2.8) before. It is just equation (0.4) of the Introduction, which perhaps seems somewhat less mysterious now than it did then. Indeed, (1.2.8) is now seen to be the condition that Λ is the matrix of a linear transformation which preserves the quadratic form of $\mathcal{M}$. In particular, if $x - x_0$ is the displacement vector between two events for which $\mathcal{Q}(x - x_0) = 0$, then both $(\Delta x^1)^2 + (\Delta x^2)^2 + (\Delta x^3)^2 - (\Delta x^4)^2$ and $(\Delta\hat{x}^1)^2 + (\Delta\hat{x}^2)^2 + (\Delta\hat{x}^3)^2 - (\Delta\hat{x}^4)^2$, where the $\Delta\hat{x}^a$ are, from (1.2.7), $\Delta\hat{x}^a = \Lambda^a{}_b\Delta x^b$, are zero. Physically, the two observers presiding over the hatted and unhatted reference frames agree that x_0 and x are "connectible by a light ray", i.e., they agree on the speed of light.

Any 4×4 matrix Λ that satisfies (1.2.8) is called a *general (homogeneous) Lorentz transformation*. At times we shall indulge in a traditional abuse of terminology and refer to the coordinate transformation (1.2.7) as a Lorentz transformation. Since the orthogonal transformations of $\mathcal{M}$ are isomorphisms and therefore invertible, the matrix Λ associated with such an orthogonal transformation must be invertible [also see (1.3.6)]. From (1.2.8) we find that $\Lambda^T\eta\Lambda = \eta$ implies $\Lambda^T\eta = \eta\Lambda^{-1}$ so that $\Lambda^{-1} = \eta^{-1}\Lambda^T\eta$ or, since $\eta^{-1} = \eta$,

$$\Lambda^{-1} = \eta\Lambda^T\eta. \tag{1.2.9}$$

Exercise 1.2.9. Show that the set of all general (homogeneous) Lorentz transformations forms a group under matrix multiplication, i.e., that it is closed under the formation of products and inverses. This group is called the *general (homogeneous) Lorentz group* and we shall denote it by $\mathcal{L}_{GH}$.

We shall denote the entries in the matrix Λ^{-1} by $\Lambda_a{}^b$ so that, by (1.2.9),

$$\begin{bmatrix} \Lambda_1{}^1 & \Lambda_2{}^1 & \Lambda_3{}^1 & \Lambda_4{}^1 \\ \Lambda_1{}^2 & \Lambda_2{}^2 & \Lambda_3{}^2 & \Lambda_4{}^2 \\ \Lambda_1{}^3 & \Lambda_2{}^3 & \Lambda_3{}^3 & \Lambda_4{}^3 \\ \Lambda_1{}^4 & \Lambda_2{}^4 & \Lambda_3{}^4 & \Lambda_4{}^4 \end{bmatrix} = \begin{bmatrix} \Lambda^1{}_1 & \Lambda^2{}_1 & \Lambda^3{}_1 & -\Lambda^4{}_1 \\ \Lambda^1{}_2 & \Lambda^2{}_2 & \Lambda^3{}_2 & -\Lambda^4{}_2 \\ \Lambda^1{}_3 & \Lambda^2{}_3 & \Lambda^3{}_3 & -\Lambda^4{}_3 \\ -\Lambda^1{}_4 & -\Lambda^2{}_4 & -\Lambda^3{}_4 & \Lambda^4{}_4 \end{bmatrix}. \tag{1.2.10}$$

Exercise 1.2.10. Show that

$$\Lambda_a{}^b = \eta_{ac}\eta^{bd}\Lambda^c{}_d, \qquad a, b = 1, 2, 3, 4, \tag{1.2.11}$$

and similarly

$$\Lambda^a{}_b = \eta^{ac}\eta_{bd}\Lambda_c{}^d, \qquad a, b = 1, 2, 3, 4. \tag{1.2.12}$$

Since we have seen (Exercise 1.2.9) that Λ^{-1} is in $\mathcal{L}_{GH}$ whenever Λ is it must also satisfy conditions analogous to (1.2.5) and (1.2.6), namely,

$$\Lambda_a{}^c\Lambda_b{}^d\eta^{ab} = \eta^{cd}, \qquad c, d = 1, 2, 3, 4, \tag{1.2.13}$$

and

$$\Lambda_a{}^c\Lambda_b{}^d\eta_{cd} = \eta_{ab}, \qquad a, b = 1, 2, 3, 4. \tag{1.2.14}$$

The analogues of (1.2.3) and (1.2.7) are

$$\hat{e}_u = \Lambda_u{}^a e_a, \qquad u = 1, 2, 3, 4, \tag{1.2.15}$$

and

$$x^b = \Lambda_a{}^b\hat{x}^a, \qquad b = 1, 2, 3, 4. \tag{1.2.16}$$

1.3 The Lorentz Group

Observe that by setting $c = d = 4$ in (1.2.5) one obtains $(\Lambda^4{}_4)^2 = 1 + (\Lambda^1{}_4)^2 + (\Lambda^2{}_4)^2 + (\Lambda^3{}_4)^2$ so that, in particular, $(\Lambda^4{}_4)^2 \geq 1$. Consequently,

$$\Lambda^4{}_4 \geq 1 \quad \text{or} \quad \Lambda^4{}_4 \leq -1 \tag{1.3.1}$$

An element Λ of $\mathcal{L}_{GH}$ is said to be *orthochronous* if $\Lambda^4{}_4 \geq 1$ and *nonorthochronous* if $\Lambda^4{}_4 \leq -1$. Nonorthochronous Lorentz transformations have certain unsavory characteristics which we now wish to expose. First, however, the following extremely important preliminary.

Theorem 1.3.1. *Suppose that v is timelike and w is either timelike or null and nonzero. Let $\{e_a\}$ be an orthonormal basis for $\mathcal{M}$ with $v = v^a e_a$ and $w = w^a e_a$. Then either*

(a) $v^4 w^4 > 0$, in which case $g(v, w) < 0$, or

(b) $v^4 w^4 < 0$, in which case $g(v, w) > 0$.

Proof: By assumption we have $g(v, v) = (v^1)^2 + (v^2)^2 + (v^3)^2 - (v^4)^2 < 0$ and $(w^1)^2 + (w^2)^2 + (w^3)^2 - (w^4)^2 \leq 0$ so $(v^4 w^4)^2 > ((v^1)^2 + (v^2)^2 + (v^3)^2)((w^1)^2 + (w^2)^2 + (w^3)^2) \geq (v^1 w^1 + v^2 w^2 + v^3 w^3)^2$, the second inequality following from the Schwartz Inequality for $\mathbb{R}^3$ (see Exercise 1.2.2). Thus, we find that

$$| v^4 w^4 | > | v^1 w^1 + v^2 w^2 + v^3 w^3 |,$$

so, in particular, $v^4 w^4 \neq 0$ and, moreover, $g(v, w) \neq 0$. Suppose that $v^4 w^4 > 0$. Then $v^4 w^4 = |v^4 w^4| > |v^1 w^1 + v^2 w^2 + v^3 w^3| \geq v^1 w^1 + v^2 w^2 + v^3 w^3$ and so $v^1 w^1 + v^2 w^2 + v^3 w^3 - v^4 w^4 < 0$, i.e., $g(v, w) < 0$. On the other hand, if $v^4 w^4 < 0$, then $g(v, -w) < 0$ so $g(v, w) > 0$. ∎

Corollary 1.3.2. *If a nonzero vector in $\mathcal{M}$ is orthogonal to a timelike vector, then it must be spacelike.*

We denote by τ the collection of all timelike vectors in $\mathcal{M}$ and define a relation $\sim$ on τ as follows: If v and w are in τ, then $v \sim w$ if and only if $g(v, w) < 0$ (so that v^4 and w^4 have the same sign in any orthonormal basis).

Exercise 1.3.1. Verify that $\sim$ is an equivalence relation on τ with precisely two equivalence classes. That is, show that $\sim$ is

1. reflexive ($v \sim v$ for every v in τ),
2. symmetric ($v \sim w$ implies $w \sim v$),
3. transitive ($v \sim w$ and $w \sim x$ imply $v \sim x$)

and that τ is the union of two disjoint subsets τ^+ and τ^- with the property that $v \sim w$ for all v and w in τ^+, $v \sim w$ for all v and w in τ^- and $v \not\sim w$ if one of v or w is in τ^+ and the other is in τ^-.

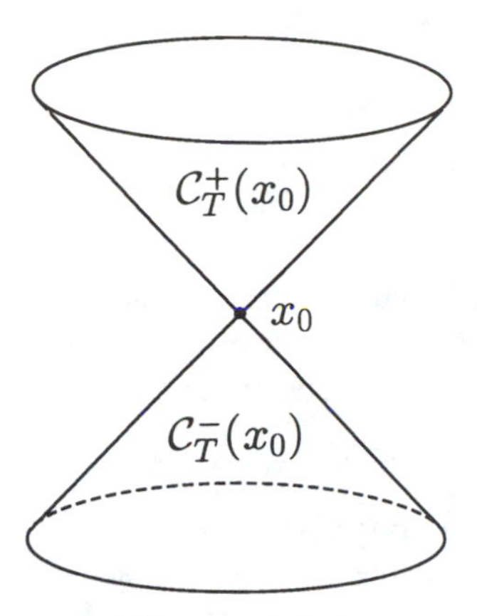

Figure 1.3.1

We think of the elements of τ^+ (and τ^-) as having the same time orientation. More specifically, we select (arbitrarily) τ^+ and refer to its elements as *future-directed* timelike vectors, whereas the vectors in τ^- we call *past-directed.*

Exercise 1.3.2. Show that τ^+ (and τ^-) are *cones,* i.e., that if v and w are in $\tau^+(\tau^-)$ and r is a *positive* real number, then rv and $v+w$ are also in $\tau^+(\tau^-)$.

For each x_0 in $\mathcal{M}$ we define the *time cone* $\mathcal{C}_T(x_0)$, *future time cone* $\mathcal{C}_T^+(x_0)$ and *past time cone* $\mathcal{C}_T^-(x_0)$ *at* x_0 by

$$\mathcal{C}_T(x_0) = \{x \in \mathcal{M} : \mathcal{Q}(x - x_0) < 0\},$$

$$\mathcal{C}_T^+(x_0) = \{x \in \mathcal{M} : x - x_0 \in \tau^+\} = \mathcal{C}_T(x_0) \cap \tau^+,$$

and

$$\mathcal{C}_T^-(x_0) = \{x \in \mathcal{M} : x - x_0 \in \tau^-\} = \mathcal{C}_T(x_0) \cap \tau^-.$$

We picture $\mathcal{C}_T(x_0)$ as the interior of the null cone $\mathcal{C}_N(x_0)$. It is the disjoint union of $\mathcal{C}_T^+(x_0)$ and $\mathcal{C}_T^-(x_0)$ and we shall adopt the convention that our pictures will always be drawn with future-directed vectors "pointing up" (see Figure 1.3.1).

We wish to extend the notion of past- and future-directed to nonzero null vectors as well. First we observe that if n is a nonzero null vector, then $n \cdot v$ has the same sign for all v in τ^+. To see this we suppose that there exist vectors v_1 and v_2 in τ^+ such that $n \cdot v_1 < 0$ and $n \cdot v_2 > 0$. We may assume that $|\, n \cdot v_1 \,| = n \cdot v_2$ since if this is not the case we can replace v_1 by $(n \cdot v_2 / \,|\, n \cdot v_1 \,|)v_1$, which is still in τ^+ by Exercise 1.3.2 and satisfies $g(n, (n \cdot v_2 / \,|\, n \cdot v_1 \,|)v_1) = (n \cdot v_2 / \,|\, n \cdot v_1 \,|)g(n, v_1) = -n \cdot v_2$. Thus, $n \cdot v_1 = -n \cdot v_2$ so $n \cdot v_1 + n \cdot v_2 = 0$ and therefore $n \cdot (v_1 + v_2) = 0$. But, again by Exercise 1.3.2, $v_1 + v_2$ is in τ^+ and so, in particular, is timelike. Since n is nonzero and null this contradicts Corollary 1.3.2. Thus, we may say that a nonzero null vector n is *future-directed* if $n \cdot v < 0$ for all v in τ^+ and *past-directed* if $n \cdot v > 0$ for all v in τ^+.

Exercise 1.3.3. Show that two nonzero null vectors n_1 and n_2 have the same time orientation (i.e., are both past-directed or both future-directed) if and only if n_1^4 and n_2^4 have the same sign relative to any orthonormal basis for $\mathcal{M}$.

For any x_0 in $\mathcal{M}$ we define the *future null cone at* x_0 by $\mathcal{C}_N^+(x_0) = \{x \in \mathcal{C}_N(x_0) : x - x_0$ is future-directed$\}$ and the *past null cone at* x_0 by $\mathcal{C}_N^-(x_0) = \{x \in \mathcal{C}_N(x_0) : x - x_0$ is past-directed$\}$. Physically, event x is in $\mathcal{C}_N^+(x_0)$ if x_0 and x respectively can be regarded as the emission and reception of a light signal. Consequently, $\mathcal{C}_N^+(x_0)$ may be thought of as the history in spacetime of a spherical electromagnetic wave (photons in all directions) whose emission event is x_0 (see Figure 1.3.2).

The disagreeable nature of nonorthochronous Lorentz transformations is that they always reverse time orientations (and so presumably relate reference frames in which someone's clock is running backwards).

Theorem 1.3.3. *Let* $\Lambda = [\Lambda^a{}_b]_{a,b=1,2,3,4}$ *be an element of* $\mathcal{L}_{GH}$ *and* $\{e_a\}_{a=1,2,3,4}$ *an orthonormal basis for* $\mathcal{M}$. *Then the following are equivalent:*

- *(a)* Λ *is orthochronous.*
- *(b)* Λ *preserves the time orientation of all nonzero null vectors, i.e., if* $v = v^a e_a$ *is a nonzero null vector, then the numbers* v^4 *and* $\hat{v}^4 = \Lambda^4{}_b v^b$ *have the same sign.*
- *(c)* Λ *preserves the time orientation of all timelike vectors.*

Proof: Let $v = v^a e_a$ be a vector which is either timelike or null and nonzero. By the Schwartz Inequality for $\mathbb{R}^3$ we have

$$(\Lambda^4{}_1 v^1 + \Lambda^4{}_2 v^2 + \Lambda^4{}_3 v^3)^2 \le \left(\sum_{i=1}^{3} (\Lambda^4{}_i)^2\right)\left(\sum_{i=1}^{3} (v^i)^2\right). \qquad (1.3.2)$$

Now, by (1.2.6) with $a = b = 4$, we have

$$(\Lambda^4{}_1)^2 + (\Lambda^4{}_2)^2 + (\Lambda^4{}_3)^2 - (\Lambda^4{}_4)^2 = -1 \qquad (1.3.3)$$

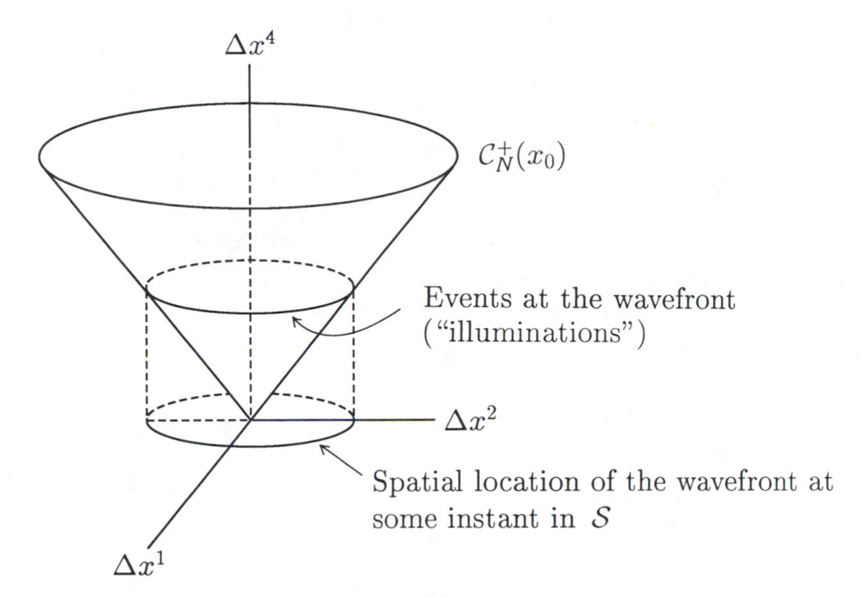

Figure 1.3.2

and so $(\Lambda^4{}_4)^2 > (\Lambda^4{}_1)^2 + (\Lambda^4{}_2)^2 + (\Lambda^4{}_3)^2$. Moreover, since v is either timelike or null, $(v^4)^2 \geq (v^1)^2 + (v^2)^2 + (v^3)^2$. Since v is nonzero, (1.3.2) therefore yields $(\Lambda^4{}_1 v^1 + \Lambda^4{}_2 v^2 + \Lambda^4{}_3 v^3)^2 < (\Lambda^4{}_4 v^4)^2$, which we may write as

$$(\Lambda^4{}_1 v^1 + \Lambda^4{}_2 v^2 + \Lambda^4{}_3 v^3 - \Lambda^4{}_4 v^4)(\Lambda^4{}_1 v^1 + \Lambda^4{}_2 v^2 + \Lambda^4{}_3 v^3 + \Lambda^4{}_4 v^4) < 0. \quad (1.3.4)$$

Define w in $\mathcal{M}$ by $w = \Lambda^4{}_1 e_1 + \Lambda^4{}_2 e_2 + \Lambda^4{}_3 e_3 + \Lambda^4{}_4 e_4$. By (1.3.3), w is timelike. Moreover, (1.3.4) can now be written

$$(v \cdot w)\hat{v}^4 < 0. \quad (1.3.5)$$

Consequently, $v \cdot w$ and $\hat{v}^4$ have opposite signs.

We now show that $\Lambda^4{}_4 \geq 1$ if and only if v^4 and $\hat{v}^4$ have the same sign. First suppose $\Lambda^4{}_4 \geq 1$. If $v^4 > 0$, then, by Theorem 1.3.1, $v \cdot w < 0$ so $\hat{v}^4 > 0$ by (1.3.5). Similarly, if $v^4 < 0$, then $v \cdot w > 0$ so $\hat{v}^4 < 0$. Thus, $\Lambda^4{}_4 \geq 1$ implies that v^4 and $\hat{v}^4$ have the same sign. In the same way, $\Lambda^4{}_4 \leq -1$ implies that v^4 and $\hat{v}^4$ have opposite signs. ■

Notice that we have actually shown that if Λ is nonorthochronous, then it necessarily reverses the time orientation of *all* timelike and nonzero null vectors. For this reason we elect to restrict our attention henceforth to the orthochronous elements of $\mathcal{L}_{GH}$. Since such a Lorentz transformation never reverses the time orientation of a timelike vector we may also limit ourselves to orthonormal bases $\{e_1, e_2, e_3, e_4\}$ with e_4 future-directed. At this point the reader may wish to return to the Introduction with a somewhat better understanding of why the condition ${\Lambda^4}_4 \geq 1$ appeared in Zeeman's Theorem.

There is yet one more restriction we would like to impose on our Lorentz transformations. Observe that taking determinants on both sides of (1.2.8) yields $(\det \Lambda^T)(\det \eta)(\det \Lambda) = \det \eta$ so that, since $\det \Lambda^T = \det \Lambda$, $(\det \Lambda)^2 = 1$ and therefore

$$\det \Lambda = 1 \quad \text{or} \quad \det \Lambda = -1. \tag{1.3.6}$$

We shall say that a Lorentz transformation Λ is *proper* if $\det \Lambda = 1$ and *improper* if $\det \Lambda = -1$.

Exercise 1.3.4. Show that an orthochronous Lorentz transformation is improper if and only if it is of the form

$$\begin{bmatrix} -1 & 0 & 0 & 0 \\ 0 & 1 & 0 & 0 \\ 0 & 0 & 1 & 0 \\ 0 & 0 & 0 & 1 \end{bmatrix} \Lambda, \tag{1.3.7}$$

where Λ is proper and orthochronous.

Notice that the matrix on the left in (1.3.7) is an orthochronous Lorentz transformation and, as a coordinate transformation, has the effect of changing the sign of the first spatial coordinate, i.e., of reversing the spatial orientation (left-handed to right-handed or right-handed to left-handed). Since there seems to be no compelling reason to make such a change we intend to restrict our attention to the set $\mathcal{L}$ of proper, orthochronous Lorentz transformations. Having done so we may further limit the orthonormal bases we consider by selecting an orientation for the spatial coordinate axes. Specifically, we define an *admissible basis* for $\mathcal{M}$ to be an orthonormal basis $\{e_1, e_2, e_3, e_4\}$ with e_4 timelike and future-directed and $\{e_1, e_2, e_3\}$ spacelike and "right-handed", i.e., satisfying $e_1 \times e_2 \cdot e_3 = 1$ (since the restriction of g to the span of $\{e_1, e_2, e_3\}$ is the usual dot product on $\mathbb{R}^3$, the cross product and dot product here are the familiar ones from vector calculus). At this point we fully identify an "admissible basis" with an "admissible frame of reference" as discussed in the Introduction. Any two such bases (frames) are related by a proper, orthochronous Lorentz transformation.

Exercise 1.3.5. Show that the set $\mathcal{L}$ of proper, orthochronous Lorentz transformations is a subgroup of $\mathcal{L}_{GH}$, i.e., that it is closed under the formation of products and inverses.

Generally, we shall refer to $\mathcal{L}$ simply as the *Lorentz group* and its elements as Lorentz transformations with the understanding that they are all proper and orthochronous. Occasionally it is convenient to enlarge the group of coordinate transformations to include spacetime translations (see the statement of Zeeman's Theorem in the Introduction), thereby obtaining the so-called *inhomogeneous Lorentz group* or *Poincaré group*. Physically, this amounts to allowing "admissible" observers to use different spacetime origins.

The Lorentz group $\mathcal{L}$ has an important subgroup $\mathcal{R}$ consisting of those $R = [R^a{}_b]$ of the form

$$R = \begin{bmatrix} & & & 0 \\ & [R^i{}_j] & & 0 \\ & & & 0 \\ 0 & 0 & 0 & 1 \end{bmatrix},$$

where $[R^i{}_j]_{i,j=1,2,3}$ is a unimodular orthogonal matrix, i.e., satisfies $\det[R^i{}_j] = 1$ and $[R^i{}_j]^T = [R^i{}_j]^{-1}$. Observe that the orthogonality conditions (1.2.5) are clearly satisfied by such an R and that, moreover, $R^4{}_4 = 1$ and $\det R = \det[R^i{}_j] = 1$ so that R is indeed in $\mathcal{L}$. The coordinate transformation associated with R corresponds physically to a rotation of the spatial coordinate axes within a given frame of reference. For this reason $\mathcal{R}$ is called the *rotation subgroup* of $\mathcal{L}$ and its elements are called *rotations* in $\mathcal{L}$.

Lemma 1.3.4. *Let* $\Lambda = [\Lambda^a{}_b]_{a,b=1,2,3,4}$ *be a proper, orthochronous Lorentz transformation. Then the following are equivalent:*

(a) Λ *is a rotation,*
(b) $\Lambda^1{}_4 = \Lambda^2{}_4 = \Lambda^3{}_4 = 0,$
(c) $\Lambda^4{}_1 = \Lambda^4{}_2 = \Lambda^4{}_3 = 0,$
(d) $\Lambda^4{}_4 = 1.$

Proof: Set $c = d = 4$ in (1.2.5) to obtain

$$(\Lambda^1{}_4)^2 + (\Lambda^2{}_4)^2 + (\Lambda^3{}_4)^2 - (\Lambda^4{}_4)^2 = -1. \quad (1.3.8)$$

Similarly, with $a = b = 4$, (1.2.6) becomes

$$(\Lambda^4{}_1)^2 + (\Lambda^4{}_2)^2 + (\Lambda^4{}_3)^2 - (\Lambda^4{}_4)^2 = -1. \quad (1.3.9)$$

The equivalence of (b), (c) and (d) now follows immediately from (1.3.8) and (1.3.9) and the fact that Λ is assumed orthochronous. Since a rotation in $\mathcal{L}$ satisfies (b), (c) and (d) by definition, all that remains is to show that if Λ satisfies one (and therefore all) of these conditions, then $\left[\Lambda^i{}_j\right]_{i,j=1,2,3}$ is a unimodular orthogonal matrix.

Exercise 1.3.6. Complete the proof. ∎

Exercise 1.3.7. Use Lemma 1.3.4 to show that $\mathcal{R}$ is a subgroup of $\mathcal{L}$, i.e., that it is closed under the formation of inverses and products.

Exercise 1.3.8. Show that an element of $\mathcal{L}$ has the same fourth row as $[\Lambda^a{}_b]_{a,b=1,2,3,4}$ if and only if it can be obtained from $[\Lambda^a{}_b]$ by multiplying on the left by some rotation in $\mathcal{L}$. Similarly, an element of $\mathcal{L}$ has the same fourth column as $[\Lambda^a{}_b]$ if and only if it can be obtained from $[\Lambda^a{}_b]$ by multiplying on the right by an element of $\mathcal{R}$.

There are 16 parameters in every Lorentz transformation, although, by virtue of the relations (1.2.5), these are not all independent. We now derive simple physical interpretations for some of these parameters. Thus, we consider two admissible bases $\{e_a\}$ and $\{\hat{e}_a\}$ and the corresponding admissible frames of reference $\mathcal{S}$ and $\hat{\mathcal{S}}$. Any two events on the worldline of a point which can be interpreted physically as being at rest in $\hat{\mathcal{S}}$ have coordinates in $\hat{\mathcal{S}}$ which satisfy $\Delta\hat{x}^1 = \Delta\hat{x}^2 = \Delta\hat{x}^3 = 0$ and $\Delta\hat{x}^4 =$ the time separation of the two events as measured in $\hat{\mathcal{S}}$. From (1.2.16) we find that the corresponding coordinate differences in $\mathcal{S}$ are

$$\Delta x^b = \Lambda_a{}^b \Delta\hat{x}^a = \Lambda_4{}^b \Delta\hat{x}^4. \tag{1.3.10}$$

From (1.3.10) and the fact that $\Lambda^4{}_4$ and $\Lambda_4{}^4$ are nonzero it follows that the ratios

$$\frac{\Delta x^i}{\Delta x^4} = \frac{\Lambda_4{}^i}{\Lambda_4{}^4} = -\frac{\Lambda^4{}_i}{\Lambda^4{}_4}, \qquad i = 1,2,3,$$

are constant and independent of the particular point at rest in $\hat{\mathcal{S}}$ we choose to examine. Physically, these ratios are interpreted as the components of the ordinary *velocity 3-vector* of $\hat{\mathcal{S}}$ relative to $\mathcal{S}$:

$$\vec{u} = u^1 e_1 + u^2 e_2 + u^3 e_3, \text{ where } u^i = \frac{\Lambda_4{}^i}{\Lambda_4{}^4} = -\frac{\Lambda^4{}_i}{\Lambda^4{}_4}, \quad i = 1,2,3 \tag{1.3.11}$$

(notice that we use the term "3-vector" and the familiar vector notation to distinguish such highly observer-dependent spatial vectors whose physical interpretations are not invariant under Lorentz transformations, but which are familiar from physics). Similarly, the *velocity 3-vector* of $\mathcal{S}$ relative to $\hat{\mathcal{S}}$ is

$$\vec{\hat{u}} = \hat{u}^1\hat{e}_1 + \hat{u}^2\hat{e}_2 + \hat{u}^3\hat{e}_3, \quad \text{where } \hat{u}^i = \frac{\Lambda^i{}_4}{\Lambda^4{}_4} = -\frac{\Lambda_i{}^4}{\Lambda_4{}^4}, \quad i = 1,2,3. \quad (1.3.12)$$

Next observe that $\sum_{i=1}^3 (\Delta x^i/\Delta x^4)^2 = (\Lambda^4{}_4)^{-2} \sum_{i=1}^3 (\Lambda^4{}_i)^2 = (\Lambda^4{}_4)^{-2} \cdot \left[(\Lambda^4{}_4)^2 - 1\right]$. Similarly, $\sum_{i=1}^3 (\Delta\hat{x}^i/\Delta\hat{x}^4)^2 = (\Lambda^4{}_4)^{-2}[(\Lambda^4{}_4)^2 - 1]$. Physically, we interpret these equalities as asserting that the velocity of $\hat{\mathcal{S}}$ relative to $\mathcal{S}$ and the velocity of $\mathcal{S}$ relative to $\hat{\mathcal{S}}$ have the same constant magnitude which we shall denote by β. Thus, $\beta^2 = 1 - (\Lambda^4{}_4)^{-2}$, so, in particular, $0 \le \beta^2 < 1$ and $\beta = 0$ if and only if Λ is a rotation (Lemma 1.3.4). Solving for $\Lambda^4{}_4$ (and taking the positive square root since Λ is assumed orthochronous) yields

$$\Lambda^4{}_4 = (1 - \beta^2)^{-\frac{1}{2}} \; (= \Lambda_4{}^4). \quad (1.3.13)$$

The quantity $(1-\beta^2)^{-1/2}$ will occur frequently and is often designated γ. Assuming that Λ is not a rotation we may write $\vec{u}$ as

$$\vec{u} = \beta\vec{d} = \beta(d^1e_1 + d^2e_2 + d^3e_3), \qquad d^i = u^i/\beta, \quad (1.3.14)$$

where $\vec{d}$ is the *direction 3-vector* of $\hat{\mathcal{S}}$ relative to $\mathcal{S}$ and the d^i are interpreted as the direction cosines of the directed line segment in $\sum$ along which the observer in $\mathcal{S}$ sees $\hat{\sum}$ moving. Similarly,

$$\vec{\hat{u}} = \beta\vec{\hat{d}} = \beta(\hat{d}^1\hat{e}_1 + \hat{d}^2\hat{e}_2 + \hat{d}^3\hat{e}_3), \qquad \hat{d}^i = \hat{u}^i/\beta. \quad (1.3.15)$$

Exercise 1.3.9. Show that the d^i are the components of the normalized projection of $\hat{e}_4$ onto the subspace spanned by $\{e_1, e_2, e_3\}$, i.e., that

$$d^i = \left(\sum_{j=1}^3 (\hat{e}_4 \cdot e_j)^2\right)^{-\frac{1}{2}} (\hat{e}_4 \cdot e_i), \qquad i = 1,2,3, \quad (1.3.16)$$

and similarly

$$\hat{d}^i = \left(\sum_{j=1}^3 (e_4 \cdot \hat{e}_j)^2\right)^{-\frac{1}{2}} (e_4 \cdot \hat{e}_i), \qquad i = 1,2,3. \quad (1.3.17)$$

Exercise 1.3.10. Show that $\hat{e}_4 = \gamma(\beta \vec{d} + e_4)$ and, similarly, $e_4 = \gamma(\beta \vec{\hat{d}} + \hat{e}_4)$ and notice that it follows from these that $e_4 \cdot \hat{e}_4 = \hat{e}_4 \cdot e_4 = -\gamma$.

Comparing (1.3.11) and (1.3.14) and using (1.3.13) we obtain

$$\Lambda_4{}^i = -\Lambda^4{}_i = \beta(1-\beta^2)^{-\frac{1}{2}}\, d^i, \qquad i = 1,2,3, \tag{1.3.18}$$

and similarly

$$\Lambda^i{}_4 = -\Lambda_i{}^4 = \beta(1-\beta^2)^{-\frac{1}{2}}\, \hat{d}^i, \qquad i = 1,2,3. \tag{1.3.19}$$

Equations (1.3.13), (1.3.18) and (1.3.19) give the last row and column of Λ in terms of physically measurable quantities and even at this stage a number of interesting kinematic consequences become apparent. Indeed, from (1.2.7) we obtain

$$\Delta\hat{x}^4 = -\beta\gamma(d^1\Delta x^1 + d^2\Delta x^2 + d^3\Delta x^3) + \gamma\Delta x^4 \tag{1.3.20}$$

for any two events. Let us consider the special case of two events on the worldline of a point *at rest in* $\mathcal{S}$. Then $\Delta x^1 = \Delta x^2 = \Delta x^3 = 0$ so (1.3.20) becomes

$$\Delta\hat{x}^4 = \gamma\Delta x^4 = \frac{1}{\sqrt{1-\beta^2}}\,\Delta x^4. \tag{1.3.21}$$

In particular, $\Delta\hat{x}^4 = \Delta x^4$ if and only if Λ is a rotation. Any relative motion of $\mathcal{S}$ and $\hat{\mathcal{S}}$ gives rise to a *time dilation* effect according to which $\Delta\hat{x}^4 > \Delta x^4$. Since our two events can be interpreted as two readings on one of the clocks at rest in $\mathcal{S}$, an observer in $\hat{\mathcal{S}}$ will conclude that the clocks in $\mathcal{S}$ are running slow (even though they are, by assumption, identical).

Exercise 1.3.11. Show that this time dilation effect is entirely symmetrical, i.e., that for two events with $\Delta\hat{x}^1 = \Delta\hat{x}^2 = \Delta\hat{x}^3 = 0$,

$$\Delta x^4 = \gamma\Delta\hat{x}^4 = \frac{1}{\sqrt{1-\beta^2}}\,\Delta\hat{x}^4. \tag{1.3.22}$$

We shall return to this phenomenon of time dilation in much greater detail after we have introduced a geometrical construction for picturing it. Nevertheless, we should point out at the outset that it is in no sense an illusion; it is quite "real" and can manifest itself in observable phenomena. One such instance occurs in the study of cosmic rays ("showers" of various types of elementary particles from space which impact the earth). Certain types of mesons that are encountered in cosmic radiation are so short-lived (at rest) that even if they could travel at the speed of light (which they cannot) the time required to traverse our atmosphere would be some ten times their normal life span. They should not be able to reach the earth, but they do. Time dilation, in a sense, "keeps them young". The meson's notion of time is not the same as ours. What seems a normal lifetime to the meson appears much longer to us. It is well to keep in mind also that we have been rather vague about what we mean by a "clock". Essentially any phenomenon involving observable change (successive readings on a Timex, vibrations of an atom, the lifetime of a meson, or a human being) is a "clock" and is therefore subject to the effects of time dilation. Of course, the effects will be negligibly small unless β is quite close to 1 (the speed of light). On the other hand, as $\beta \to 1$, (1.3.21) shows that $\Delta \hat{x}^4 \to \infty$ so that as speeds approach that of light the effects become infinitely great.

Another special case of (1.3.20) is also of interest. Let us suppose that our two events are judged *simultaneous in* $\mathcal{S}$, i.e., that $\Delta x^4 = 0$. Then

$$\Delta \hat{x}^4 = -\beta\gamma(d^1 \Delta x^1 + d^2 \Delta x^2 + d^3 \Delta x^3). \tag{1.3.23}$$

Again assuming that $\beta \neq 0$ we find that, in general, $\Delta \hat{x}^4$ will *not* be zero, i.e., that the two events will not be judged simultaneous in $\hat{\mathcal{S}}$. Indeed, $\mathcal{S}$ and $\hat{\mathcal{S}}$ will agree on the simultaneity of these two events if and only if the spatial locations of the events in $\sum$ bear a very special relation to the direction in $\sum$ along which $\hat{\sum}$ is moving, namely,

$$d^1 \Delta x^1 + d^2 \Delta x^2 + d^3 \Delta x^3 = 0 \tag{1.3.24}$$

(the displacement vector in $\sum$ between the locations of the two events is either zero or nonzero and perpendicular to the direction of $\hat{\sum}$'s motion in $\sum$). Otherwise, $\Delta \hat{x}^4 \neq 0$ and we have an instance of what is called the *relativity of simultaneity*. Notice, incidentally, that such disagreement can arise only for spatially separated events. More precisely, if in some admissible frame $\mathcal{S}$ two events x and x_0 are simultaneous and occur at the same spatial location, then $\Delta x^a = 0$ for $a = 1, 2, 3, 4$ so $x - x_0 = 0$. Since the Lorentz transformations are linear it follows that $\Delta \hat{x}^a = 0$ for $a = 1, 2, 3, 4$, i.e., the events are also simultaneous and occur at the same spatial location in $\hat{\mathcal{S}}$. Again, we will return to this phenomenon in much greater detail shortly.

It will be useful at this point to isolate a certain subgroup of the Lorentz group $\mathcal{L}$ which contains all of the physically interesting information about Lorentz transformations, but has much of the unimportant detail pruned away. We do this in the obvious way by assuming that the spatial axes of $\mathcal{S}$ and $\hat{\mathcal{S}}$ have a particularly simple relative orientation. Specifically, we consider the special case in which the direction cosines d^i and $\hat{d}^i$ are given by $d^1 = 1$, $\hat{d}^1 = -1$ and $d^2 = \hat{d}^2 = d^3 = \hat{d}^3 = 0$. Thus, the direction vectors are $\vec{d} = e_1$ and $\vec{\hat{d}} = -\hat{e}_1$. Physically, this corresponds to the situation in which an observer in $\mathcal{S}$ sees $\hat{\sum}$ moving in the direction of the positive x^1-axis and an observer in $\hat{\mathcal{S}}$ sees $\sum$ moving in the direction of the negative $\hat{x}^1$-axis. Since the origins of the spatial coordinate systems of $\mathcal{S}$ and $\hat{\mathcal{S}}$ coincided at $x^4 = \hat{x}^4 = 0$, we picture the motion of these two systems as being along their common x^1-, $\hat{x}^1$-axis. Now, from (1.3.13), (1.3.18) and (1.3.19) we find that the Lorentz transformation matrix Λ must have the form

$$\Lambda = \begin{bmatrix} \Lambda^1{}_1 & \Lambda^1{}_2 & \Lambda^1{}_3 & -\beta\gamma \\ \Lambda^2{}_1 & \Lambda^2{}_2 & \Lambda^2{}_3 & 0 \\ \Lambda^3{}_1 & \Lambda^3{}_2 & \Lambda^3{}_3 & 0 \\ -\beta\gamma & 0 & 0 & \gamma \end{bmatrix}.$$

Exercise 1.3.12. Use the orthogonality conditions (1.2.5) and (1.2.6) to show that Λ must take the form

$$\Lambda = \begin{bmatrix} \gamma & 0 & 0 & -\beta\gamma \\ 0 & \Lambda^2{}_2 & \Lambda^2{}_3 & 0 \\ 0 & \Lambda^3{}_2 & \Lambda^3{}_3 & 0 \\ -\beta\gamma & 0 & 0 & \gamma \end{bmatrix}, \tag{1.3.25}$$

where $[\Lambda^i{}_j]_{i,j=2,3}$ is a 2×2 unimodular orthogonal matrix, i.e., a rotation of the plane $\mathbb{R}^2$.

To discover the differences between these various elements of $\mathcal{L}$ we consider first the simplest possible choice for the 2×2 unimodular orthogonal matrix $[\Lambda^i{}_j]_{i,j=2,3}$, i.e., the identity matrix. The corresponding Lorentz transformation is

$$\Lambda = \begin{bmatrix} \gamma & 0 & 0 & -\beta\gamma \\ 0 & 1 & 0 & 0 \\ 0 & 0 & 1 & 0 \\ -\beta\gamma & 0 & 0 & \gamma \end{bmatrix} \tag{1.3.26}$$

and the associated coordinate transformation is

$$\begin{aligned} \hat{x}^1 &= (1-\beta^2)^{-\frac{1}{2}}\, x^1 - \beta(1-\beta^2)^{-\frac{1}{2}}\, x^4, \\ \hat{x}^2 &= x^2, \\ \hat{x}^3 &= x^3, \\ \hat{x}^4 &= -\beta(1-\beta^2)^{-\frac{1}{2}}\, x^1 + (1-\beta^2)^{-\frac{1}{2}}\, x^4. \end{aligned} \tag{1.3.27}$$

By virtue of the equalities $\hat{x}^2 = x^2$ and $\hat{x}^3 = x^3$ we view the physical relationship between $\sum$ and $\hat{\sum}$ as shown in Figure 1.3.3. Frames of reference

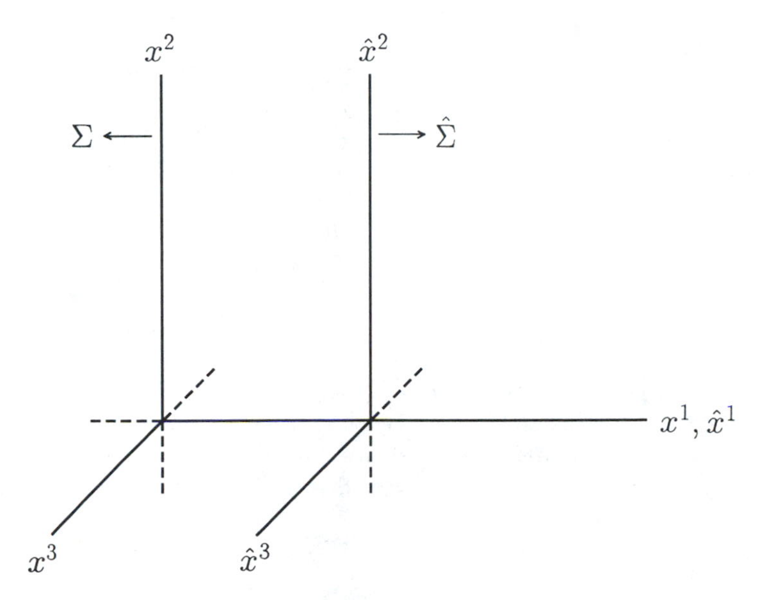

Figure 1.3.3

with spatial axes related in the manner shown in Figure 1.3.3 are said to be in *standard configuration.* Now it should be clear that any Lorentz transformation of the form (1.3.25) will correspond to the physical situation in which the $\hat{x}^2$- and $\hat{x}^3$-axes of $\hat{\mathcal{S}}$ are rotated in their own plane from the position shown in Figure 1.3.3.

By (1.2.10) the inverse of the Lorentz transformation Λ defined by (1.3.26) is

$$\Lambda^{-1} = \begin{bmatrix} \gamma & 0 & 0 & \beta\gamma \\ 0 & 1 & 0 & 0 \\ 0 & 0 & 1 & 0 \\ \beta\gamma & 0 & 0 & \gamma \end{bmatrix} \tag{1.3.28}$$

and the corresponding coordinate transformation is

$$\begin{aligned} x^1 &= (1-\beta^2)^{-\frac{1}{2}}\,\hat{x}^1 + \beta(1-\beta^2)^{-\frac{1}{2}}\,\hat{x}^4, \\ x^2 &= \hat{x}^2, \\ x^3 &= \hat{x}^3, \\ x^4 &= \beta(1-\beta^2)^{-\frac{1}{2}}\,\hat{x}^1 + (1-\beta^2)^{-\frac{1}{2}}\,\hat{x}^4. \end{aligned} \tag{1.3.29}$$

Any Lorentz transformation of the form (1.3.26) or (1.3.28), i.e., with $\Lambda^2{}_4 = \Lambda^3{}_4 = \Lambda^4{}_2 = \Lambda^4{}_3 = 0$ and $[\Lambda^i{}_j]_{i,j=2,3}$ equal to the 2×2 identity matrix, is called *a special Lorentz transformation.* Since Λ and Λ^{-1} differ only in the signs of the (1,4) and (4,1) entries it is customary, when discussing special Lorentz transformations, to allow $-1 < \beta < 1$. By choosing $\beta > 0$ when $\Lambda^1{}_4 < 0$ and $\beta < 0$ when $\Lambda^1{}_4 > 0$ all special Lorentz transformations can be written in the form (1.3.26) and we shall henceforth adopt this convention. For each real number β with $-1 < \beta < 1$ we therefore define $\gamma = \gamma(\beta) = (1-\beta^2)^{-1/2}$ and

$$\Lambda(\beta) = \begin{bmatrix} \gamma & 0 & 0 & -\beta\gamma \\ 0 & 1 & 0 & 0 \\ 0 & 0 & 1 & 0 \\ -\beta\gamma & 0 & 0 & \gamma \end{bmatrix}.$$

The matrix $\Lambda(\beta)$ is often called *a boost in the x^1-direction.*

Exercise 1.3.13. Define matrices which represent boosts in the x^2- and x^3-directions. One can define a boost in an arbitrary direction by first rotating, say, the positive x^1-axis into that direction and then applying $\Lambda(\beta)$.

Exercise 1.3.14. Suppose $-1 < \beta_1 \leq \beta_2 < 1$. Show that

(a) $\left|\dfrac{\beta_1+\beta_2}{1+\beta_1\beta_2}\right| < 1$. *Hint:* Show that if a is a constant, then the function $f(x) = \dfrac{x+a}{1+ax}$ is increasing on $-1 \leq x \leq 1$.

(b)
$$\Lambda(\beta_1)\Lambda(\beta_2) = \Lambda\left(\frac{\beta_1+\beta_2}{1+\beta_1\beta_2}\right). \tag{1.3.30}$$

It follows from Exercise 1.3.14 that the composition of two boosts in the x^1-direction is another boost in the x^1-direction. Since $\Lambda^{-1}(\beta) = \Lambda(-\beta)$ the collection of all such special Lorentz transformations forms a subgroup of $\mathcal{L}$. We point out, however, that the composition of two boosts in two different directions is, in general, *not* equivalent to a single boost in any direction.

By referring the three special Lorentz transformations $\Lambda(\beta_1)$, $\Lambda(\beta_2)$ and $\Lambda(\beta_1)\Lambda(\beta_2)$ to the corresponding admissible frames of reference one arrives at the following physical interpretation of (1.3.30): If the speed of $\hat{\mathcal{S}}$ relative to $\mathcal{S}$ is β_1 and the speed of $\hat{\hat{\mathcal{S}}}$ relative to $\hat{\mathcal{S}}$ is β_2, then the speed of $\hat{\hat{\mathcal{S}}}$ relative to $\mathcal{S}$ is not $\beta_1+\beta_2$ as one might expect, but rather

$$\frac{\beta_1+\beta_2}{1+\beta_1\beta_2},$$

which is always *less* than $\beta_1+\beta_2$ provided $\beta_1\beta_2 \neq 0$. Equation (1.3.30) is generally known as the *relativistic addition of velocities formula.* It, together with part (a) of Exercise 1.3.14, confirms the suspicion, already indicated by the behavior of (1.3.21) as $\beta \to 1$, that the relative speed of two admissible frames of reference is always less than that of light (that is,

1). Since any material object can be regarded as at rest in some admissible frame we conclude that such an object cannot attain (or exceed) the speed of light relative to an admissible frame.

Despite this "nonadditivity" of speeds in relativity it is often convenient to measure speeds with an alternative "velocity parameter" θ that *is* additive. An analogous situation occurs in plane Euclidean geometry where one has the option of describing the relative orientation of two Cartesian coordinate systems by means of angles (which are additive) or slopes (which are not). What we would like then is a measure θ of relative velocities with the property that if θ_1 is the velocity parameter of $\hat{\mathcal{S}}$ relative to $\mathcal{S}$ and θ_2 is the velocity parameter of $\hat{\hat{\mathcal{S}}}$ relative to $\hat{\mathcal{S}}$, then the velocity parameter of $\hat{\hat{\mathcal{S}}}$ relative to $\mathcal{S}$ is $\theta_1 + \theta_2$. Since θ is to measure relative speed, β will be some one-to-one function of θ, say, $\beta = f(\theta)$. Additivity and (1.3.30) require that f satisfy the functional equation

$$f(\theta_1 + \theta_2) = \frac{f(\theta_1) + f(\theta_2)}{1 + f(\theta_1)f(\theta_2)}. \tag{1.3.31}$$

Being reminiscent of the sum formula for the hyperbolic tangent, (1.3.31) suggests the change of variable

$$\beta = \tanh\theta \quad \text{or} \quad \theta = \tanh^{-1}\beta. \tag{1.3.32}$$

Observe that $\tanh^{-1}$ is a one-to-one differentiable function of $(-1, 1)$ onto $\mathbb{R}$ with the property that $\beta \to \pm 1$ implies $\theta \to \pm\infty$, i.e., the speed of light has infinite velocity parameter. If this change of variable seems to have been pulled out of the air it may be comforting to have a uniqueness theorem.

Exercise 1.3.15. Show that there is exactly one differentiable function $\beta = f(\theta)$ on $\mathbb{R}$ (namely, tanh) which satisfies (1.3.31) and the requirement that, for small speeds, β and θ are nearly equal, i.e., that

$$\lim_{\theta \to 0} \frac{f(\theta)}{\theta} = 1.$$

Hint: Show that such an f necessarily satisfies the initial value problem $f'(\theta) = 1 - (f(\theta))^2$, $f(0) = 0$ and appeal to the standard Uniqueness Theorem for solutions to such problems. Solve the problem to show that $f(\theta) = \tanh\theta$.

Exercise 1.3.16. Show that if $\beta = \tanh\theta$, then the *hyperbolic form* of the Lorentz transformation $\Lambda(\beta)$ is

$$L(\theta) = \begin{bmatrix} \cosh\theta & 0 & 0 & -\sinh\theta \\ 0 & 1 & 0 & 0 \\ 0 & 0 & 1 & 0 \\ -\sinh\theta & 0 & 0 & \cosh\theta \end{bmatrix}.$$

Earlier we suggested that all of the physically interesting behavior of proper, orthochronous Lorentz transformations is exhibited by the special Lorentz transformations. What we had in mind is the following theorem which asserts that any element of $\mathcal{L}$ differs from some $L(\theta)$ only by at most two rotations. This result will also be important in Section 1.7.

Theorem 1.3.5. *Let $\Lambda = [\Lambda^a{}_b]_{a,b=1,2,3,4}$ be a proper, orthochronous Lorentz transformation. Then there exists a real number θ and two rotations R_1 and R_2 in $\mathcal{R}$ such that $\Lambda = R_1 L(\theta) R_2$.*

Proof: Suppose first that $\Lambda^1{}_4 = \Lambda^2{}_4 = \Lambda^3{}_4 = 0$. Then, by Lemma 1.3.4, Λ is itself a rotation and so we may take $R_1 = \Lambda$, $\theta = 0$ and R_2 to be the 4×4 identity matrix. Consequently, we may assume that the vector $(\Lambda^1{}_4, \Lambda^2{}_4, \Lambda^3{}_4)$ in $\mathbb{R}^3$ is nonzero. Dividing by its magnitude in $\mathbb{R}^3$ gives a vector $\vec{u}_1 = (\alpha_1, \alpha_2, \alpha_3)$ of unit length in $\mathbb{R}^3$. Let $\vec{u}_2 = (\beta_1, \beta_2, \beta_3)$ and $\vec{u}_3 = (\gamma_1, \gamma_2, \gamma_3)$ be vectors in $\mathbb{R}^3$ such that $\{\vec{u}_1, \vec{u}_2, \vec{u}_3\}$ is an orthonormal basis for $\mathbb{R}^3$. Then

$$\begin{bmatrix} \alpha_1 & \alpha_2 & \alpha_3 \\ \beta_1 & \beta_2 & \beta_3 \\ \gamma_1 & \gamma_2 & \gamma_3 \end{bmatrix}$$

is an orthogonal matrix in $\mathbb{R}^3$ which, by a suitable ordering of the basis $\{\vec{u}_1, \vec{u}_2, \vec{u}_3\}$, we may assume unimodular, i.e., to have determinant 1. Thus, the matrix

$$R_1{}' = \begin{bmatrix} \alpha_1 & \alpha_2 & \alpha_3 & 0 \\ \beta_1 & \beta_2 & \beta_3 & 0 \\ \gamma_1 & \gamma_2 & \gamma_3 & 0 \\ 0 & 0 & 0 & 1 \end{bmatrix}$$

is a rotation in $\mathcal{R}$ and so $R_1{}'\Lambda$ is in $\mathcal{L}$. Now, since $\vec{u}_2$ and $\vec{u}_3$ are orthogonal in $\mathbb{R}^3$, the product $R_1{}'\Lambda$ must be of the form

$$R_1{}'\Lambda = \begin{bmatrix} a_{11} & a_{12} & a_{13} & a_{14} \\ a_{21} & a_{22} & a_{23} & 0 \\ a_{31} & a_{32} & a_{33} & 0 \\ \Lambda^4{}_1 & \Lambda^4{}_2 & \Lambda^4{}_3 & \Lambda^4{}_4 \end{bmatrix},$$

where $a_{14} = \alpha_1\Lambda^1{}_4 + \alpha_2\Lambda^2{}_4 + \alpha_3\Lambda^3{}_4 = ((\Lambda^1{}_4)^2 + (\Lambda^2{}_4)^2 + (\Lambda^3{}_4)^2)^{\frac{1}{2}} > 0$.

Next consider the vectors $\vec{v}_2 = (a_{21}, a_{22}, a_{23})$ and $\vec{v}_3 = (a_{31}, a_{32}, a_{33})$

in $\mathbb{R}^3$. Since $R_1{}'\Lambda$ is in $\mathcal{L}$, $\vec{v}_2$ and $\vec{v}_3$ are orthogonal unit vectors in $\mathbb{R}^3$. Select $\vec{v}_1 = (c_1, c_2, c_3)$ in $\mathbb{R}^3$ so that $\{\vec{v}_1, \vec{v}_2, \vec{v}_3\}$ is an orthonormal basis for $\mathbb{R}^3$. As for $R_1{}'$ above we may relabel if necessary and assume that

$$R_2{}' = \begin{bmatrix} c_1 & a_{21} & a_{31} & 0 \\ c_2 & a_{22} & a_{32} & 0 \\ c_3 & a_{23} & a_{33} & 0 \\ 0 & 0 & 0 & 1 \end{bmatrix}$$

is a rotation in $\mathcal{R}$. Thus, $B = R_1{}'\Lambda R_2{}'$ is also in $\mathcal{L}$.

Exercise 1.3.17. Use the available orthogonality conditions (the fact that $R_1{}'\Lambda$ and $R_2{}'$ are in $\mathcal{L}$) to show that

$$B = \begin{bmatrix} b_{11} & 0 & 0 & a_{14} \\ 0 & 1 & 0 & 0 \\ 0 & 0 & 1 & 0 \\ b_{41} & 0 & 0 & \Lambda^4{}_4 \end{bmatrix},$$

where $b_{11} = a_{11}c_1 + a_{12}c_2 + a_{13}c_3$ and $b_{41} = \Lambda^4{}_1 c_1 + \Lambda^4{}_2 c_2 + \Lambda^4{}_3 c_3$.

Thus, from the fact that B is in $\mathcal{L}$ we obtain

$$b_{11}a_{14} - b_{41}\Lambda^4{}_4 = 0, \tag{1.3.33}$$

$$b_{11}^2 - b_{41}^2 = 1, \tag{1.3.34}$$

$$a_{14}^2 - (\Lambda^4{}_4)^2 = -1. \tag{1.3.35}$$

Exercise 1.3.18. Use (1.3.33), (1.3.34) and (1.3.35) to show that neither b_{11} nor b_{41} is zero.

Thus, (1.3.33) is equivalent to $\Lambda^4{}_4/b_{11} = a_{14}/b_{41} = k$ for some k, i.e., $\Lambda^4{}_4 = kb_{11}$ and $a_{14} = kb_{41}$. Substituting these into (1.3.35) gives $k^2(b_{11}^2 - b_{41}^2) = 1$. By (1.3.34), $k^2 = 1$, i.e., $k = \pm 1$. But $k = -1$ would imply $\det B = -1$, whereas we must have $\det B = 1$ since B is in $\mathcal{L}$. Thus, $k = 1$ so

$$B = \begin{bmatrix} \Lambda^4{}_4 & 0 & 0 & a_{14} \\ 0 & 1 & 0 & 0 \\ 0 & 0 & 1 & 0 \\ a_{14} & 0 & 0 & \Lambda^4{}_4 \end{bmatrix}.$$

Now, it follows from (1.3.35) that $\Lambda^4{}_4 + a_{14} = (\Lambda^4{}_4 - a_{14})^{-1}$ so $\ln(\Lambda^4{}_4 - a_{14}) = -\ln(\Lambda^4{}_4 + a_{14})$. Define θ by

$$\theta = -\ln(\Lambda^4{}_4 + a_{14}) = \ln(\Lambda^4{}_4 - a_{14}).$$

Then $e^\theta = \Lambda^4{}_4 - a_{14}$ and $e^{-\theta} = \Lambda^4{}_4 + a_{14}$ so $\cosh\theta = \Lambda^4{}_4$ and $\sinh\theta = -a_{14}$. Consequently, $B = L(\theta)$. Since $B = R_1{}'\Lambda R_2{}' = L(\theta)$, we find that if $R_1 = (R_1{}')^{-1}$ and $R_2 = (R_2{}')^{-1}$ then $\Lambda = R_1 L(\theta) R_2$ as required. ■

The physical interpretion of Theorem 1.3.5 goes something like this: The Lorentz transformation from $\mathcal{S}$ to $\hat{\mathcal{S}}$ can be accomplished by (1) rotating the axes of $\mathcal{S}$ so that the x^1-axis coincides with the line along which the relative motion of $\hat{\sum}$ and $\sum$ takes place (positive x^1-direction coinciding with the direction of motion of $\hat{\sum}$ relative to $\sum$), (2) "boosting" to a new frame whose spatial axes are parallel to the rotated axes of $\mathcal{S}$ and at rest relative to $\hat{\sum}$ [via $L(\theta)$] and (3) rotating these spatial axes until they coincide with those of $\hat{\mathcal{S}}$. In many elementary situations the rotational part of this is unimportant and it suffices to restrict one's attention to special Lorentz transformations.

The special Lorentz transformations (1.3.27) and (1.3.29) correspond to a physical situation in which two of the three spatial coordinates are the same in both frames of reference. By suppressing these two it is possible to produce a simple, and extremely useful, 2-dimensional geometrical representation of $\mathcal{M}$ and of the effect of a Lorentz transformation. We begin by labeling two perpendicular lines in the plane "x^1" and "x^4". One should take care, however, not to attribute any physical significance to the perpendicularity of these lines. It is merely a matter of convenience and, in particular, is *not* to be identified with orthogonality in $\mathcal{M}$. Each event then has coordinates relative to e_1 and e_4 which can be obtained by projecting parallel to the opposite axis. The $\hat{x}^4$-axis is to be identified with the set of all events with $\hat{x}^1 = 0$, i.e., with $x^1 = \beta x^4$ [$= (\tanh\theta)x^4$] and we consequently picture the $\hat{x}^4$-axis as coinciding with this line. Similarly, the $\hat{x}^1$-axis is taken to lie along the line $\hat{x}^4 = 0$, i.e., $x^4 = \beta x^1$. In Figure 1.3.4 we have drawn these axes together with one branch of each of the hyperbolas $(x^1)^2 - (x^4)^2 = 1$ and $(x^1)^2 - (x^4)^2 = -1$.

Since the transformation (1.3.27) leaves invariant the quadratic form on $\mathcal{M}$ and since $\hat{x}^2 = x^2$ and $\hat{x}^3 = x^3$, it follows that the hyperbolas $(x^1)^2 - (x^4)^2 = 1$ and $(x^1)^2 - (x^4)^2 = -1$ coincide with the curves $(\hat{x}^1)^2 - (\hat{x}^4)^2 = 1$ and $(\hat{x}^1)^2 - (\hat{x}^4)^2 = -1$ respectively. From this it is clear that picturing the $\hat{x}^1$- and $\hat{x}^4$-axes as we have has distorted the picture [e.g., the point of intersection of $(x^1)^2 - (x^4)^2 = 1$ with the $\hat{x}^1$-axis must have hatted coordinates $(\hat{x}^1, \hat{x}^4) = (1,0)$] and necessitates a change of scale on these axes. To determine precisely what this change of scale should be we observe that one unit of length on the $\hat{x}^1$-axis must be represented by a segment whose Euclidean length in the picture is the Euclidean distance from the origin to the point $(\hat{x}^1, \hat{x}^4) = (1,0)$. This point has unhatted coordinates $(x^1, x^4) = ((1-\beta^2)^{-\frac{1}{2}}, \beta(1-\beta^2)^{-\frac{1}{2}})$ [by (1.3.29)] and the Euclidean distance from this point to the origin is, by the distance formula, $(1+\beta^2)^{\frac{1}{2}}(1-\beta^2)^{-\frac{1}{2}}$. A similar argument shows that

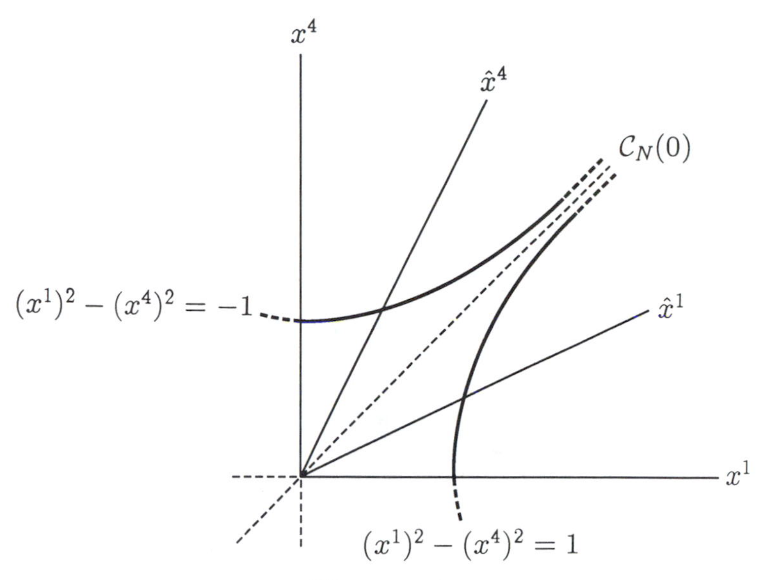

Figure 1.3.4

one unit of time on the $\hat{x}^4$-axis must also be represented by a segment of Euclidean length $(1+\beta^2)^{\frac{1}{2}}(1-\beta^2)^{-\frac{1}{2}}$. However, before we can legitimately calibrate these axes with this unit we must verify that all of the hyperbolas $(x^1)^2 - (x^4)^2 = \pm k^2$ $(k > 0)$ intersect the $\hat{x}^1$- and $\hat{x}^4$-axes a Euclidean distance $k(1+\beta^2)^{\frac{1}{2}}(1-\beta^2)^{-\frac{1}{2}}$ from the origin [the calibration must be consistent with the invariance of these hyperbolas under (1.3.27)].

Exercise 1.3.19. Verify this.

With this we have justified the calibration of the axes shown in Figure 1.3.5.

Exercise 1.3.20. Show that with this calibration of the $\hat{x}^1$- and $\hat{x}^4$-axes the hatted coordinates of any event can be obtained geometrically by projecting parallel to the opposite axis.

From this it is clear that the dotted lines in Figure 1.3.5 parallel to the $\hat{x}^1$- and $\hat{x}^4$-axes and through the points $(\hat{x}^1, \hat{x}^4) = (0,1)$ and $(\hat{x}^1, \hat{x}^4) = (1,0)$ are the lines $\hat{x}^4 = 1$ and $\hat{x}^1 = 1$ respectively.

Exercise 1.3.21. Show that, for any k, the line $\hat{x}^4 = k$ intersects the hyperbola $(x^1)^2 - (x^4)^2 = -k^2$ only at the point $(\hat{x}^1, \hat{x}^4) = (0,k)$, where it is, in fact, the tangent line. Similarly, $\hat{x}^1 = k$ is tangent to $(x^1)^2 - (x^4)^2 = k^2$ at $(\hat{x}^1, \hat{x}^4) = (k,0)$ and intersects this hyperbola only at that point.

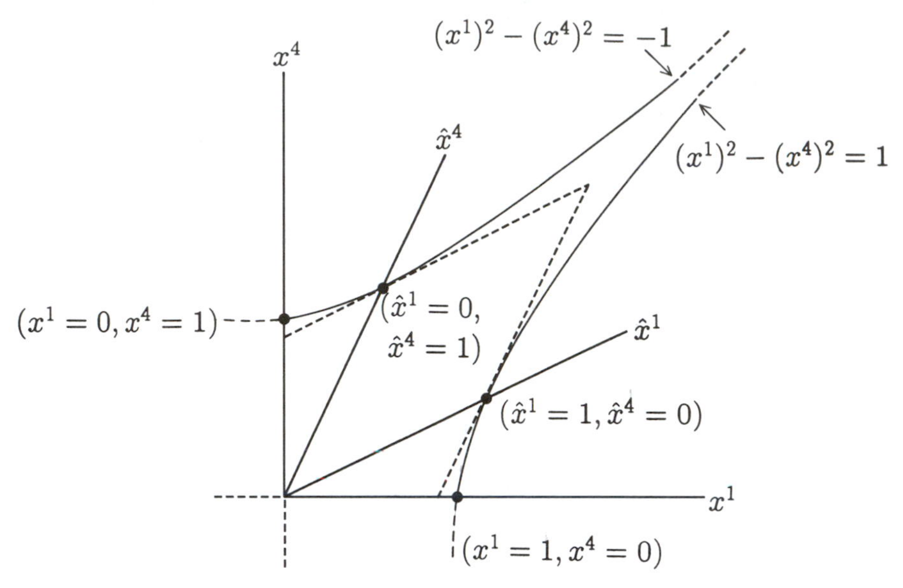

Figure 1.3.5

Next we would like to illustrate the utility of these 2-dimensional *Minkowski diagrams,* as they are called, by examining in detail the basic kinematic effects of special relativity (two of which we have already encountered). Perhaps the most fundamental of these is the so-called *relativity of simultaneity* which asserts that two admissible observers will, in general, disagree as to whether or not a given pair of spatially separated events were simultaneous. That this is the case was already clear in (1.3.23) which gives the time difference in $\hat{\mathcal{S}}$ between two events judged simultaneous in $\mathcal{S}$. Since, in a Minkowski diagram, lines of simultaneity (x^4=constant or $\hat{x}^4$=constant) are lines parallel to the respective spatial axes (Exercise 1.3.20) and since the line through two given events cannot be parallel to both the x^1- and $\hat{x}^1$-axes (unless $\beta = 0$), the geometrical representation is particularly persuasive (see Figure 1.3.6).

Notice, however, that some information is lost in such diagrams. In particular, the two lines of simultaneity in Figure 1.3.6 intersect in what appears to be a single point. But our diagram intentionally suppresses two spatial dimensions so the "lines" of simultaneity actually represent "instantaneous 3-spaces" which intersect in an entire plane of events and *both* observers judge all of these events to be simultaneous [recall (1.3.24)]. One can visualize at least an entire line of such events by mentally reinserting one of the missing spatial dimensions with an axis perpendicular to the

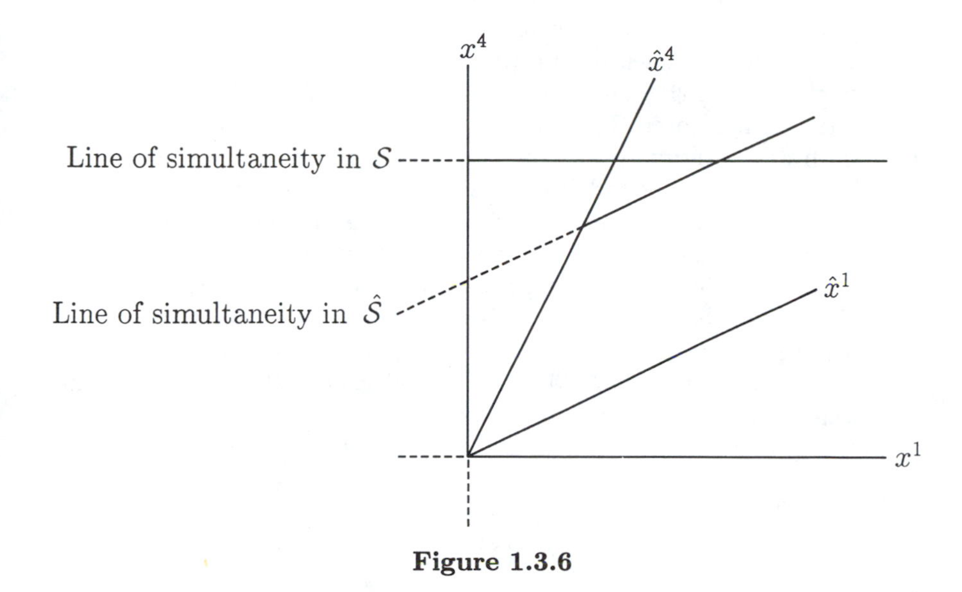

Figure 1.3.6

sheet of paper on which Figure 1.3.6 is drawn. The lines of simultaneity become planes of simultaneity which intersect in a "line of agreement" for $\mathcal{S}$ and $\hat{\mathcal{S}}$.

And so, it all seems quite simple. Too simple perhaps. One cannot escape the feeling that something must be wrong. Two events are given (for dramatic effect, two explosions). Surely the events either are, or are not, simultaneous and there is no room for disagreement. It seems inconceivable that two equally competent observers could arrive at different conclusions. And it is difficult to conceive, but only, we claim, because very few of us have ever met "another" admissible observer. We are, for the most part, all confined to the same frame of reference and, as is often the case in human affairs, our experience is too narrow, our view too parochial to comprehend other possiblities. We shall try to remedy this situation by moving the events far away from our all-too-comfortable earthly reference frame. Before getting started, however, we recommend that the reader return to the Introduction to review the procedure outlined there for synchronizing clocks as well as the properties of light signals enumerated there. In addition, it will be important to keep in mind that "simultaneity" becomes questionable only for spatially separated events. All observers agree that two given events either are, or are not, "simultaneous at the same spatial location".

Thus we consider two events (explosions) E_1 and E_2 occurring deep in space (to avoid the psychological inclination to adopt any large body nearby as a "standard of rest"). We suppose that E_1 and E_2 are observed in two admissible frames $\mathcal{S}$ and $\hat{\mathcal{S}}$ whose spatial axes are in standard configuration (Figure 1.3.3). Let us also suppose that when the explosions take place they permanently "mark" the locations at which they occur in each frame and, at the same time, emit light rays in all directions whose arrival times are recorded by local "assistants" at each spatial point within the two frames. Naturally, an observer in a given frame of reference will say that the events E_1 and E_2 are simultaneous if two such assistants, each of whom is in the immediate vicinity of one of the events, record times x_1^4 and x_2^4 for these events which, when compared later, are found to be equal. It is useful, however, to rephrase this notion of simultaneity in terms of readings taken at a single point. To do so we let $2d$ denote the distance between the spatial locations of E_1 and E_2 as determined in the given frame of reference and let M denote the midpoint of the line segment in that frame which joins these two locations:

Figure 1.3.7

Since $x_1^4 = x_2^4$ if and only if $x_1^4 + d = x_2^4 + d$ and since $x_1^4 + d$ is, *by definition,* the time of arrival at M of a light signal emitted with E_1 and, similarly, $x_2^4 + d$ is the arrival time at M of a light signal emitted with E_2 we conclude that *E_1 and E_2 are simultaneous in the given frame of reference if and only if light signals emitted with these events arrive simultaneously at the midpoint of the line segment joining the spatial locations of E_1 and E_2 within that frame.*

Now let us denote by A and $\hat{A}$ the spatial locations of E_1 in $\mathcal{S}$ and $\hat{\mathcal{S}}$ respectively and by B and $\hat{B}$ the locations of E_2 in $\mathcal{S}$ and $\hat{\mathcal{S}}$. Thus, the points A and $\hat{A}$ coincide at the instant E_1 occurs (they are the points "marked" by E_1) and similarly B and $\hat{B}$ coincide when E_2 occurs. At their convenience the two observers $\mathcal{O}$ and $\hat{\mathcal{O}}$ presiding over $\mathcal{S}$ and $\hat{\mathcal{S}}$ respectively collect all of the data recorded by their assistants for analysis. Each will inspect the coordinates of the two marked points, calculate from them the coordinates of the midpoint of the line segment joining these two points in his coordinate system (denote these midpoints by M and $\hat{M}$) and inquire of his assistant located at this point whether or not the light signals emitted from the two explosions arrived simultaneously at his location. In general, of course, there is no reason to expect an affirmative answer from either, but let us just suppose that in this particular case one of the observers, say $\mathcal{O}$, finds that the light signals from the two explosions did indeed arrive simultaneously at the midpoint of the line segment joining the spatial locations of the explosions in $\sum$. According to the criteria we have established, $\mathcal{O}$ will therefore conclude that E_1 and E_2 were simultaneous so that, from his point of view, A and $\hat{A}$, M and $\hat{M}$ and B and $\hat{B}$ all coincide "at the same time".

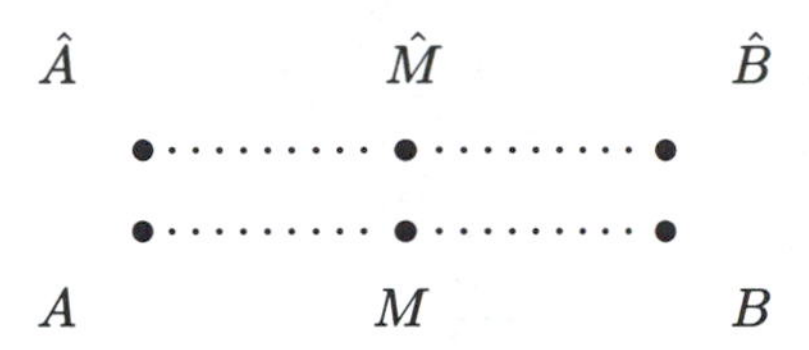

Figure 1.3.8

Continuing to analyze the situation as it is viewed from $\mathcal{S}$ we observe that, by virtue of the finite speed at which the light signals propagate, a nonzero time interval is required for these signals to reach M and that, during this time interval, M and $\hat{M}$ separate so that the signals cannot meet simultaneously at $\hat{M}$.

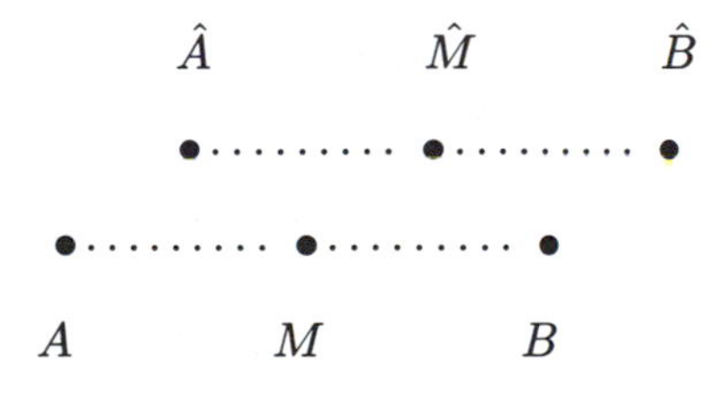

Figure 1.3.9

Indeed, if the motion is as indicated in Figures 1.3.8 and 1.3.9, the light from E_2 will clearly reach $\hat{M}$ before the light from E_1. Although we have reached this conclusion by examining the situation from the point of view of $\mathcal{O}$, any other admissible observer will necessarily concur since we have assumed that all such observers agree on the temporal order of any two events on the worldline of a photon (consider a photon emitted at E_2 and the two events on its worldline corresponding to its encounters with $\hat{M}$ and the light signal emitted at E_1). In particular, $\hat{\mathcal{O}}$ must conclude that E_2 occurred before E_1 and consequently that these two events are *not* simultaneous. When $\mathcal{O}$ and $\hat{\mathcal{O}}$ next meet they compare their observations of the two explosions and discover, much to their chagrin, that they disagree as to whether or not these two events were simultaneous. Having given the matter some thought, $\mathcal{O}$ believes that he has resolved the difficulty. The two events were indeed simultaneous as he had claimed, but they did not appear so to $\hat{\mathcal{O}}$ *because* $\hat{\mathcal{O}}$ *was moving* (running toward the light signal from E_2 and away from that of E_1). To this $\hat{\mathcal{O}}$ responds without hesitation "I wasn't moving — you were! The explosions were not simultaneous, but only appeared so to you because of your motion toward E_1 and away from E_2". This apparent impasse could, of course, be easily overcome if one could determine with some assurance which of the two observers was "really moving". But it is precisely this determination which the Relativity Principle disallows: One can attach no objective meaning to the phrase "really at rest". The conclusion is inescapable: It makes no more sense to ask if the events were "really simultaneous" than it does to ask if $\mathcal{O}$ was "really at rest". "Simultaneity", like "motion" is a purely relative term. If two events are simultaneous in one admissible frame of reference they will, in general, not be simultaneous in another such frame.

The relativity of simultaneity is not easy to come to terms with, but it is essential that one do so. Without it even the most basic contentions of relativity appear riddled with logical inconsistencies.

Exercise 1.3.22. Observer $\hat{\mathcal{O}}$ is moving to the right at constant speed β relative to observer $\mathcal{O}$ (along their common x^1-, $\hat{x}^1$-axes with origins coinciding at $x^4 = \hat{x}^4 = 0$). At the instant $\mathcal{O}$ and $\hat{\mathcal{O}}$ pass each other a flashbulb emits a spherical electromagnetic wavefront. $\mathcal{O}$ observes this spherical wavefront moving away from him with speed 1. After x_0^4 meters of time the wavefront will have reached points a distance x_0^4 meters from him. According to $\mathcal{O}$, at the instant the light has reached point A in Figure 1.3.10 it has also reached point B. However, $\hat{\mathcal{O}}$ regards himself as at rest with $\mathcal{O}$ moving so he will also observe a spherical wavefront moving away from him with speed 1. But as the light travels to A, $\hat{\mathcal{O}}$ has moved a short distance to the right of $\mathcal{O}$ so that the spherical wavefront observed by $\hat{\mathcal{O}}$ is not concentric with that observed by $\mathcal{O}$. In particular, when the light arrives at A, $\hat{\mathcal{O}}$ will contend that it also reaches (not B yet, but) C. They cannot both be right. Resolve the "paradox". *Hint:* There is an error in Figure 1.3.10. Compare it with Figure 1.3.11 after you have filled in the blanks.

To be denied the absolute, universal notion of simultaneity which the rather limited scope of our day-to-day experience has led us to accept uncritically is a serious matter. Disconcerting enough in its own right, this relativity of simultaneity also necessitates a profound reevaluation of the most basic concepts with which we describe the world. For example, since our observers $\mathcal{O}$ and $\hat{\mathcal{O}}$ need not agree on the time lapse between two events even when one of them measures it to be zero, one could scarcely

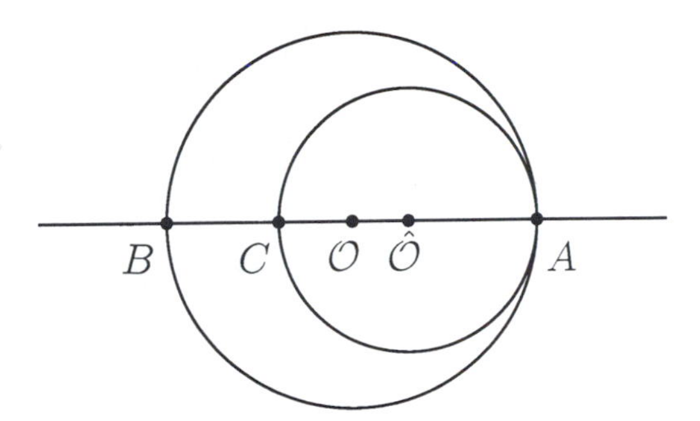

Figure 1.3.10

expect them to agree on the elapsed time between two arbitrarily given events. And, indeed, we have already seen in (1.3.20) that Δx^4 and $\Delta \hat{x}^4$ are generally not equal. This effect, known as *time dilation,* has a particularly nice geometrical representation in a Minkowski diagram (see Figure 1.3.12). E_1 (resp., E_2) can be identified physically with the appearance of the reading "1" on the clock at the origin of $\mathcal{S}$ (resp., $\hat{\mathcal{S}}$). In $\mathcal{S}$, E_1 is simultaneous with E_3 which corresponds to a reading strictly less than 1 on the clock at the origin in $\hat{\mathcal{S}}$. Since the clocks at the origins of $\mathcal{S}$ and $\hat{\mathcal{S}}$ agreed at $x^4 = \hat{x}^4 = 0$, $\mathcal{O}$ concludes that $\hat{\mathcal{O}}$'s clock is running slow. Indeed, (1.3.21) and (1.3.22) show that each observes the other's time dilated by the same constant factor $\gamma = (1-\beta^2)^{-\frac{1}{2}}$. The moral of the story, perhaps a bit too tersely stated, is that "moving clocks run slow".

Exercise 1.3.23. Pions are subatomic particles which decay spontaneously and have a half-life (at rest) of 1.8×10^{-8} sec $(= 5.4\text{m})$. A beam of pions is accelerated to a speed of $\beta = 0.99$. One would expect that the beam would drop to one-half its original intensity after travelling a distance of $(0.99)(5.4\text{m}) = 5.3\text{m}$. However, it is found experimentally that the beam reaches one-half intensity after travelling approximately 38m. Explain! *Hint:* Let $\mathcal{S}$ denote the laboratory frame of reference, $\hat{\mathcal{S}}$ the rest frame of the pions and assume that $\mathcal{S}$ and $\hat{\mathcal{S}}$ are related by (1.3.27) and (1.3.29). Draw a Minkowski diagram which represents the situation.

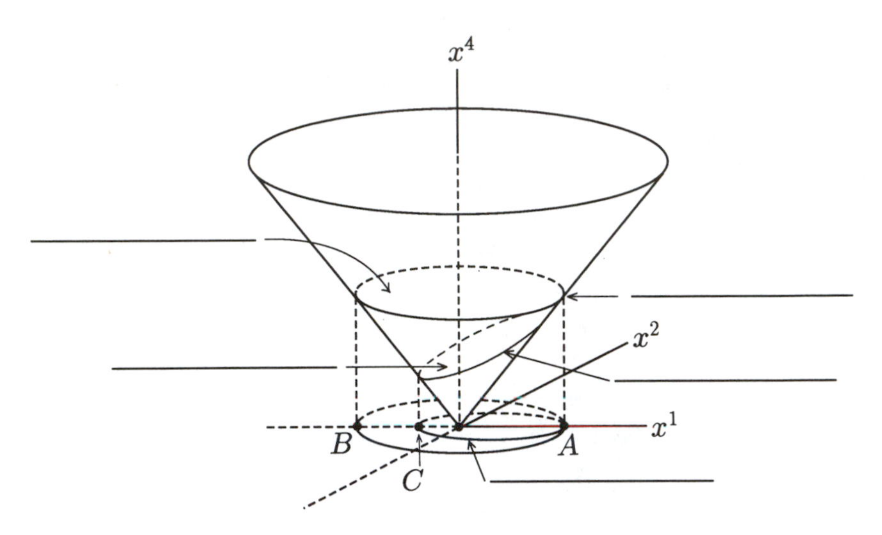

Figure 1.3.11

Return for a moment to Figure 1.3.12 and, in particular, to the line $\hat{x}^4 = 1$. Each point on this line can be identified with the appearance of the reading "1" on a clock that is stationary at some point in $\hat{\sum}$. These all occur "simultaneously" for $\hat{\mathcal{O}}$ because his clocks have been synchronized. However, each of these events occurs at a different "time" in $\mathcal{S}$ so $\mathcal{O}$ will disagree. Clocks at different locations in $\hat{\sum}$ read 1 at different "times" so, according to $\mathcal{O}$, they cannot be synchronized.

Here is an old, and much abused, "paradox" with its roots in the phenomenon of time dilation, or rather, in a basic misunderstanding of that phenomenon. Suppose that, at $(0,0,0,0)$, two identical twins part company. One remains at rest in the admissible frame in which he was born. The other is transported away at some constant speed to a distant point in space where he turns around and returns at the same constant speed to rejoin his brother. At the reunion the stationary twin finds that he is considerably older than his more adventurous brother. Not surprising; after all, moving clocks run slow. However, is it not true that, from the point of view of the "rocket" twin, it is the "stationary" brother who has been moving and must, therefore, be the younger of the two?

The error concealed in this argument, of course, is that it hinges upon a supposed symmetry between the two twins which simply does not exist. If the stationary twin does, in fact, remain at rest in an admissible frame, then his brother certainly does not. Indeed, to turn around and re-

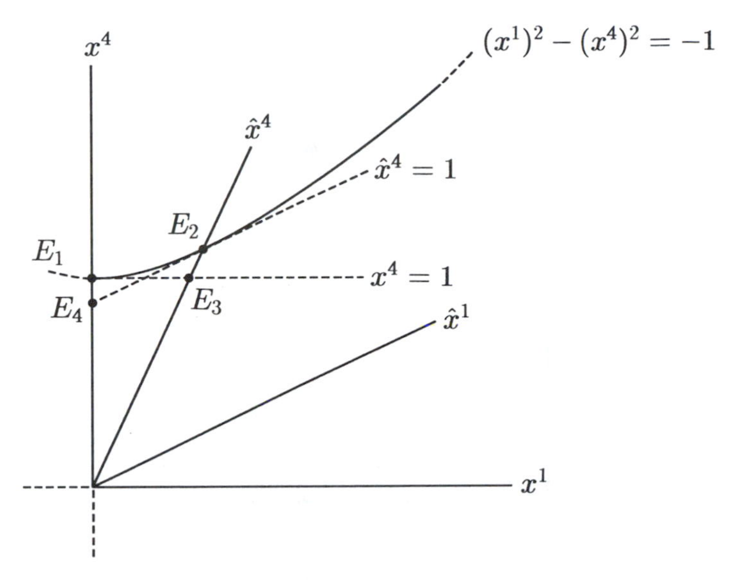

Figure 1.3.12

turn midway through his journey he must "transfer" from one admissible frame to another and, in practice, such a transfer would require *accelerations* (slow down, turn around, speed up) and these accelerations would be experienced only by the traveller and not by his brother. Nothing we have done thus far equips us to deal with these accelerations and so we can come to no conclusions about their physical effects (we will pursue this further in Section 1.4). That they do have physical effects, however, can be surmised even now by idealizing the situation a bit. Let us replace our two twins with three admissible frames: $\mathcal{S}$ (stationary twin), $\hat{\mathcal{S}}$ (rocket twin on his outward journey) and $\hat{\hat{\mathcal{S}}}$ (rocket twin on his return journey). What this amounts to is the assumption that the two individuals involved compare ages in passing (without stopping to discuss it) at the beginning and end of the trip and that, at the turnaround point, the traveller "jumps" instantaneously from one admissible frame to another (he cannot do that, of course, but it seems reasonable that, with a sufficiently durable observer, we could approximate such a jump arbitrarily well by a "large" acceleration over a "small" time interval). Figure 1.3.13 represents the outward journey from O to the turnaround event T.

Notice that, in $\hat{\mathcal{S}}$, T is simultaneous with the event P on the worldline of the stay-at-home. In $\mathcal{S}$, P is simultaneous with some earlier event on the worldline of the traveller. Each sees the other's time dilated. Figure 1.3.14 represents the return journey. Notice that, in $\hat{\hat{\mathcal{S}}}$, T is simultaneous with

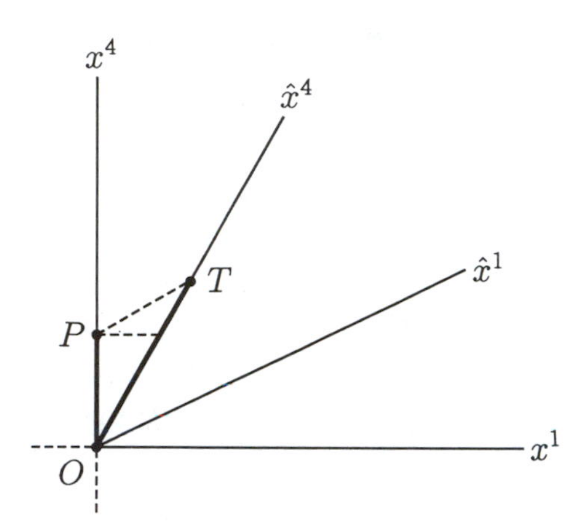

Figure 1.3.13

(not P, but) the event Q on the worldline of the stationary twin, whereas, in $\mathcal{S}$, Q is simultaneous with some later event on the traveller's worldline. Each sees the other's time dilated. Now, put the two pictures together in Figure 1.3.15 and notice that in "jumping" from $\hat{\mathcal{S}}$ to $\hat{\hat{\mathcal{S}}}$, our rocket twin has also jumped over the entire interval from P to Q on the worldline of his brother; an interval over which his brother ages, but he does not. The lesson to be learned is that, while all motion is indeed relative, it is not all physically equivalent.

Exercise 1.3.24. Account, in a sentence or two, for the "missing" time in Figure 1.3.15. *Hint:* $\dfrac{2\beta}{1+\beta^2} > \beta$ for $0 < \beta < 1$.

There is one last kinematic consequence of the relativity of simultaneity, as interesting, as important and as surprising as time dilation. To trace its origins we return once again to the explosions E_1 and E_2, observed by $\mathcal{S}$ and $\hat{\mathcal{S}}$ and discussed on pages 35-38. Recall that the points A in $\sum$ and $\hat{A}$ in $\hat{\sum}$ coincided when E_1 occurred, whereas B in $\sum$ and $\hat{B}$ in $\hat{\sum}$ coincided when E_2 occurred. Since the two events were simultaneous in $\mathcal{S}$, the observer $\mathcal{O}$ will conclude that A coincides with $\hat{A}$ at the same instant that B coincides with $\hat{B}$ and, in particular, that the segments AB and $\hat{A}\hat{B}$ have the *same length* (see Figure 1.3.8). However, in $\hat{\mathcal{S}}$, E_2 occurred before E_1 so B coincides with $\hat{B}$ before A coincides with $\hat{A}$

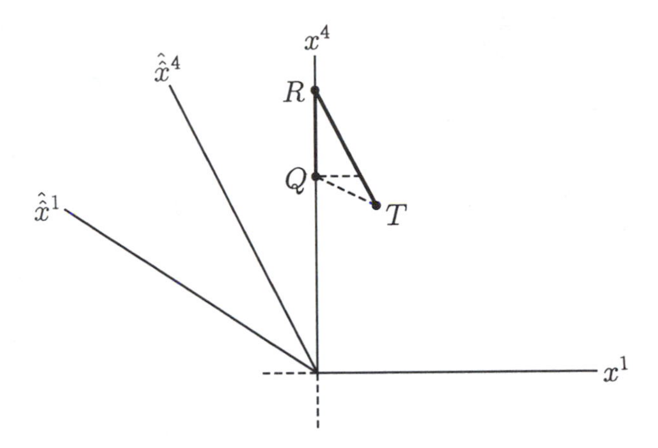

Figure 1.3.14

and $\hat{O}$ must conclude that the length of $\hat{A}\hat{B}$ is *greater* than the length of AB. More generally, two objects (say, measuring rods) in relative motion are considered to be equal in length if, when they pass each other, their respective endpoints A, $\hat{A}$ and B, $\hat{B}$ coincide simultaneously. But, "simultaneously" according to whom? Here we have two events (the coincidence of A and $\hat{A}$ and the coincidence of B and $\hat{B}$) and we have seen that if one admissible observer claims that they are simultaneous (i.e., that the lengths AB and $\hat{A}\hat{B}$ are equal), then another will, in general, disagree and we have no reason to prefer the judgment of one such observer to that of another (Relativity Principle). "Length", we must conclude, cannot be regarded as an objective attribute of the rods, but is rather simply the result of a specific measurement which we can no longer go on believing must be the same for all observers. Notice also that these conclusions have nothing whatever to do with the material construction of the measuring rods (in particular, their "rigidity") since, in the case of the two explosions, for example, there need not be any material connection between the two events. This phenomenon is known as *length contraction* (or *Lorentz contraction*) and we shall now look into the quantitative side of it.

To simplify the calculations and to make available an illuminating Minkowski diagram we shall restrict our discussion to frames of reference whose spatial axes are in standard configuration (see Figure 1.3.3) and whose coordinates are therefore related by (1.3.27) and (1.3.29). For the picture let us consider a "rigid" rod resting along the $\hat{x}^1$-axis of $\hat{\mathcal{S}}$ with

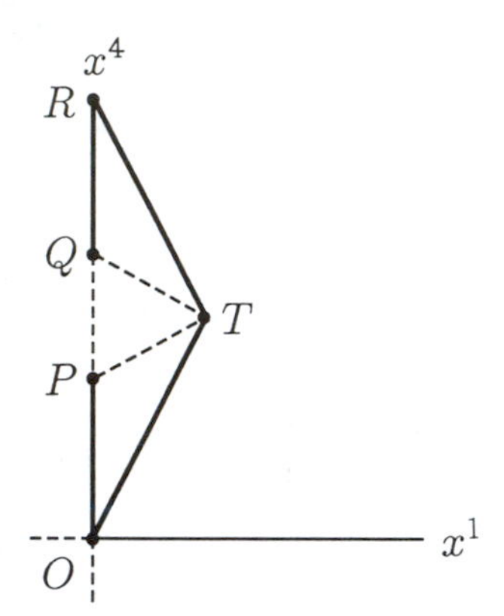

Figure 1.3.15

ends fixed at $\hat{x}^1 = 0$ and $\hat{x}^1 = 1$. Thus, the length of the rod as measured in $\hat{\mathcal{S}}$ is 1. The worldlines of the left and right ends of the rod are the $\hat{x}^1$-axis and the line $\hat{x}^1 = 1$ respectively. Geometrically, the measured length of the rod in $\mathcal{S}$ is the Euclidean length of the segment joining two points on these worldlines *at the same instant in* $\mathcal{S}$ ("locate the ends of the rod *simultaneously* and compute the length from their coordinates at this instant"). Since the Euclidean length of such a segment is clearly the same as the x^1-coordinate of the point P in Figure 1.3.16 and since this is clearly less than 1, length contraction is visually apparent.

For the calculation we will be somewhat more general and consider a rod lying along the $\hat{x}^1$-axis of $\hat{\mathcal{S}}$ between $\hat{x}_0^1$ and $\hat{x}_1^1$ with $\hat{x}_0^1 < \hat{x}_1^1$ so that its measured length in $\hat{\mathcal{S}}$ is $\Delta\hat{x}^1 = \hat{x}_1^1 - \hat{x}_0^1$. The worldline of the rod's left- (resp., right-) hand endpoint has $\hat{\mathcal{S}}$-coordinates $(\hat{x}_0^1, 0, 0, \hat{x}^4)$ [resp., $(\hat{x}_1^1, 0, 0, \hat{x}^4)$], with $-\infty < \hat{x}^4 < \infty$. $\mathcal{S}$ will measure the length of this rod by locating its endpoints "simultaneously", i.e., by finding one event on each of these worldlines with the same x^4 (*not* $\hat{x}^4$). But, for any *fixed* x^4, the transformation equations (1.3.27) give

$$\hat{x}_0^1 = (1-\beta^2)^{-\frac{1}{2}}\, x_0^1 - \beta(1-\beta^2)^{-\frac{1}{2}}\, x^4,$$

$$\hat{x}_1^1 = (1-\beta^2)^{-\frac{1}{2}}\, x_1^1 - \beta(1-\beta^2)^{-\frac{1}{2}}\, x^4,$$

so that $\Delta\hat{x}^1 = (1-\beta^2)^{-\frac{1}{2}}\, \Delta x^1$ and therefore

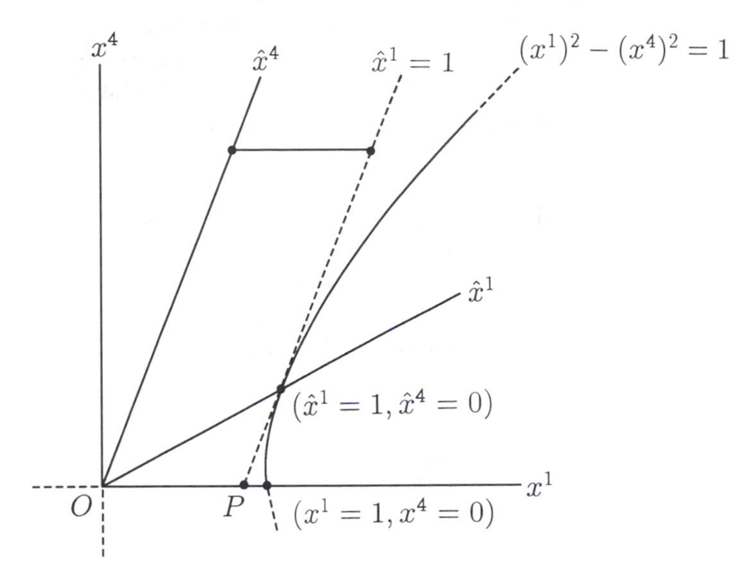

Figure 1.3.16

$$\Delta x^1 = \left(1 - \beta^2\right)^{\frac{1}{2}} \Delta \hat{x}^1. \tag{1.3.36}$$

Since $(1-\beta^2)^{\frac{1}{2}} < 1$ we find that the measured length of the rod in $\mathcal{S}$ is less than its measured length in $\hat{\mathcal{S}}$ by a factor of $\sqrt{1-\beta^2}$. By reversing the roles of $\mathcal{S}$ and $\hat{\mathcal{S}}$ we again find that this effect is entirely symmetrical.

Exercise 1.3.25. Return to Exercise 1.3.23 and offer another explanation based, not on time dilation, but on length contraction.

As it is with time dilation, the correct physical interpretation of the Lorentz contraction often requires rather subtle and delicate argument.

Exercise 1.3.26. Imagine a barn which, at rest, measures 8 meters in length. A (very fast) runner carries a pole of rest length 16 meters toward the barn at such a high speed that, for an observer at rest with the barn, it appears Lorentz contracted to 8 meters and therefore fits inside the barn. This observer slams the front door shut at the instant the back of the pole enters the front of the barn and so encloses the pole entirely within the barn. But is it not true that the runner sees the barn Lorentz contracted to 4 meters so that the 16 meter pole could never fit entirely within it? Resolve the difficulty! *Hint:* Let $\mathcal{S}$ and $\hat{\mathcal{S}}$ respectively denote the rest frames of the barn and the pole and assume that these frames are related by (1.3.27) and (1.3.29). Calculate β. Suppose the front of the pole enters the front of the barn at $(0,0,0,0)$. Now consider the two events at which the front of the pole hits the back of the barn and the back of the pole enters the front of the barn. Finally, think about the maximum speed at which the signal to stop can be communicated from the front to the back of the pole.

The underlying message of Exercise 1.3.26 would seem to be that the classical notion of a perfectly "rigid" body has no place in relativity, even as an idealization. The pole *must* compress since otherwise the signal to halt would proceed from the front to the back instantaneously and, in particular, the situation described in the exercise would, indeed, be "paradoxical", i.e., represent a logical inconsistency.

1.4 Timelike Vectors and Curves

Let us now consider in somewhat more detail a pair of events x_0 and x for which $x - x_0$ is timelike, i.e., $\mathcal{Q}(x - x_0) < 0$. Relative to any admissible basis $\{e_a\}$ we have $(\Delta x^1)^2 + (\Delta x^2)^2 + (\Delta x^3)^2 < (\Delta x^4)^2$. Clearly then, $\Delta x^4 \neq 0$ and we may assume without loss of generality that $\Delta x^4 > 0$, i.e., that $x - x_0$ is future-directed. Thus, we obtain

$$\frac{((\Delta x^1)^2 + (\Delta x^2)^2 + (\Delta x^3)^2)^{\frac{1}{2}}}{\Delta x^4} < 1.$$

Physically, it is therefore clear that if one were to move with speed

$$\frac{((\Delta x^1)^2 + (\Delta x^2)^2 + (\Delta x^3)^2)^{\frac{1}{2}}}{\Delta x^4}$$

relative to the frame $\mathcal{S}$ corresponding to $\{e_a\}$ along the line in $\sum$ from $(x_0^1,\ x_0^2,\ x_0^3)$ to $(x^1,\ x^2,\ x^3)$ and if one were present at x_0, then one would also experience x, i.e., that there is an admissible frame of reference $\hat{\mathcal{S}}$ in which x_0 and x occur at the same spatial point, one after the other. Specifically, we now prove that if one chooses $\beta = ((\Delta x^1)^2 + (\Delta x^2)^2 + (\Delta x^3)^2)^{1/2} / \Delta x^4$ and lets d^1, d^2 and d^3 be the direction cosines in $\sum$ of the directed line segment from $(x_0^1,\ x_0^2,\ x_0^3)$ to $(x^1,\ x^2,\ x^3)$, then the basis $\{\hat{e}_a\}$ for $\mathcal{M}$ obtained from $\{e_a\}$ by performing any Lorentz transformation whose fourth row is $\Lambda^4{}_i = -\beta(1-\beta^2)^{-1/2}\, d^i,\ i = 1,2,3,$ and $\Lambda^4{}_4 = (1-\beta^2)^{-1/2} = \gamma$, has the property that $\Delta\hat{x}^1 = \Delta\hat{x}^2 = \Delta\hat{x}^3 = 0$.

Exercise 1.4.1. There will, in general, be many Lorentz transformations with this fourth row. Show that defining the remaining entries by $\Lambda^i{}_4 = -\beta\gamma d^i,\ i = 1,2,3,$ and $\Lambda^i{}_j = (\gamma - 1)d^i d^j + \delta^i{}_j,\ i,j = 1,2,3$ ($\delta^i{}_j$ being the Kronecker delta) gives an element of $\mathcal{L}$.

To prove this we compute $\Delta\hat{x}^4 = \Lambda^4{}_b \Delta x^b$. To simplify the calculations we let $\Delta\vec{x} = ((\Delta x^1)^2 + (\Delta x^2)^2 + (\Delta x^3)^2)^{1/2}$. We may clearly assume that $\Delta\vec{x} \neq 0$ since otherwise there is nothing to prove. Thus, $\beta^2 = \Delta\vec{x}^2/(\Delta x^4)^2$, $\gamma = \Delta x^4/\sqrt{-\mathcal{Q}(x-x_0)}$, $\beta\gamma = \Delta\vec{x}/\sqrt{-\mathcal{Q}(x-x_0)}$ and $d^i = \Delta x^i/\Delta\vec{x}$ for $i = 1,2,3$. From (1.3.20) we therefore obtain

$$\begin{aligned}\Delta\hat{x}^4 &= -\frac{\Delta\vec{x}}{\sqrt{-\mathcal{Q}(x-x_0)}}(\Delta\vec{x}) + \frac{(\Delta x^4)^2}{\sqrt{-\mathcal{Q}(x-x_0)}} \\ &= \sqrt{-\mathcal{Q}(x-x_0)}.\end{aligned}$$

Consequently, $\mathcal{Q}(x-x_0) = -(\Delta\hat{x}^4)^2$. But, computing $\mathcal{Q}(x-x_0)$ relative to the basis $\{\hat{e}_a\}$ we find that $\mathcal{Q}(x-x_0) = (\Delta\hat{x}^1)^2+(\Delta\hat{x}^2)^2+(\Delta\hat{x}^3)^2-(\Delta\hat{x}^4)^2$ so we must have $(\Delta\hat{x}^1)^2+(\Delta\hat{x}^2)^2+(\Delta\hat{x}^3)^2 = 0$, i.e., $\Delta\hat{x}^1 = \Delta\hat{x}^2 = \Delta\hat{x}^3 = 0$ as required.

For any timelike vector v in $\mathcal{M}$ we define the *duration* $\tau(v)$ of v by $\tau(v) = \sqrt{-\mathcal{Q}(v)}$. If v is the displacement vector $v = x - x_0$ between two events x_0 and x, then, as we have just shown, $\tau(x-x_0)$ *is to be interpreted physically as the time separation of* x_0 *and* x *in any admissible frame of reference in which both events occur at the same spatial location.*

A subset of $\mathcal{M}$ of the form $\{x_0 + t(x-x_0) : t \in \mathbb{R}\}$, where $x - x_0$ is timelike, is called a *timelike straight line* in $\mathcal{M}$. A timelike straight line which passes through the origin is called a *time axis.* We show that the name is justified by proving that *if* T *is a time axis, then there exists an admissible basis* $\{\hat{e}_a\}$ *for* $\mathcal{M}$ *such that the subspace of* $\mathcal{M}$ *spanned by* $\hat{e}_4$ *is* T. To see this we select an event $\tilde{e}_4$ on T with $\tilde{e}_4 \cdot \tilde{e}_4 = -1$ and let Span $\{\tilde{e}_4\}$ be the linear span of $\tilde{e}_4$ in $\mathcal{M}$. Next let Span $\{\tilde{e}_4\}^\perp$ be the orthogonal complement of Span $\{\tilde{e}_4\}$ in $\mathcal{M}$. By Exercise 1.1.2, Span $\{\tilde{e}_4\}^\perp$ is also a subspace of $\mathcal{M}$. We claim that $\mathcal{M} = \text{Span}\{\tilde{e}_4\} \oplus \text{Span}\{\tilde{e}_4\}^\perp$ (recall that a vector space V is the direct sum of two subspaces W_1 and W_2 of V, written $V = W_1 \oplus W_2$, if $W_1 \cap W_2 = \{0\}$ and if every vector in V can be written as the sum of a vector in W_1 and a vector in W_2). Since every nonzero vector in Span $\{\tilde{e}_4\}$ is timelike, whereas, by Corollary 1.3.2, every nonzero vector in Span $\{\tilde{e}_4\}^\perp$ is spacelike, it is clear that these two subspaces intersect only in the zero vector. Next we let v denote an arbitrary vector in $\mathcal{M}$ and consider the vector $w = v + (v \cdot \tilde{e}_4)\tilde{e}_4$ in $\mathcal{M}$. Since $w \cdot \tilde{e}_4 = v \cdot \tilde{e}_4 + (v \cdot \tilde{e}_4)(\tilde{e}_4 \cdot \tilde{e}_4) = 0$ we find that w is in Span $\{\tilde{e}_4\}^\perp$. Thus, the expression $v = -(v \cdot \tilde{e}_4)\tilde{e}_4 + w$ completes the proof that $\mathcal{M} = \text{Span}\{\tilde{e}_4\} \oplus \text{Span}\{\tilde{e}_4\}^\perp$. Now, the restriction of the $\mathcal{M}$-inner product to Span $\{\tilde{e}_4\}^\perp$ is positive definite so, by Theorem 1.1.1, we may select three vectors $\tilde{e}_1$, $\tilde{e}_2$ and $\tilde{e}_3$ in Span $\{\tilde{e}_4\}^\perp$ such that $\tilde{e}_i \cdot \tilde{e}_j = \delta_{ij}$ for $i,j = 1,2,3$. Thus, $\{\tilde{e}_1, \tilde{e}_2, \tilde{e}_3, \tilde{e}_4\}$ is an orthonormal basis for $\mathcal{M}$. Now let us fix an admissible basis $\{e_a\}$ for $\mathcal{M}$. There is a unique orthogonal transformation of $\mathcal{M}$ that carries e_a onto $\tilde{e}_a$ for each $a = 1,2,3,4$. If the corresponding Lorentz transformation is either improper or nonorthochronous or both we may multiply $\tilde{e}_1$ or $\tilde{e}_4$ or both by -1 to obtain an admissible basis $\{\hat{e}_a\}$ for $\mathcal{M}$ with $\text{Span}\{\hat{e}_4\} = T$ and so the proof is complete. Any time axis is therefore the x^4-axis of some admissible coordinatization of $\mathcal{M}$ and so may be identified with the worldline of some admissible observer. Since any timelike straight line is parallel to some time axis we view such a straight line as the worldline of a point at rest in the corresponding admissible frame (say, the worldline of one of the "assistants" to our observer).

Exercise 1.4.2. Show that if T is a time axis and x and x_0 are two events, then $x - x_0$ is orthogonal to T if and only if x and x_0 are simultaneous in any reference frame whose x^4-axis is T.

Exercise 1.4.3. Show that if $x - x_0$ is timelike and s is an arbitrary non-negative real number, then there is an admissible frame of reference in which the spatial separation of x and x_0 is s. Show also that the time separation of x and x_0 can assume any real value greater than or equal to $\tau(x - x_0)$. *Hint:* Begin with a basis $\{e_a\}$ in which $\Delta x^1 = \Delta x^2 = \Delta x^3 = 0$ and $\Delta x^4 = \tau(x - x_0)$. Now perform the special Lorentz transformation (1.3.27), where $-1 < \beta < 1$ is arbitrary.

According to Exercise 1.4.3, $\tau(x - x_0)$ is a lower bound for the temporal separation of x_0 and x and, for this reason, it is often called the *proper time separation* of x_0 and x; when no reference to the specific events under consideration is required $\tau(x - x_0)$ is generally denoted $\Delta\tau$.

Any timelike vector v lies along some time axis so $\tau(v)$ can be regarded as a sort of "temporal length" of v (the time separation of its tail and tip as recorded by an observer who experiences both). It is a rather unusual notion of length, however, since the analogues of the basic inequalities one is accustomed to dealing with for Euclidean lengths are generally reversed.

Theorem 1.4.1. *(Reversed Schwartz Inequality) If v and w are timelike vectors in $\mathcal{M}$, then*

$$(v \cdot w)^2 \geq v^2 w^2 \tag{1.4.1}$$

and equality holds if and only if v and w are linearly dependent.

Proof: Consider the vector $u = av - bw$, where $a = v \cdot w$ and $b = v \cdot v = v^2$. Observe that $u \cdot v = av^2 - bv \cdot w = v^2(v \cdot w) - v^2(v \cdot w) = 0$. Since v is timelike, Corollary 1.3.2 implies that u is either zero or spacelike. Thus, $0 \leq u^2 = a^2v^2 + b^2w^2 - 2abv \cdot w$, with equality holding only if $u = 0$.

Consequently, $2abv \cdot w \le a^2v^2 + b^2w^2$, i.e.,

$$\begin{aligned} 2v^2\,(v\cdot w)^2 &\le v^2(v\cdot w)^2 + (v^2)^2\,w^2, \\ 2(v\cdot w)^2 &\ge (v\cdot w)^2 + v^2w^2 \quad (\text{since } v^2 < 0), \\ (v\cdot w)^2 &\ge v^2w^2, \end{aligned}$$

and equality holds only if $u = 0$. But $u = 0$ implies $av - bw = 0$ which, since $a = v \cdot w \ne 0$ by Theorem 1.3.1, implies that v and w are linearly dependent. Conversely, if v and w are linearly dependent, then one is a multiple of the other and equality clearly holds in (1.4.1). ∎

Theorem 1.4.2. *(Reversed Triangle Inequality) Let v and w be timelike vectors with the same time orientation (i.e., $v \cdot w < 0$). Then*

$$\tau(v+w) \ge \tau(v) + \tau(w) \tag{1.4.2}$$

and equality holds if and only if v and w are linearly dependent.

Proof: By Theorem 1.4.1, $(v\cdot w)^2 \ge v^2w^2 = (-v^2)(-w^2)$ so $|\, v\cdot w \,| \ge \sqrt{-v^2}\,\sqrt{-w^2}$. But $(v\cdot w) < 0$ so we must have $v\cdot w \le -\sqrt{-v^2}\,\sqrt{-w^2}$ and therefore

$$-2v\cdot w \ge 2\sqrt{-v^2}\,\sqrt{-w^2}\,. \tag{1.4.3}$$

Now, by Exercise 1.3.2, $v + w$ is timelike. Moreover, $-(v+w)^2 = -v^2 - 2v\cdot w - w^2 \ge -v^2 + 2\sqrt{-v^2}\,\sqrt{-w^2} - w^2$ by (1.4.3). Thus,

$$\begin{aligned} -(v+w)^2 &\ge (\sqrt{-v^2} + \sqrt{-w^2}\,)^2, \\ \sqrt{-(v+w)^2} &\ge \sqrt{-v^2} + \sqrt{-w^2}, \\ \sqrt{-\mathcal{Q}(v+w)} &\ge \sqrt{-\mathcal{Q}(v)} + \sqrt{-\mathcal{Q}(w)}, \\ \tau(v+w) &\ge \tau(v) + \tau(w), \end{aligned}$$

as required. If equality holds in (1.4.2), then, by reversing the preceding steps, we obtain

$$-2v\cdot w = 2\sqrt{-v^2}\,\sqrt{-w^2}$$

and therefore $(v\cdot w)^2 = v^2w^2$ so, by Theorem 1.4.1, v and w are linearly dependent. ∎

To extend Theorem 1.4.2 to arbitrary finite sums of similarly oriented timelike vectors (and for other purposes as well) we require:

Lemma 1.4.3. *The sum of any finite number of vectors in $\mathcal{M}$ all of which are timelike or null and all future-directed (resp., past-directed) is timelike and future-directed (resp., past-directed) except when all of the vectors are null and parallel, in which case the sum is null and future-directed (resp., past-directed).*

Proof: It suffices to prove the result for future-directed vectors since the corresponding result for past-directed vectors will then follow by changing signs. Moreover, it is clear that any sum of future-directed vectors is, indeed, future-directed.

First we observe that if v_1 and v_2 are timelike and future-directed, then $v_1 \cdot v_1 < 0$, $v_2 \cdot v_2 < 0$ and $v_1 \cdot v_2 < 0$ so $(v_1 + v_2) \cdot (v_1 + v_2) = v_1 \cdot v_1 + 2v_1 \cdot v_2 + v_2 \cdot v_2 < 0$ and therefore $v_1 + v_2$ is timelike.

Exercise 1.4.4. Show that if v_1 is timelike, v_2 is null and both are future-directed, then $v_1 + v_2$ is timelike and future-directed.

Next suppose that v_1 and v_2 are null and future-directed. We show that $v_1 + v_2$ is timelike unless v_1 and v_2 are parallel (in which case, it is obviously null). To this end we note that $(v_1 + v_2) \cdot (v_1 + v_2) = 2v_1 \cdot v_2$. By Theorem 1.2.1, $v_1 \cdot v_2 = 0$ if and only if v_1 and v_2 are parallel. Suppose then that v_1 and v_2 are not parallel. Fix an admissible basis $\{e_a\}$ for $\mathcal{M}$ and let $v_1 = v_1^a e_a$ and $v_2 = v_2^a e_a$. For each $n = 1, 2, 3, \ldots$, define w_n in $\mathcal{M}$ by $w_n = v_1^1 e_1 + v_1^2 e_2 + v_1^3 e_3 + (v_1^4 + \frac{1}{n}) e_4$. Then each w_n is timelike and future-directed. By Theorem 1.3.1, $0 > w_n \cdot v_2 = v_1 \cdot v_2 - \frac{1}{n} v_2^4$, i.e., $v_1 \cdot v_2 < \frac{1}{n} v_2^4$ for every n. Thus, $v_1 \cdot v_2 \leq 0$. But $v_1 \cdot v_2 \neq 0$ by assumption so $v_1 \cdot v_2 < 0$ and therefore $(v_1 + v_2) \cdot (v_1 + v_2) < 0$ as required.

Exercise 1.4.5. Complete the proof by induction. ∎

Corollary 1.4.4. *Let $v_1, \ldots, v_n$ be timelike vectors, all with the same time orientation. Then*

$$\tau(v_1 + v_2 + \cdots + v_n) \geq \tau(v_1) + \tau(v_2) + \cdots + \tau(v_n) \tag{1.4.4}$$

and equality holds if and only if $v_1, v_2, \ldots, v_n$ are all parallel.

Proof: Inequality (1.4.4) is clear from Theorem 1.4.2 and Lemma 1.4.3. We show, by induction on n, that equality in (1.4.4) implies that $v_1, \ldots, v_n$ are all parallel. For $n = 2$ this is just Theorem 1.4.2. Thus, we assume that the statement is true for sets of n vectors and consider a set $v_1, \ldots, v_n, v_{n+1}$ of timelike vectors which are, say, future-directed and for which

$$\tau(v_1 + \cdots + v_n + v_{n+1}) = \tau(v_1) + \cdots + \tau(v_n) + \tau(v_{n+1}).$$

$v_1 + \cdots + v_n$ is timelike and future-directed so, again by Theorem 1.4.2,

$$\tau(v_1 + \cdots + v_n) + \tau(v_{n+1}) \leq \tau(v_1) + \cdots + \tau(v_n) + \tau(v_{n+1}).$$

We claim that, in fact, equality must hold here. Indeed, otherwise we have $\tau(v_1 + \cdots + v_n) < \tau(v_1) + \cdots + \tau(v_n)$ and so (Theorem 1.4.2 again) $\tau(v_1 + \cdots + v_{n-1}) < \tau(v_1) + \cdots + \tau(v_{n-1})$. Continuing the process we eventually conclude that $\tau(v_1) < \tau(v_1)$ which is a contradiction. Thus,

$$\tau(v_1 + \cdots + v_n) = \tau(v_1) + \cdots + \tau(v_n)$$

and the induction hypothesis implies that $v_1, \ldots, v_n$ are all parallel. Let $v = v_1 + \cdots + v_n$. Then v is timelike and future-directed. Thus, $\tau(v + v_{n+1}) = \tau(v) + \tau(v_{n+1})$ and one more application of Theorem 1.4.2 implies that v_{n+1} is parallel to v and therefore to all of $v_1, \ldots, v_n$ and the proof is complete. ■

Corollary 1.4.5. *Let v and w be two nonparallel null vectors. Then v and w have the same time orientation if and only if $v \cdot w < 0$.*

Proof: Suppose first that v and w have the same time orientation. By Lemma 1.4.3, $v + w$ is timelike so $0 > (v + w) \cdot (v + w) = 2v \cdot w$ so $v \cdot w < 0$. Conversely, if v and w have opposite time orientation, then v and $-w$ have the same time orientation so $v \cdot (-w) < 0$ and therefore $v \cdot w > 0$. ■

The reason that the sense of the inequality in Theorem 1.4.2 is "reversed" becomes particularly transparent by choosing a coordinate system relative to which $v = (v^1, v^2, v^3, v^4)$, $w = (w^1, w^2, w^3, w^4)$ and $v + w = (0,\ 0,\ 0,\ v^4 + w^4)$ (this simply amounts to taking the time axis through $v + w$ as the x^4-axis). For then $\tau(v) = ((v^4)^2 - (v^1)^2 - (v^2)^2 - (v^3)^2)^{\frac{1}{2}} < v^4$ and $\tau(w) < w^4$, but $\tau(v + w) = v^4 + w^4$.

A timelike straight line is regarded as the worldline of a material particle that is "free" in the sense of Newtonian mechanics and consequently is at rest in some admissible frame of reference. Not all material particles of interest have this property (e.g., the "rocket twin"). To model these in $\mathcal{M}$ we will require a few preliminaries. Let $I \subseteq \mathbb{R}$ be an open interval. A map $\alpha : I \to \mathcal{M}$ is a *curve* in $\mathcal{M}$. Relative to any admissible basis $\{e_a\}$ for $\mathcal{M}$ we can write $\alpha(t) = x^a(t)\, e_a$ for each t in I. We will assume that α is *smooth,* i.e., that each component function $x^a(t)$ is infinitely differentiable and that α's *velocity vector*

$$\alpha'(t) = \frac{dx^a}{dt}\, e_a$$

is nonzero for each t in I.

Exercise 1.4.6. Show that this definition of smoothness does not depend on the choice of admissible basis. *Hint:* Let $\{\hat{e}_a\}$ be another admissible basis, L the orthogonal transformation that carries e_a onto $\hat{e}_a$ for $a = 1, 2, 3, 4$ and $[\Lambda^a{}_b]$ the corresponding element of $\mathcal{L}$. If $\alpha(t) = \hat{x}^a(t)\,\hat{e}_a$, then $\hat{x}^a(t) = \Lambda^a{}_b\, x^b(t)$ so $\frac{d\hat{x}^a}{dt} = \Lambda^a{}_b \frac{dx^b}{dt}$. Keep in mind that $[\Lambda^a{}_b]$ is nonsingular.

A curve $\alpha : I \to \mathcal{M}$ is said to be *spacelike, timelike* or *null* respectively if its velocity vector $\alpha'(t)$ has that character for every t in I, that is, if $\alpha'(t) \cdot \alpha'(t)$ is > 0, < 0 or $= 0$ respectively for each t. A timelike or null curve α is *future-directed* (resp., *past-directed*) if $\alpha'(t)$ is future-directed (resp., past-directed) for each t. A future-directed timelike curve is called a *timelike worldline* or *worldline of a material particle*. We extend all of these definitions to the case in which I contains either or both of its endpoints by requiring that $\alpha : I \to \mathcal{M}$ be extendible to an open interval containing I. More precisely, if I is an (not necessarily open) interval in $\mathbb{R}$, then $\alpha : I \to \mathcal{M}$ is *smooth, spacelike, ...* if there exists an open interval $\tilde{I}$ containing I and a curve $\tilde{\alpha} : \tilde{I} \to \mathcal{M}$ which is smooth, spacelike, ... and satisfies $\tilde{\alpha}(t) = \alpha(t)$ for each t in I. Generally, we will drop the tilda and use the same symbol for α and its extension.

If $\alpha : I \to \mathcal{M}$ is a curve and $J \subseteq \mathbb{R}$ is another interval and $h : J \to I$, $t = h(s)$, is an infinitely differentiable function with $h'(s) > 0$ for each s in J, then the curve $\beta = \alpha \circ h : J \to \mathcal{M}$ is called a *reparametrization* of α.

Exercise 1.4.7. Show that $\beta'(s) = h'(s)\alpha'(h(s))$ and conclude that all of the definitions we have given are independent of parametrization.

We arrive at a particularly convenient parametrization of a timelike worldline in the following way: If $\alpha : [a, b] \to \mathcal{M}$ is a timelike worldline in $\mathcal{M}$ we define the *proper time length* of α by

$$L(\alpha) = \int_a^b |\,\alpha'(t) \cdot \alpha'(t)\,|^{\frac{1}{2}}\, dt = \int_a^b \sqrt{-\eta_{ab}\,\frac{dx^a}{dt}\,\frac{dx^b}{dt}}\; dt.$$

Exercise 1.4.8. Show that the definition of $L(\alpha)$ is independent of parametrization.

As the appropriate physical interpretation of $L(\alpha)$ we take

The Clock Hypothesis: *If $\alpha : [a, b] \to \mathcal{M}$ is a timelike worldline in $\mathcal{M}$, then $L(\alpha)$ is interpreted as the time lapse between the events $\alpha(a)$ and $\alpha(b)$ as measured by an ideal standard clock carried along by the particle whose worldline is represented by α.*

The motivation for the Clock Hypothesis is at the same time "obvious" and subtle. For it we shall require the following theorem which asserts that two events can be experienced by a single admissible observer if and only if some (not necessarily free) material particle has both on its worldline.

Theorem 1.4.6. *Let p and q be two points in $\mathcal{M}$. Then $p - q$ is timelike and future-directed if and only if there exists a smooth, future-directed timelike curve $\alpha\colon [a,b] \to \mathcal{M}$ such that $\alpha(a) = q$ and $\alpha(b) = p$.*

We postpone the proof for a moment to show its relevance to the Clock Hypothesis. We partition the interval $[a,b]$ into subintervals by $a = t_0 < t_1 < \cdots < t_{n-1} < t_n = b$. Then, by Theorem 1.4.6, each of the displacement vectors $v_i = \alpha(t_i) - \alpha(t_{i-1})$ is timelike and future-directed. $\tau(v_i)$ is then interpreted as the time lapse between $\alpha(t_{i-1})$ and $\alpha(t_i)$ as measured by an admissible observer who is present at both events. If the "material particle" whose worldline is represented by α has constant velocity between the events $\alpha(t_{i-1})$ and $\alpha(t_i)$, then $\tau(v_i)$ would be the time lapse between these events as measured by a clock carried along by the particle. Relative to any admissible frame,

$$\tau(v_i) = \sqrt{-\eta_{ab}\Delta x_i^a \Delta x_i^b} = \sqrt{-\eta_{ab}\,\frac{\Delta x_i^a}{\Delta t_i}\,\frac{\Delta x_i^b}{\Delta t_i}}\,\Delta t_i .$$

By choosing Δt_i sufficiently small, Δx_i^4 can be made small (by continuity of α) and, since the speed of the particle relative to our frame of reference is "nearly" constant over "small" x^4-time intervals, $\tau(v_i)$ should be a good approximation to the time lapse between $\alpha(t_{i-1})$ and $\alpha(t_i)$ measured by the material particle. Consequently, the sum

$$\sum_{i=1}^{n} \sqrt{-\eta_{ab}\,\frac{\Delta x_i^a}{\Delta t_i}\,\frac{\Delta x_i^b}{\Delta t_i}}\;\Delta t_i \tag{1.4.5}$$

approximates the time lapse between $\alpha(a)$ and $\alpha(b)$ that this particle measures. The approximations become better as the Δt_i approach 0 and, in the limit, the sum (1.4.5) approaches the definition of $L(\alpha)$.

The argument seems persuasive enough, but it clearly rests on an assumption about the behavior of ideal clocks that we had not previously made explicit, namely, that acceleration as such has no effect on their rates, i.e., that the "instantaneous rate" of such a clock depends only on its instantaneous speed and not on the rate at which this speed is changing. Justifying such an assumption is a nontrivial matter. One must perform experiments with various types of clocks subjected to real accelerations and, in the end, will no doubt be forced to a more modest proposal ("The Clock Hypothesis is valid for such and such a clock over such and such a range of accelerations").

For the proof of Theorem 1.4.6 we will require the following preliminary result.

Lemma 1.4.7. *Let $\alpha : (A, B) \to \mathcal{M}$ be smooth, timelike and future-directed and fix a t_0 in (A, B). Then there exists an $\varepsilon > 0$ such that $(t_0 - \varepsilon, t_0 + \varepsilon)$ is contained in (A, B), $\alpha(t)$ is in the past time cone at $\alpha(t_0)$ for every t in $(t_0 - \varepsilon, t_0)$ and $\alpha(t)$ is in the future time cone at $\alpha(t_0)$ for every t in $(t_0, t_0 + \varepsilon)$.*

Proof: We prove that there exists an $\varepsilon_1 > 0$ such that $\alpha(t)$ is in $\mathcal{C}_T^+(\alpha(t_0))$ for each t in $(t_0, t_0 + \varepsilon_1)$. The argument to produce an $\varepsilon_2 > 0$ with $\alpha(t)$ in $\mathcal{C}_T^-(\alpha(t_0))$ for each t in $(t_0 - \varepsilon_2, t_0)$ is similar. Taking ε to be the smaller of ε_1 and ε_2 proves the lemma.

Fix an admissible basis $\{e_a\}$ and write $\alpha(t) = x^a(t)e_a$ for $A < t < B$. Now suppose that no such ε_1 exists. Then one can produce a sequence $t_1 > t_2 > \cdots > t_0$ in $(t_0,\ B)$ such that $\lim_{n\to\infty} t_n = t_0$ and such that one of the following is true:

(I) $\mathcal{Q}(\alpha(t_n) - \alpha(t_0)) \geq 0$ for all n [i.e., $\alpha(t_n) - \alpha(t_0)$ is spacelike or null for every n], or

(II) $\mathcal{Q}(\alpha(t_n) - \alpha(t_0)) < 0$, but $\alpha(t_n) - \alpha(t_0)$ is past-directed for every n [i.e., $\alpha(t_n)$ is in $\mathcal{C}_T^-(\alpha(t_0))$ for every n].

We show first that (I) is impossible. Suppose to the contrary that such a sequence does exist. Then

$$\mathcal{Q}\left(\frac{\alpha(t_n) - \alpha(t_0)}{t_n - t_0}\right) \geq 0$$

for all n so

$$\mathcal{Q}\left(\frac{x^1(t_n) - x^1(t_0)}{t_n - t_0}, \ldots, \frac{x^4(t_n) - x^4(t_0)}{t_n - t_0}\right) \geq 0.$$

Thus,

$$\lim_{n\to\infty} \mathcal{Q}\left(\frac{x^1(t_n) - x^1(t_0)}{t_n - t_0}, \ldots, \frac{x^4(t_n) - x^4(t_0)}{t_n - t_0}\right) \geq 0,$$

$$\mathcal{Q}\left(\lim_{n\to\infty} \frac{x^1(t_n) - x^1(t_0)}{t_n - t_0}, \ldots, \lim_{n\to\infty} \frac{x^4(t_n) - x^4(t_0)}{t_n - t_0}\right) \geq 0,$$

$$\mathcal{Q}\left(\frac{dx^1}{dt}(t_0), \ldots, \frac{dx^4}{dt}(t_0)\right) \geq 0,$$

$$\mathcal{Q}(\alpha'(t_0)) \geq 0,$$

and this contradicts the fact that $\alpha'(t_0)$ is timelike.

Exercise 1.4.9. Apply a similar argument to $g(\alpha(t_n) - \alpha(t_0),\ \alpha'(t_0))$ to show that (II) is impossible.

We therefore infer the existence of the ε_1 as required and the proof is complete. ■

Proof of Theorem 1.4.6: The necessity is clear. To prove the sufficiency we denote by α also a smooth, future-directed timelike extension of α to some interval (A, B) containing $[a, b]$. By Lemma 1.4.7, there exists an $\varepsilon_1 > 0$ with $(a,\ a + \varepsilon_1) \subseteq (A, B)$ and such that $\alpha(t)$ is in $\mathcal{C}_T^+(q)$ for each t in $(a,\ a+\varepsilon_1)$. Let t_0 be the supremum of all such ε_1. Since $b < B$ it will suffice to show that $t_0 = B$ and for this we assume to the contrary that $A < t_0 < B$.

According to Lemma 1.4.7 there exists an $\varepsilon > 0$ such that $(t_0 - \varepsilon, t_0 + \varepsilon) \subseteq (A, B)$, $\alpha(t) \in \mathcal{C}_T^-(\alpha(t_0))$ for t in $(t_0 - \varepsilon,\ t_0)$ and $\alpha(t) \in \mathcal{C}_T^+(\alpha(t_0))$ for t in $(t_0,\ t_0 + \varepsilon)$. Observe that if $\alpha(t_0)$ were itself in $\mathcal{C}_T^+(q)$, then for any t in $(t_0, t_0 + \varepsilon)$, $(\alpha(t_0) - q) + (\alpha(t) - \alpha(t_0)) = \alpha(t) - q$ would be future-directed and timelike by Lemma 1.4.3 and this contradicts the definition of t_0. On the other hand, if $\alpha(t_0)$ were outside the null cone at q, then for some t's in $(t_0 - \varepsilon,\ t_0)$, $\alpha(t)$ would be outside the null cone at q and this is impossible since, again by the definition of t_0, any such $\alpha(t)$ is in $\mathcal{C}_T^+(q)$. The only remaining possibility is that $\alpha(t_0)$ is on the null cone at q. But then the past time cone at $\alpha(t_0)$ is disjoint from the future time cone at q and any t in $(t_0 - \varepsilon,\ t_0)$ gives a contradiction. We conclude that t_0 must be equal to B and the proof is complete. ■

As promised we now deliver what is for most purposes the most useful parametrization of a timelike worldline $\alpha\colon I \to \mathcal{M}$. First let us appeal to Exercises 1.4.7 and 1.4.8 and translate the domain of α in the real line if necessary to assume that it contains 0. Now define the *proper time function* $\tau(t)$ on I by

$$\tau = \tau(t) = \int_0^t |\ \alpha'(u) \cdot \alpha'(u)\ |^{\frac{1}{2}}\ du.$$

Thus, $\frac{d\tau}{dt} = |\ \alpha'(t) \cdot \alpha'(t)\ |^{1/2}$ which is positive and infinitely differentiable since α is timelike. The inverse $t = h(\tau)$ therefore exists and $\frac{dh}{d\tau} = (\frac{d\tau}{dt})^{-1} > 0$ so we conclude that τ is a legitimate parameter along α (physically, we are simply parametrizing α by time readings actually recorded along α). We shall abuse our notation somewhat and use the same name for α and its coordinate functions relative to an admissible basis when they are parametrized by τ rather than t:

$$\alpha(\tau) = x^a(\tau) e_a\,. \tag{1.4.6}$$

Exercise 1.4.10. Define $\alpha: \mathbb{R} \to \mathcal{M}$ by $\alpha(t) = x_0 + t(x - x_0)$, where $\mathcal{Q}(x - x_0) < 0$ and t is in $\mathbb{R}$. Show that $\tau = \tau(x - x_0)t$ and write down the proper time parametrization of α.

The velocity vector $\alpha'(\tau) = \frac{dx^a}{d\tau} e_a$ of α is called the *world velocity* (or *4-velocity*) of α and denoted $U = U^a e_a$. Just as the familiar arc length parametrization of a curve in $\mathbb{R}^3$ has unit speed, so the world velocity of a timelike worldline is always a unit timelike vector.

Exercise 1.4.11. Show that

$$U \cdot U = -1 \tag{1.4.7}$$

at each point along α.

The second proper time derivative $\alpha''(\tau) = \frac{d^2x^a}{d\tau^2} e_a$ of α is called the *world acceleration* (or *4-acceleration*) of α and denoted $A = A^a e_a$. It is always orthogonal to U and so, in particular, must be spacelike if it is nonzero.

Exercise 1.4.12. Show that

$$U \cdot A = 0 \tag{1.4.8}$$

at each point along α. *Hint:* Differentiate (1.4.7) with respect to τ.

The world velocity and acceleration of a timelike worldline are, as we shall see, crucial to an understanding of the dynamics of the particle whose worldline is represented by α. A given admissible observer, however, is more likely to parametrize a particle's worldline by his time x^4 than by τ and so will require procedures for calculating U and A from this parametrization. First observe that since $\alpha(\tau) = (x^1(\tau), \ldots, x^4(\tau))$ is smooth, $x^4(\tau)$ is infinitely differentiable. Since α is future-directed, $\frac{dx^4}{d\tau}$ is positive so the inverse $\tau = h(x^4)$ exists and $h'(x^4) = (\frac{dx^4}{d\tau})^{-1}$ is positive. Thus, x^4 is a legitimate parameter for α. Moreover,

$$\begin{aligned}
\frac{d\tau}{dx^4} &= \left|\alpha'(x^4) \cdot \alpha'(x^4)\right|^{\frac{1}{2}} \\
&= \sqrt{1 - \left[(\frac{dx^1}{dx^4})^2 + (\frac{dx^2}{dx^4})^2 + (\frac{dx^3}{dx^4})^2 \right]} \\
&= \sqrt{1 - \beta^2(x^4)}\,,
\end{aligned}$$

where we have denoted by $\beta(x^4)$ the usual instantaneous speed of the particle whose worldline is α relative to the frame $\mathcal{S}(x^1, x^2, x^3, x^4)$. Thus,

$$\frac{dx^4}{d\tau} = (1 - \beta^2(x^4))^{-\frac{1}{2}}$$

which we denote by $\gamma = \gamma(x^4)$. Now, we compute

$$U^i = \frac{dx^i}{d\tau} = \frac{dx^i}{dx^4}\frac{dx^4}{d\tau} = \gamma\frac{dx^i}{dx^4}, \quad i = 1, 2, 3,$$

and

$$U^4 = \gamma,$$

so

$$U = U^a e_a = \gamma\frac{dx^1}{dx^4}\, e_1 + \gamma\frac{dx^2}{dx^4}\, e_2 + \gamma\frac{dx^3}{dx^4}\, e_3 + \gamma e_4$$

which it is often more convenient to write as

$$\left(U^1, U^2, U^3, U^4\right) = \gamma\left(\frac{dx^1}{dx^4}, \frac{dx^2}{dx^4}, \frac{dx^3}{dx^4}, 1\right) \tag{1.4.9}$$

or, even more compactly as

$$\left(U^1, U^2, U^3, U^4\right) = \gamma(\vec{u}, 1), \tag{1.4.10}$$

where $\vec{u}$ is the ordinary velocity 3-vector of α in $\mathcal{S}$. Similarly, one computes

$$A^i = \gamma\frac{d}{dx^4}\left(\gamma\frac{dx^i}{dx^4}\right), \quad i = 1, 2, 3,$$

and

$$A^4 = \gamma\frac{d}{dx^4}(\gamma),$$

so that

$$\left(A^1, A^2, A^3, A^4\right) = \gamma\frac{d}{dx^4}(\gamma\vec{u}, \gamma). \tag{1.4.11}$$

Exercise 1.4.13. Using (in this exercise only) a dot to indicate differentiation with respect to x^4 and $E\colon \mathbb{R}^3 \times \mathbb{R}^3 \to \mathbb{R}$ for the usual positive definite inner product on $\mathbb{R}^3$, prove each of the following in an arbitrary admissible frame of reference $\mathcal{S}$:

(a) $\dot{\gamma} = \gamma^3\beta\dot{\beta}$.

(b) $E(\vec{u}, \vec{u}) = |\vec{u}|^2 = \beta^2$.

(c) $E(\vec{u}, \dot{\vec{u}}) = E(\vec{u}, \vec{a}) = \beta\dot{\beta}$ ($\vec{a} = \dot{\vec{u}}$ is the usual 3-acceleration in $\mathcal{S}$).

(d) $g(A, A) = \gamma^4 E(\vec{a}, \vec{a}) + \gamma^6\beta^2(\dot{\beta})^2 = \gamma^4|\vec{a}|^2 + \gamma^6\beta^2(\dot{\beta})^2$.

At each fixed point $\alpha(\tau_0)$ along the length of a timelike worldline α, $U(\tau_0)$ is a future-directed unit timelike vector and so may be taken as the timelike vector e_4 in some admissible basis for $\mathcal{M}$. Relative to such a basis, $U(\tau_0) = (0,0,0,1)$. Letting $x_0^4 = x^4(\tau_0)$ we find from (1.4.9) that

$$\left(\frac{dx^i}{dx^4}\right)_{x^4=x_0^4} = 0, \quad i = 1,2,3,$$

and so $\beta(x_0^4) = 0$ and $\gamma(x_0^4) = 1$. The reference frame corresponding to such a basis is therefore thought of as being "momentarily ($x^4 = x_0^4$) at rest" relative to the particle whose worldline is α. Any such frame of reference is called an *instantaneous rest frame for* α *at* $\alpha(\tau_0)$. Notice that Exercise 1.4.12(d) gives

$$g(A,A) = |\vec{a}|^2 \tag{1.4.12}$$

in an instantaneous rest frame. Since $g(A,A)$ is invariant under Lorentz transformations we find that all admissible observers will agree, at each point along α, on the magnitude of the 3-acceleration of α relative to its instantaneous rest frames.

As an illustration of these ideas we will examine in some detail the following situation. A futuristic explorer plans a journey to a distant part of the universe. For the sake of comfort he will maintain a constant acceleration of $1g$ (one "earth gravity") relative to his instantaneous rest frames (assuming that he neither diets nor overindulges his "weight" will remain the same as on earth throughout the trip). We begin by calculating the explorer's worldline $\alpha(\tau)$. As usual we denote by $U(\tau)$ and $A(\tau)$ the world velocity and world acceleration of α respectively. Thus, (1.4.7), (1.4.8) and (1.4.12) give

$$U \cdot U = -1, \tag{1.4.13}$$

$$U \cdot A = 0, \tag{1.4.14}$$

$$A \cdot A = g^2 \quad \text{(a constant)}. \tag{1.4.15}$$

We examine the situation from an admissible frame of reference in which the explorer's motion is along the positive x^1-axis. Thus, $U^2 = U^3 = A^2 = A^3 = 0$ and (1.4.13), (1.4.14) and (1.4.15) become

$$(U^1)^2 - (U^4)^2 = -1, \tag{1.4.16}$$

$$U^1 A^1 - U^4 A^4 = 0, \tag{1.4.17}$$

$$(A^1)^2 - (A^4)^2 = g^2. \tag{1.4.18}$$

Exercise 1.4.14. Solve these last three equations for A^1 and A^4 to obtain $A^1 = gU^4$ and $A^4 = gU^1$.

The result of Exercise 1.4.14 is a system of ordinary differential equations for U^1 and U^4. Specifically, we have

$$\frac{dU^1}{d\tau} = gU^4 \tag{1.4.19}$$

and

$$\frac{dU^4}{d\tau} = gU^1. \tag{1.4.20}$$

Differentiate (1.4.19) with respect to τ and substitute into (1.4.20) to obtain

$$\frac{d^2U^1}{d\tau^2} = g^2U^1. \tag{1.4.21}$$

The general solution to (1.4.21) can be written

$$U^1 = U^1(\tau) = a \sinh g\tau + b \cosh g\tau.$$

Assuming that the explorer accelerates from rest at $\tau = 0$ ($U^1(0) = 0$, $A^1(0) = g$) one obtains

$$U^1(\tau) = \sinh g\tau. \tag{1.4.22}$$

Equation (1.4.19) now gives

$$U^4(\tau) = \cosh g\tau. \tag{1.4.23}$$

Integrating (1.4.22) and (1.4.23) and assuming, for convenience, that $x^1(0) = 1/g$ and $x^4(0) = 0$, one obtains

$$\begin{cases} x^1 = \dfrac{1}{g} \cosh g\tau, \\[2ex] x^4 = \dfrac{1}{g} \sinh g\tau. \end{cases} \tag{1.4.24}$$

Observe that (1.4.24) implies that $(x^1)^2 - (x^4)^2 = 1/g^2$ so that our explorer's worldline lies on a hyperbola in the 2-dimensional representation of $\mathcal{M}$ (see Figure 1.4.1).

Exercise 1.4.15. Assume that the explorer's point of departure (at $x^1 = 1/g$) was the earth, which is at rest in the frame of reference under consideration. How far from the earth (as measured in the earth's frame) will the explorer be after

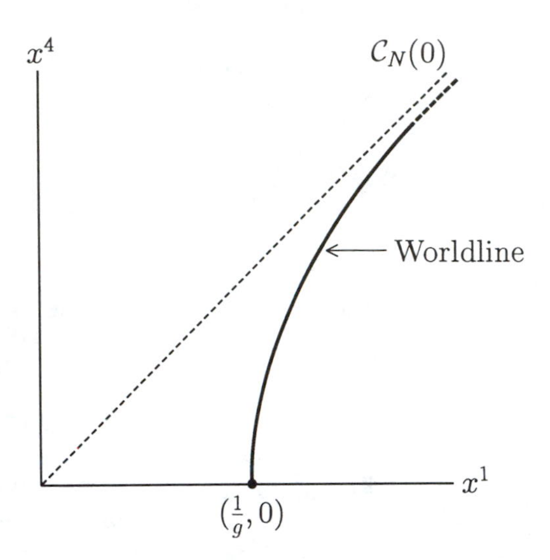

Figure 1.4.1

(a) 40 years as measured on earth? (How much time will have elapsed on the rocket?) *Answers:* 39 light years (4.38 years).

(b) 40 years as measured on the rocket? (How much time will have elapsed on earth?) *Answers:* 10^{17} light years (10^{17} years).

Hint: It will simplify the arithmetic to measure in light years rather than meters. Then $g \approx 1$ (light year)$^{-1}$.

We conclude this section with a theorem which asserts quite generally that an accelerated observer such as the explorer in the preceding discussion of hyperbolic motion or the "rocket twin" in the twin paradox must always experience a time dilation not experienced by those of us who remain at rest in an admissible frame.

Theorem 1.4.8. *Let* $\alpha\colon [a,b] \to \mathcal{M}$ *be a timelike worldline in* $\mathcal{M}$ *from* $\alpha(a) = q$ *to* $\alpha(b) = p$. *Then*

$$L(\alpha) \leq \tau(p - q) \tag{1.4.25}$$

and equality holds if and only if α *is a parametrization of a timelike straight line joining* q *and* p.

Proof: By Theorem 1.4.6, $p - q$ is timelike and future-directed so we may select a basis $\{e_a\}$ with $q = x_0^1\, e_1 + x_0^2\, e_2 + x_0^3\, e_3 + x_q^4\, e_4$, $p = x_0^1\, e_1 +$

$x_0^2 e_2 + x_0^3 e_3 + x_p^4 e_4$ and $\tau(p-q) = x_p^4 - x_q^4 = \Delta x^4$. Now parametrize α by x^4. Then

$$\begin{aligned} L(\alpha) &= \int_{x_q^4}^{x_p^4} \sqrt{1 - \left[(\frac{dx^1}{dx^4})^2 + (\frac{dx^2}{dx^4})^2 + (\frac{dx^3}{dx^4})^2 \right]}\, dx^4 \\ &\leq \int_{x_q^4}^{x_p^4} dx^4 = \Delta x^4 = \tau(p-q). \end{aligned}$$

Moreover, equality holds if and only if $\frac{dx^i}{dx^4} = 0$ for $i = 1,2,3$, that is, if and only if x^i is constant for $i = 1,2,3$ and this is the case if and only if $\alpha(x^4) = x_0^1 e_1 + x_0^2 e_2 + x_0^3 e_3 + x^4 e_4$ for $x_q^4 \leq x^4 \leq x_p^4$ as required. ■

1.5 Spacelike Vectors

Now we turn to spacelike separations, i.e., we consider two events x and x_0 for which $\mathcal{Q}(x - x_0) > 0$. Relative to any admissible basis we have $(\Delta x^1)^2 + (\Delta x^2)^2 + (\Delta x^3)^2 > (\Delta x^4)^2$ so that $x - x_0$ lies *outside* the null cone at x_0 and there is obviously no admissible basis in which the spatial separation of the two events is zero, i.e., there is no admissible observer who can experience both events (to do so he would have to travel faster than the speed of light). However, an argument analogous to that given at the beginning of Section 1.4 will show that there is a frame in which x and x_0 are simultaneous.

Exercise 1.5.1. Show that if $\mathcal{Q}(x - x_0) > 0$, then there is an admissible basis $\{\hat{e}_a\}$ for $\mathcal{M}$ relative to which $\Delta \hat{x}^4 = 0$. *Hint:* With $\{e_a\}$ arbitrary, take $\beta = \frac{\Delta x^4}{\Delta \vec{x}}$ and $d^i = \frac{\Delta x^i}{\Delta \vec{x}}$ and proceed as at the beginning of Section 1.4.

Exercise 1.5.2. Show that if $\mathcal{Q}(x - x_0) > 0$ and s is an arbitrary real number (positive, negative or zero), then there is an admissible basis for $\mathcal{M}$ relative to which the temporal separation Δx^4 of x and x_0 is s (so that admissible observers will, in general, not even agree on the *temporal order* of x and x_0).

Since $((\Delta x^1)^2 + (\Delta x^2)^2 + (\Delta x^3)^2)^{\frac{1}{2}} = \sqrt{(\Delta x^4)^2 + \mathcal{Q}(x - x_0)}$ in any admissible frame and since $(\Delta x^4)^2$ can assume any non-negative real value, the spatial separation of x and x_0 can assume any value greater than or equal to $\sqrt{\mathcal{Q}(x - x_0)}$; there is no frame in which the spatial separation is less than this value. For any two events x and x_0 for which $\mathcal{Q}(x - x_0) > 0$ we define the *proper spatial separation* $S(x - x_0)$ of x and x_0 by

$$S(x - x_0) = \sqrt{\mathcal{Q}(x - x_0)},$$

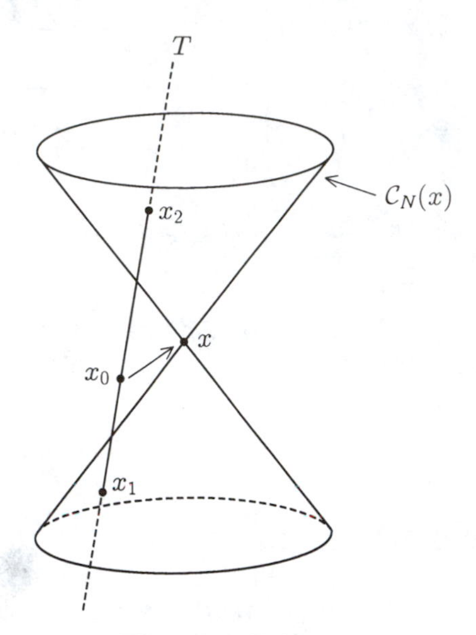

Figure 1.5.1

and regard it as the spatial separation of x and x_0 in any frame of reference in which x and x_0 are simultaneous.

Let T be an arbitrary timelike straight line containing x_0. We have seen that T can be identified with the worldline of some observer at rest in an admissible frame, but not necessarily stationed at the origin of the spatial coordinate system of this frame (we consider the special case of a time axis shortly). Let x in $\mathcal{M}$ be such that $x - x_0$ is spacelike and let x_1 and x_2 be the points of intersection of T with $\mathcal{C}_N(x)$ as shown in Figure 1.5.1. We claim that

$$S^2(x - x_0) = \tau(x_0 - x_1)\tau(x_2 - x_0) \tag{1.5.1}$$

(a result first proved by Robb [**R**]). To prove (1.5.1) we observe that, since $x - x_1$ is null,

$$0 = \mathcal{Q}(x - x_1) = \mathcal{Q}((x_0 - x_1) + (x - x_0)),$$

$$0 = -\tau^2(x_0 - x_1) + 2(x_0 - x_1) \cdot (x - x_0) + S^2(x - x_0). \tag{1.5.2}$$

Similarly, since $x_2 - x$ is null,

$$0 = -\tau^2(x_2 - x_0) - 2(x_2 - x_0) \cdot (x - x_0) + S^2(x - x_0). \tag{1.5.3}$$

There exists a constant $k > 0$ such that $x_2 - x_0 = k(x_0 - x_1)$ so $\tau^2(x_2 - x_0) = k^2\tau^2(x_0 - x_1)$. Multiplying (1.5.2) by k and adding the result to (1.5.3) therefore yields

$$-(k + k^2)\tau^2(x_0 - x_1) + (k+1)S^2(x - x_0) = 0.$$

Since $k + 1 \neq 0$ this can be written

$$\begin{aligned} S^2(x - x_0) &= k\tau^2(x_0 - x_1) \\ &= \tau(x_0 - x_1)(k\tau(x_0 - x_1)) \\ &= \tau(x_0 - x_1)\tau(x_2 - x_0) \end{aligned}$$

as required.

Suppose that the spacelike displacement vector $x - x_0$ is orthogonal to the timelike straight line T. Then (with the notation as above) $(x_0 - x_1)\cdot(x - x_0) = (x_2 - x_0)\cdot(x - x_0) = 0$ so (1.5.2) and (1.5.3) yield $S(x - x_0) = \tau(x_2 - x_0) = \tau(x_0 - x_1)$ which we prefer to write as

$$S(x - x_0) = \tfrac{1}{2}\left(\tau(x_0 - x_1) + \tau(x_2 - x_0)\right). \qquad (1.5.4)$$

In particular, this is true if T is a time axis. We have seen that, in this case, T can be identified with the worldline of an admissible observer $\mathcal{O}$ and the events x and x_0 are simultaneous in this observer's reference frame. But then $S(x - x_0)$ is the distance in this frame between x and x_0. Since x_0 lies on T we find that (1.5.4) admits the following physical interpretation: *The $\mathcal{O}$-distance of an event x from an admissible observer $\mathcal{O}$ is one-half the time lapse measured by $\mathcal{O}$ between the emission and reception of light signals connecting $\mathcal{O}$ with x.*

Exercise 1.5.3. Let x, x_0 and x_1 be events for which $x - x_0$ and $x_1 - x$ are spacelike and orthogonal. Show that

$$S^2(x_1 - x_0) = S^2(x_1 - x) + S^2(x - x_0) \qquad (1.5.5)$$

and interpret the result physically by considering a time axis T which is orthogonal to both $x - x_0$ and $x_1 - x$.

Suppose that v and w are nonzero vectors in $\mathcal{M}$ with $v \cdot w = 0$. Thus far we have shown the following: If v and w are null, then they must be parallel (Theorem 1.2.1). If v is timelike, then w must be spacelike (Corollary 1.3.2). If v and w are spacelike, then their proper spatial lengths satisfy the Pythagorean Theorem $S^2(v+w) = S^2(v) + S^2(w)$ (Exercise 1.5.3).

Exercise 1.5.4. Can a spacelike vector be orthogonal to a nonzero null vector?

1.6 Causality Relations

We begin by defining two order relations $\ll$ and $<$ on $\mathcal{M}$ as follows: For x and y in $\mathcal{M}$ we say that x *chronologically precedes* y and write $x \ll y$ if $y - x$ is timelike and future-directed, i.e., if y is in $C_T^+(x)$. We will say that x *causally precedes* y and write $x < y$ if $y - x$ is null and future-directed, i.e., if y is in $C_N^+(x)$. Both $\ll$ and $<$ are called *causality relations* because they establish a causal connection between the two events in the sense that the event x can influence the event y either by way of the propagation of some material phenomenon if $x \ll y$ or some electromagnetic effect if $x < y$.

Exercise 1.6.1. Prove that $\ll$ is *transitive*, i.e., that $x \ll y$ and $y \ll z$ implies $x \ll z$, and show by example that $<$ is not transitive.

It is an interesting, and useful, fact that each of the relations $\ll$ and $<$ can be defined in terms of the other.

Lemma 1.6.1. *For distinct points x and y in $\mathcal{M}$,*

$$x < y \text{ if and only if } \begin{cases} x \not\ll y & \text{and} \\ y \ll z & \text{implies } x \ll z\,. \end{cases}$$

Proof: First suppose $x < y$. Then $\mathcal{Q}(y - x) = 0$ so $x \not\ll y$ is clear. Moreover, if $y \ll z$, then $z - y$ is timelike and future-directed. Since $y - x$ is null and future-directed, Lemma 1.4.3 implies that $z - x = (z - y) + (y - x)$ is timelike and future-directed, i.e., $x \ll z$.

For the converse we suppose $x \not< y$ and show that either $x \ll y$ or there exists a z in $\mathcal{M}$ with $y \ll z$, but $x \not\ll z$. If $x \not< y$ and $x \not\ll y$, then $y - x$ is either timelike and past-directed, null and past-directed or spacelike. In the first case any z with $x < z$ has the property that $z - y = (z - x) + (x - y)$ is timelike and future-directed (Lemma 1.4.3 again) so $y \ll z$, but $x \not\ll z$. Finally, suppose $y - x$ is either null and past-directed or spacelike [see Figure 1.6.1 (a) and (b) respectively]. In each case we produce a z in $\mathcal{M}$ with $y \ll z$, but $x \not\ll z$ in the same way. Fix an admissible basis $\{e_a\}$ for $\mathcal{M}$ with $x = x^a e_a$ and $y = y^a e_a$. If $y - x$ is null and past-directed, then $x^4 - y^4 > 0$. If $y - x$ is spacelike we may choose $\{e_a\}$ so that $x^4 - y^4 > 0$ (Exercise 1.5.2). Now, for each $n = 1, 2, 3, \ldots$, define z_n in $\mathcal{M}$ by $z_n = y^1 e_1 + y^2 e_2 + y^3 e_3 + (y^4 + \frac{1}{n})e_4$. Then $z_n - y = \frac{1}{n} e_4$ is timelike and future-directed so $y \ll z_n$ for each n. However,

$$\begin{aligned} \mathcal{Q}(z_n - x) &= ((z_n - y) + (y - x))^2 \\ &= \mathcal{Q}(z_n - y) + 2(z_n - y) \cdot (y - x) + \mathcal{Q}(y - x) \\ &= -\tfrac{1}{n^2} + \tfrac{2}{n}(x^4 - y^4) + \mathcal{Q}(y - x) \\ &= \mathcal{Q}(y - x) + \tfrac{1}{n}\left[2(x^4 - y^4) - \tfrac{1}{n}\right]. \end{aligned}$$

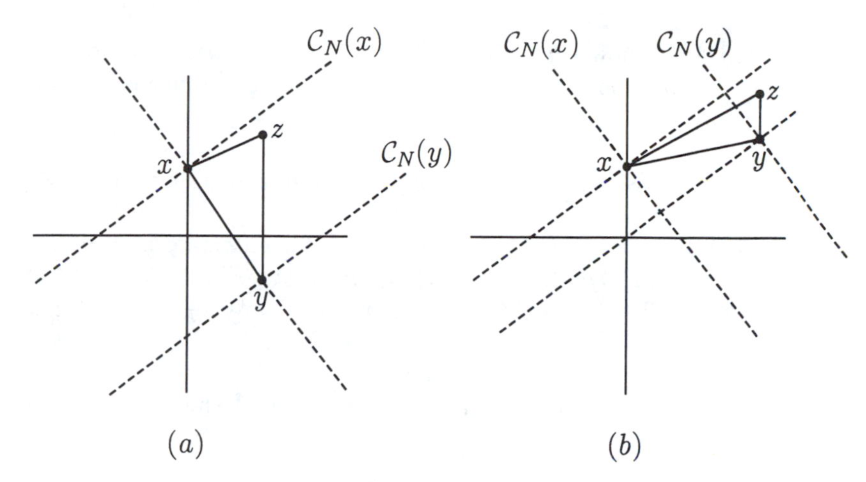

Figure 1.6.1

Since $\mathcal{Q}(y-x) \geq 0$ and $x^4 - y^4 > 0$ we can clearly choose n sufficiently large that $\mathcal{Q}(z_n - x) > 0$. For this n, $z = z_n$ satisfies $y \ll z$, but $x \not\ll z$. ■

Exercise 1.6.2. Show that, for distinct x and y in $\mathcal{M}$,

$$x \ll y \text{ if and only if } \begin{cases} x \not< y & \text{and} \\ x < z < y & \text{for some } z \text{ in } \mathcal{M}. \end{cases}$$

A map $F: \mathcal{M} \to \mathcal{M}$ is said to be a *causal automorphism* if it is one-to-one, onto and both F and F^{-1} preserve $<$, i.e., $x < y$ if and only if $F(x) < F(y)$. Note that, in particular, F is *not* assumed to be linear (or even continuous). We will eventually prove that this actually follows from the definition.

Exercise 1.6.3. Show that a one-to-one map F of $\mathcal{M}$ onto $\mathcal{M}$ is a causal automorphism if and only if both F and F^{-1} preserve $\ll$, i.e., $x \ll y$ if and only if $F(x) \ll F(y)$.

We propose next to embark upon a proof of the remarkable result of Zeeman [**Z₁**] to which we referred in the Introduction.[1] For the statement

---

[1] The proof is considerably more demanding than anything we have attempted thus far and might reasonably be omitted on a first reading.

of the theorem we define a *translation* of $\mathcal{M}$ to be a map $T\colon \mathcal{M} \to \mathcal{M}$ of the form $T(v) = v + v_0$ for some fixed v_0 in $\mathcal{M}$ and a *dilation* to be a map $K\colon \mathcal{M} \to \mathcal{M}$ such that $K(v) = kv$ for some positive real number k. An orthogonal transformation $L\colon \mathcal{M} \to \mathcal{M}$ is said to be *orthochronous* if $x \cdot Lx < 0$ for all timelike or null and nonzero x.

Exercise 1.6.4. Show that any translation, dilation, orthochronous orthogonal transformation, or any composition of such mappings is a causal automorphism.

Zeeman's Theorem asserts that we have just enumerated them all.

Theorem 1.6.2. *Let $F : \mathcal{M} \to \mathcal{M}$ be a causal automorphism of $\mathcal{M}$. Then there exists an orthochronous orthogonal transformation $L : \mathcal{M} \to \mathcal{M}$, a translation $T : \mathcal{M} \to \mathcal{M}$ and a dilation $K : \mathcal{M} \to \mathcal{M}$ such that $F = T \circ K \circ L$.*

For the proof we will require a sequence of five lemmas, the first of which, at least, is easy.

Lemma 1.6.3. *A causal automorphism $F : \mathcal{M} \to \mathcal{M}$ maps light rays to light rays. More precisely, if $x < y$ and $R_{x,y}$ is the light ray through x and y, then*

$$F(R_{x,y}) = R_{F(x),F(y)}.$$

Proof: Since both F and F^{-1} preserve $<$, F maps null cones to null cones so $F(\mathcal{C}_N(x)) = \mathcal{C}_N(F(x))$ and $F(\mathcal{C}_N(y)) = \mathcal{C}_N(F(y))$. By Theorem 1.2.2, $R_{x,y} = \mathcal{C}_N(x) \cap \mathcal{C}_N(y)$ and $R_{F(x),F(y)} = \mathcal{C}_N(F(x)) \cap \mathcal{C}_N(F(y))$. Thus,

$$\begin{aligned} F(R_{x,y}) &= F(\mathcal{C}_N(x) \cap \mathcal{C}_N(y)) \\ &= F(\mathcal{C}_N(x)) \cap F(\mathcal{C}_N(y)) \\ &= \mathcal{C}_N(F(x)) \cap \mathcal{C}_N(F(y)) \\ &= R_{F(x),F(y)} \,. \end{aligned}$$ ∎

Lemma 1.6.4. *A causal automorphism $F\colon \mathcal{M} \to \mathcal{M}$ maps parallel light rays onto parallel light rays.*

Proof: Let R_1 and R_2 be two distinct parallel light rays in $\mathcal{M}$ and P the (2-dimensional) plane containing them. Any plane in $\mathcal{M}$ is the translation of a plane through the origin which contains 0, 1 or 2 independent null vectors (depending on whether the plane is outside the null cone to each of its points, tangent to these null cones or intersects all of its time cones). Only the second two cases are relevant to P however.

Suppose first that P contains two independent null directions. Then it contains two families $\{R_\alpha\}$ and $\{S_\beta\}$ of light rays with all of the R_α parallel to R_1 and R_2 and all of the S_β parallel to some light ray which intersects both R_1 and R_2. Thus, the families $\{F(R_\alpha)\}$ and $\{F(S_\beta)\}$ are two families of light rays in $\mathcal{M}$ with the following properties:

1. No two of the $F(R_\alpha)$ intersect.

2. No two of the $F(S_\beta)$ intersect.

3. Each $F(R_\alpha)$ intersects every $F(S_\beta)$.

To show that $F(R_1)$ and $F(R_2)$ are parallel it will suffice (since they do not intersect) to show them coplanar. Suppose not. Then $F(R_1)$ and $F(R_2)$ lie in some 3-dimensional affine subspace R^3 of $\mathcal{M}$. Since each $F(S_\beta)$ intersects both $F(R_1)$ and $F(R_2)$, it too must lie in R^3. Thus, by #3 above, all of the $F(R_\alpha)$ are contained in R^3. We claim that, as a result, no $F(R_\alpha)$ can be coplanar with either $F(R_1)$ or $F(R_2)$ (unless $\alpha = 1$ or $\alpha = 2$). For suppose to the contrary that some $F(R_\alpha)$ were coplanar with, say, $F(R_1)$. Every $F(S_\beta)$ intersects both $F(R_\alpha)$ and $F(R_1)$ so it too must lie in this plane. Since $F(R_2)$ does not (by assumption) lie in this plane it can intersect the plane in at most one point. Thus, $F(R_2)$ intersects at most one $F(S_\beta)$ and this contradicts #3 above. Consequently, we may select an $F(R_3)$ such that no two of $\{F(R_1), F(R_2), F(R_3)\}$ are coplanar. Since $\{F(S_\beta)\}$ is then the family of straight lines in R^3 intersecting all of $\{F(R_1), F(R_2), F(R_3)\}$ it is the family of generators (rulings) for a hyperboloid of one sheet in R^3 (this old, and none-too-well-known, result in analytic geometry is proved on pages 105-106 of [**Sa**]). In the same way one shows that $\{F(R_\alpha)\}$ is the other family of rulings for this hyperboloid. But then each $F(R_\alpha)$ would be parallel to some $F(S_\beta)$ and this again contradicts #3 above.

Finally, we consider the case in which P contains only one independent null direction (and so is tangent to each of its null cones). Any point in $\mathcal{M}$ has through it a light ray parallel to both R_1 and R_2. Since the tangent space to the null cone at each point of R_1 is (only) 3-dimensional and since the same is true of R_2 we may select a light ray R_3 parallel to both R_1 and R_2 and not in either of these tangent spaces. Thus, the argument given above applies to R_1 and R_3 as well as R_2 and R_3. Consequently, $F(R_1)$ and $F(R_2)$ are both parallel to $F(R_3)$ and so are parallel to each other. ∎

Let $R_{x,y} = \{x + r(y - x) : r \in \mathbb{R}\}$ be a light ray and $F(R_{x,y}) = \{F(x) + s(F(y) - F(x)) : s \in \mathbb{R}\}$ its image under F. We regard s as a function of r: $s = f(r)$. Our next objective is to show that f is linear, i.e., that $f(r+t) = f(r) + f(t)$ and $f(tr) = tf(r)$ for all r and t in $\mathbb{R}$. First though, a few preliminaries. A map $g \colon R_{x,y} \to R_{x,y}$ is called a *translation* of $R_{x,y}$ if there exists a fixed t in $\mathbb{R}$ such that

$$g\,(x + r(y - x)) = x + (r + t)(y - x)$$

for all r in $\mathbb{R}$. We shall say that a translation g of a light ray R *lifts* to $F(R)$ if there is a translation $e\colon F(R) \to F(R)$ such that the diagram

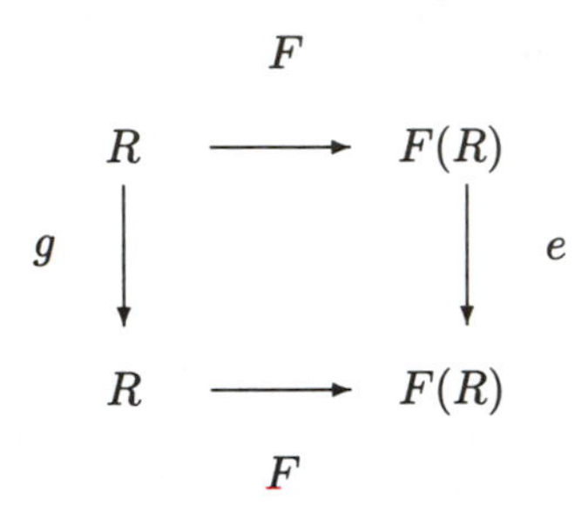

commutes, i.e., such that $F \circ g = e \circ F$. We show next that, in fact, every translation of R lifts to $F(R)$.

Lemma 1.6.5. *Let R be a light ray, $g : R \to R$ a translation of R and $F : \mathcal{M} \to \mathcal{M}$ a causal automorphism. Then g lifts to a translation $e\colon F(R) \to F(R)$ of $F(R)$.*

Proof: For the proof we will construct a family of translations of R which clearly do lift and then prove that this family exhausts all the translations of R.

Select a light ray R_1 parallel to R and such that the plane of R and R_1 contains two independent null directions. This plane therefore contains a family $\{S_\beta\}$ of parallel light rays all of which meet R and R_1. The family $\{S_\beta\}$ therefore determines an obvious parallel displacement map g_1 of R onto R_1 (see Figure 1.6.2). Since F carries parallel light rays to parallel light rays there is a parallel displacement e_1 of $F(R)$ onto $F(R_1)$ for which the diagram

$$\begin{array}{ccc} R & \xrightarrow{\;F\;} & F(R) \\ {\scriptstyle g_1}\big\downarrow & & \big\downarrow{\scriptstyle e_1} \\ R_1 & \xrightarrow[\;F\;]{} & F(R_1) \end{array}$$

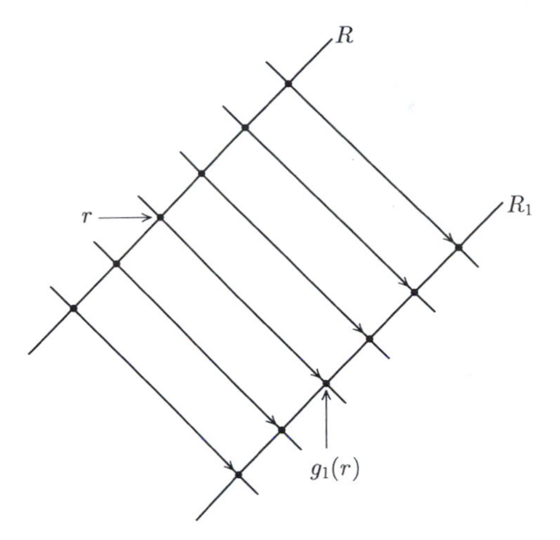

Figure 1.6.2

commutes. Now choose a light ray R_2 parallel to R_1 (and therefore to R) such that the planes of R_1 and R_2 and of R and R_2 both contain two independent null directions. Construct $g_2,\ e_2$ and $g_3,\ e_3$ as above so that all of the following diagrams commute.

$$
\begin{array}{ccc}
R & \xrightarrow{F} & F(R) \\
g_1 \downarrow & & \downarrow e_1 \\
R_1 & \xrightarrow{F} & F(R_1) \\
g_2 \downarrow & & \downarrow e_2 \\
R_2 & \xrightarrow{F} & F(R_2) \\
g_3 \downarrow & & \downarrow e_3 \\
R & \xrightarrow{F} & F(R)
\end{array}
$$

Now compose to get

$$\begin{array}{ccc} & F & \\ R & \longrightarrow & F(R) \\ g = g_3 \circ g_2 \circ g_1 \Big\downarrow & & \Big\downarrow e = e_3 \circ e_2 \circ e_1 \\ & F & \\ R & \longrightarrow & F(R) \end{array}$$

Observe that if R, R_1 and R_2 were all coplanar, then g and e would both necessarily be the identity. As it is g and e, being compositions of parallel displacements, are translations of R and $F(R)$ respectively. Consequently, any translation g of R constructed in this way as a composition of three such parallel displacements lifts to $F(R)$.

We claim now that the proof will be complete if we can show that for some particular light ray $\tilde{R}$ every translation of $\tilde{R}$ is realizable as such a composition. Indeed, if this has been proved for some $\tilde{R}$ we show that it is also true for R as follows: Select some composition G of a translation and an orthochronous orthogonal transformation that carries R onto $\tilde{R}$ (convince yourself that this can be done, or see Theorem 1.7.2). Since G is affine, a translation g of R gives rise to a translation $\tilde{g} = G \circ g \circ G^{-1}$ of $\tilde{R}$. Now represent $\tilde{g}$ as a composition $\tilde{g} = \tilde{g}_3 \circ \tilde{g}_2 \circ \tilde{g}_1$ of parallel displacements as indicated above. Then $g = G^{-1} \circ \tilde{g}_3 \circ \tilde{g}_2 \circ \tilde{g}_1 \circ G = (G^{-1} \circ \tilde{g}_3 \circ G) \circ (G^{-1} \circ \tilde{g}_2 \circ G) \circ (G^{-1} \circ \tilde{g}_3 \circ G)$. Moreover, since G and G^{-1} are causal automorphisms and so preserve parallel light rays by Lemma 1.6.4, we have produced a decomposition

$$R = G^{-1}(\tilde{R}) \xrightarrow[g_1]{} G^{-1}(\tilde{R}_1) \xrightarrow[g_2]{} G^{-1}(\tilde{R}_2) \xrightarrow[g_3]{} G^{-1}(\tilde{R}) = R$$

of g into a composition of parallel displacements $g_i = G^{-1} \circ \tilde{g}_i \circ G$ as required.

The particular light ray we choose to focus our attention on is obtained as follows: Fix an admissible basis $\{e_a\}$ and take $\tilde{R}$ to be the light ray through $x = (0,0,0,0)$ and $y = (0,0,1,1)$. Now consider a translation $\tilde{g}$ of $\tilde{R}$ defined by $\tilde{g}(x + r(y-x)) = \tilde{g}(0,0,r,r) = (0,0,r+t,r+t)$. In particular, $\tilde{g}$ carries $x = (0,0,0,0)$ to $\tilde{g}(x) = (0,0,t,t)$. Let $x_1 = (0,-t,0,-t)$ and $x_2 = (0,0,0,2t)$ and take $\tilde{R}_1$ and $\tilde{R}_2$ to be the light rays parallel to $\tilde{R}$ and through x_1 and x_2 respectively. We claim that the required parallel displacements $\tilde{g}_1$, $\tilde{g}_2$ and $\tilde{g}_3$ are defined and moreover that

$$x \xrightarrow[\tilde{g}_1]{} x_1 \xrightarrow[\tilde{g}_2]{} x_2 \xrightarrow[\tilde{g}_3]{} \tilde{g}(x) \qquad (1.6.1)$$

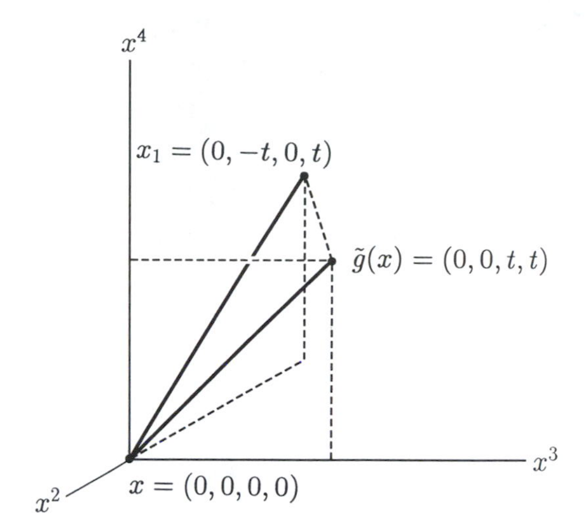

Figure 1.6.3

so that $\tilde{g}(x) = (\tilde{g}_3 \circ \tilde{g}_2 \circ \tilde{g}_1)(x)$. Since $\tilde{g}_3 \circ \tilde{g}_2 \circ \tilde{g}_1$ is a translation of $\tilde{R}$ that agrees with $\tilde{g}$ at $x = (0,0,0,0)$ it follows that $\tilde{g} = \tilde{g}_3 \circ \tilde{g}_2 \circ \tilde{g}_1$. All the verifications in (1.6.1) are the same so we illustrate by showing that $\tilde{g}_1(x) = x_1$ (see Figure 1.6.3). Note that the plane of $\tilde{R}$ and $\tilde{R}_1$ can contain at most two families of parallel light rays. The light rays parallel to $\tilde{R}$ (and $\tilde{R}_1$) form one such family. Since the line joining x and x_1 is also null and not parallel to $\tilde{R}$ it must be in the second family. Thus, $\tilde{g}_1$ exists and, obviously, $\tilde{g}_1(x) = x_1$. ∎

With Lemma 1.6.5 we can show that a causal automorphism is linear on each light ray. More precisely, we prove:

Lemma 1.6.6. *Let* $R = \{x + r(y - x) : x < y,\ r \in \mathbb{R}\}$ *be a light ray,* $F : \mathcal{M} \to \mathcal{M}$ *a causal automorphism and* $F(R) = \{F(x) + s(F(y) - F(x)) : s \in \mathbb{R}\}$ *the image of* R *under* F*. Then, regarding* s *as a function of* r*, say,* $s = f(r)$*, we have* $f(r+t) = f(r) + f(t)$ *and* $f(tr) = tf(r)$ *for all* r *and* t *in* $\mathbb{R}$.

Proof: Observe first that $f(0) = 0$. Now, fix a t in $\mathbb{R}$. We wish to show that, for any r in $\mathbb{R}$, $f(r+t) = f(r) + f(t)$, i.e., that

$$F(x + (r+t)(y - x)) = F(x) + (f(r) + f(t))(F(y) - F(x)). \quad (1.6.2)$$

Let $g : R \to R$ denote the translation of R by t, i.e., $g(x + r(y-x)) = x + (r+t)(y-x)$. By Lemma 1.6.5, there exists a translation $e : F(R) \to F(R)$

of $F(R)$ such that $F \circ g = e \circ F$. Suppose that e is the translation of $F(R)$ by $u = u(t)$, i.e., that $e(F(x) + s(F(y) - F(x))) = F(x) + (s + u(t))(F(y) - F(x))$. Then

$$\begin{aligned} F(x+(r+t)(y-x)) &= F(g(x+r(y-x))) \\ &= F \circ g(x+r(y-x)) \\ &= e \circ F(x+r(y-x)) \\ &= e(F(x)+f(r)(F(y)-F(x))) \\ &= F(x)+[f(r)+u(t)](F(y)-F(x)) \end{aligned}$$

so that $f(r+t) = f(r) + u(t)$ for any r. Setting $r = 0$ gives $f(t) = f(0) + u(t) = u(t)$ so we obtain $f(r+t) = f(r) + f(t)$ as required.

In particular, $f(2r) = f(r+r) = f(r) + f(r) = 2f(r)$ and, by induction, $f(nr) = nf(r)$ for $n = 0, 1, 2, \ldots$. Moreover, $f(r) = f(-r+2r) = f(-r) + 2f(r)$ so $f(-r) = -f(r)$ and, again by induction, $f(nr) = nf(r)$ for $n = 0, \pm 1, \pm 2, \ldots$. If m is also an integer and n is a nonzero integer, $nf(\frac{m}{n}r) = f(mr) = mf(r)$ so $f(\frac{m}{n}r) = \frac{m}{n}f(r)$. Thus, $f(tr) = tf(r)$ for any rational number t. Finally, observe that, since F preserves $<$ in $\mathcal{M}$, f preserves $<$ in $\mathbb{R}$ and is therefore continuous on $\mathbb{R}$. Since any real number t is the limit of a sequence of rational numbers we find that $f(tr) = tf(r)$ for any t in $\mathbb{R}$ and the proof is complete. ■

We conclude from Lemma 1.6.6 that if $R_{x,y} = \{x + r(y-x) : r \in \mathbb{R}\}$ is a light ray and F is a causal automorphism, then there exists a nonzero constant k, called the *expansion factor of F on $R_{x,y}$*, such that $F(R_{x,y}) = \{F(x) + kr(F(y) - F(x)) : r \in \mathbb{R}\}$.

Lemma 1.6.7. *Let $F : \mathcal{M} \to \mathcal{M}$ be a causal automorphism. Then F is an affine mapping, i.e., its composition with some translation of $\mathcal{M}$ (perhaps the identity) is a linear transformation.*

Proof: By first composing with a translation if necessary we may assume that $F(0) = 0$ and so the problem is to show that F is linear (the composition of a causal automorphism and a translation is clearly another causal automorphism).

Select a basis $\{v_1, v_2, v_3, v_4\}$ for $\mathcal{M}$ consisting of null vectors (Exercise 1.2.1). Define a map $G : \mathcal{M} \to \mathcal{M}$ by

$$G(y) = G\left(\sum_{i=1}^{4} y^i v_i\right) = \sum_{i=1}^{4} y^i F(v_i)$$

for each $y = \sum_{i=1}^{4} y^i v_i$ (for the remainder of this proof we temporarily suspend the summation convention and use a $\sum$ whenever a summation is intended). G is obviously linear and we shall prove that F is linear by

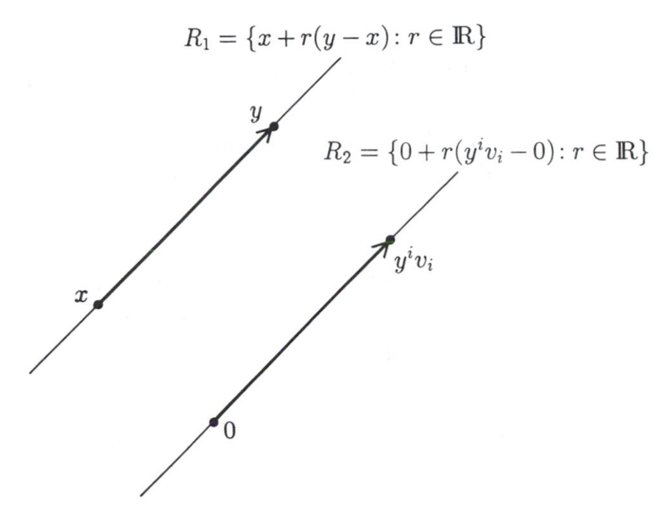

Figure 1.6.4

showing that, in fact, $F = G$. For each $i = 1, 2, 3, 4$ we let M_i denote the subspace of $\mathcal{M}$ spanned by $\{v_j : j \leq i\}$. Thus, M_1 is a light ray and M_4 is all of $\mathcal{M}$. We prove $F \mid M_i = G \mid M_i$ for all $i = 1, 2, 3, 4$. For $i = 1$ this is clear since $F(v_1) = G(v_1)$ and, by Lemma 1.6.6, F is linear on M_1. Now assume that $i = 2, 3$ or 4 and that $F \mid M_{i-1} = G \mid M_{i-1}$. We show from this that $F \mid M_i = G \mid M_i$ as follows: Any y in M_i can be uniquely represented as $y = x + y^i v_i$, where x is in M_{i-1} and there is *no sum* over i in $y^i v_i$. Thus, $y - x = y^i v_i$ is null since v_i is null. We consider two light rays, the first (R_1) through x and y and the second (R_2) through 0 and $y^i v_i$ (see Figure 1.6.4). R_1 and R_2 are parallel so $F(R_1)$ and $F(R_2)$ are parallel by Lemma 1.6.4. Moreover, if the expansion factors of F on R_1 and R_2 were not the same, then the images under F of the light rays intersecting both R_1 and R_2 would not be parallel (in case there is no such family use a third light ray R_3 as in the proof of Lemma 1.6.4). Consequently, there exists a nonzero real number k such that

$$F(R_1) = \{F(x) + kr(F(y) - F(x)) : r \in \mathbb{R}\}$$

and

$$\begin{aligned} F(R_2) &= \{F(0) + kr(F(y^i v_i) - F(0)) : r \in \mathbb{R}\} \\ &= \{0 + kr(F(y^i v_i) - 0) : r \in \mathbb{R}\}. \end{aligned}$$

Since $F(R_1)$ and $F(R_2)$ are parallel and $r = 0$ gives 0 on $F(R_2)$ and $F(x)$ on $F(R_1)$, translation of $F(R_2)$ by $F(x)$ gives $F(R_1)$. For $r = 1$

this gives

$$F(x) + \left[0 + k(F(y^i v_i) - 0)\right] = F(x) + k(F(y) - F(x)),$$

that is,

$$F(y^i v_i) = F(y) - F(x).$$

Thus,

$$\begin{aligned} F(y) &= F(x) + F(y^i v_i) \\ &= G(x) + F(y^i v_i) \quad \text{since } x \in \mathcal{M}_{i-1} \\ &= G(x) + y^i F(v_i) \quad \text{by Lemma 1.6.6} \\ &= G(x) + y^i G(v_i) \\ &= G(x) + G(y^i v_i) \quad \text{by linearity of } G \\ &= G(x + y^i v_i) \\ &= G(y) \end{aligned}$$

and the proof is complete. ∎

Finally, we are prepared for:

Proof of Theorem 1.6.2: According to Lemma 1.6.7 there is a translation, which we write $T^{-1} : \mathcal{M} \to \mathcal{M}$, such that $T^{-1} \circ F$ is linear. To complete the proof we need only produce a positive constant $\frac{1}{k}$ such that $\frac{1}{k} T^{-1} \circ F$ preserves the quadratic form on $\mathcal{M}$. For then, by Lemma 1.2.3, $\frac{1}{k} T^{-1} \circ F$ is a (necessarily orthochronous) orthogonal transformation L. Denoting by K the dilation $K(v) = kv$, $\frac{1}{k} T^{-1} \circ F = L$ therefore gives $F = T \circ K \circ L$ as required.

Since both $T^{-1} \circ F$ and its inverse take 0 to 0 and preserve $<, T^{-1} \circ F$ must carry the null cone $\mathcal{C}_N(0)$ onto itself, i.e., $\mathcal{Q}(x) = 0$ if and only if $\mathcal{Q}(T^{-1} \circ F(x)) = 0$. Since $T^{-1} \circ F$ is linear, both $\mathcal{Q}(x)$ and $\mathcal{Q}(T^{-1} \circ F(x))$ are quadratic forms and, as we have just observed, they have the same kernel, i.e., vanish for the same x's. But two quadratic forms with the same kernel differ at most by a multiplicative constant (Theorem 14.10 of [**K**]) so there exists a constant k' such that $\mathcal{Q}(x) = k' \mathcal{Q}(T^{-1} \circ F(x))$ for all x. But $T^{-1} \circ F$ is a causal automorphism and so preserves the upper time cone. In particular, $\mathcal{Q}(x) < 0$ if and only if $\mathcal{Q}(T^{-1} \circ F(x)) < 0$, so k' must be positive. Letting $k = (k')^{-1/2}$ we therefore have $\mathcal{Q}(x) = \mathcal{Q}(\frac{1}{k} T^{-1} \circ F(x))$ so $\frac{1}{k} T^{-1} \circ F$ preserves the quadratic form on $\mathcal{M}$ and the proof is complete. ∎

1.7 Spin Transformations and the Lorentz Group

In this section we develop a new and very powerful technique for the construction and investigation of Lorentz transformations. The principal tool is a certain homomorphism (called the "spinor map") from the group of 2×2 complex matrices with determinant 1 onto the Lorentz group $\mathcal{L}$. With it we uncover a remarkable connection between Lorentz transformations and the familiar fractional linear transformations of complex analysis. This, in turn, has some rather startling things to say about the Lorentz group and the phenomenon of length contraction.

We begin by establishing some notation. $\mathbb{C}^{2\times 2}$ denotes the set of all 2×2 matrices

$$A = [a_{ij}] = \begin{bmatrix} a_{11} & a_{12} \\ a_{21} & a_{22} \end{bmatrix}$$

with complex entries. Using an overbar to designate complex conjugation, the *conjugate transpose* A^{CT} of A is defined by

$$A^{CT} = \begin{bmatrix} \bar{a}_{11} & \bar{a}_{21} \\ \bar{a}_{12} & \bar{a}_{22} \end{bmatrix}.$$

An H in $\mathbb{C}^{2\times 2}$ is said to be *Hermitian* if $H^{CT} = H$ and we denote by $\mathcal{H}_2$ the set of all such.

Exercise 1.7.1. Show that any Hermitian H in $\mathbb{C}^{2\times 2}$ is uniquely expressible in the form

$$H = \begin{bmatrix} x^3 + x^4 & x^1 + ix^2 \\ x^1 - ix^2 & -x^3 + x^4 \end{bmatrix}, \tag{1.7.1}$$

where x^a, $a = 1, 2, 3, 4$, are real. Show, moreover, that the representation (1.7.1) is equivalent to

$$H = x^1\sigma_1 + x^2\sigma_2 + x^3\sigma_3 + x^4\sigma_4, \tag{1.7.2}$$

where σ_i, $i = 1, 2, 3$, are the *Pauli spin matrices*

$$\sigma_1 = \begin{bmatrix} 0 & 1 \\ 1 & 0 \end{bmatrix}, \quad \sigma_2 = \begin{bmatrix} 0 & i \\ -i & 0 \end{bmatrix}, \quad \sigma_3 = \begin{bmatrix} 1 & 0 \\ 0 & -1 \end{bmatrix}$$

and σ_4 is the 2×2 identity matrix.

We denote by $SL(2, \mathbb{C})$ the set of all A in $\mathbb{C}^{2\times 2}$ with determinant 1. $SL(2, \mathbb{C})$ is called the *special linear group* of order 2 and is, indeed, a group of matrices, that is, closed under the formation of products and inverses. Elements of $SL(2, \mathbb{C})$ are often called *spin transformations.* Each A in $SL(2, \mathbb{C})$ gives rise to a mapping $M_A : \mathcal{H}_2 \rightarrow \mathcal{H}_2$ defined by

$$M_A(H) = AHA^{CT}$$

for every H in $\mathcal{H}_2(M_A(H)$ is in $\mathcal{H}_2$ since $(AHA^{CT})^{CT} = (A^{CT})^{CT} \cdot (AH)^{CT} = AH^{CT}A^{CT} = AHA^{CT}$). Moreover, $\det M_A(H) = \det(AHA^{CT}) = (\det A)(\det H)(\det A^{CT}) = \det H$. But $M_A(H)$ can be uniquely written in the form

$$M_A(H) = \begin{bmatrix} \hat{x}^3 + \hat{x}^4 & \hat{x}^1 + i\hat{x}^2 \\ \hat{x}^1 - i\hat{x}^2 & -\hat{x}^3 + \hat{x}^4 \end{bmatrix} \tag{1.7.3}$$

for some real numbers $\hat{x}^a$, $a = 1, 2, 3, 4$. Computing the determinants in (1.7.1) and (1.7.2) therefore gives

$$(\hat{x}^1)^2 + (\hat{x}^2)^2 + (\hat{x}^3)^2 - (\hat{x}^4)^2 = (x^1)^2 + (x^2)^2 + (x^3)^2 - (x^4)^2. \tag{1.7.4}$$

Thus, the mapping $[x^a] \to [\hat{x}^a]$ defined by

$$\begin{bmatrix} \hat{x}^3 + \hat{x}^4 & \hat{x}^1 + i\hat{x}^2 \\ \hat{x}^1 - i\hat{x}^2 & -\hat{x}^3 + \hat{x}^4 \end{bmatrix} = A \begin{bmatrix} x^3 + x^4 & x^1 + ix^2 \\ x^1 - ix^2 & -x^3 + x^4 \end{bmatrix} A^{CT}, \tag{1.7.5}$$

which is clearly linear, preserves the quadratic form $\eta_{ab}x^a x^b$. According to Lemma 1.2.3, the matrix of this map is therefore a general, homogeneous Lorentz transformation. We intend to construct this matrix explicitly from the entries of

$$A = \begin{bmatrix} \alpha & \beta \\ \gamma & \delta \end{bmatrix}.$$

Letting $h_{11} = x^3 + x^4$, $h_{12} = x^1 + ix^2$, $h_{21} = x^1 - ix^2$, $h_{22} = -x^3 + x^4$ (and $\hat{h}_{11} = \hat{x}^3 + \hat{x}^4$, etc.) we have

$$\begin{bmatrix} h_{11} \\ h_{12} \\ h_{21} \\ h_{22} \end{bmatrix} = \begin{bmatrix} 0 & 0 & 1 & 1 \\ 1 & i & 0 & 0 \\ 1 & -i & 0 & 0 \\ 0 & 0 & -1 & 1 \end{bmatrix} \begin{bmatrix} x^1 \\ x^2 \\ x^3 \\ x^4 \end{bmatrix}$$

which we will write more compactly as

$$[\,h_{ij}\,] = G\,[\,x^i\,]$$

and similarly for $\left[\hat{h}_{ij}\right]$. Moreover, it is easy to check that

$$G^{-1} = \frac{1}{2} \begin{bmatrix} 0 & 1 & 1 & 0 \\ 0 & -i & i & 0 \\ 1 & 0 & 0 & -1 \\ 1 & 0 & 0 & 1 \end{bmatrix}.$$

Exercise 1.7.2. Write out the product

$$AHA^{CT} = \begin{bmatrix} \alpha & \beta \\ \gamma & \delta \end{bmatrix} \begin{bmatrix} h_{11} & h_{12} \\ h_{21} & h_{22} \end{bmatrix} \begin{bmatrix} \bar{\alpha} & \bar{\gamma} \\ \bar{\beta} & \bar{\delta} \end{bmatrix}$$

explicitly and show that $M_A(H) = AHA^{CT}$ is equivalent to

$$\begin{bmatrix} \hat{h}_{11} \\ \hat{h}_{12} \\ \hat{h}_{21} \\ \hat{h}_{22} \end{bmatrix} = \begin{bmatrix} \alpha\bar{\alpha} & \alpha\bar{\beta} & \bar{\alpha}\beta & \beta\bar{\beta} \\ \alpha\bar{\gamma} & \alpha\bar{\delta} & \beta\bar{\gamma} & \beta\bar{\delta} \\ \bar{\alpha}\gamma & \bar{\beta}\gamma & \bar{\alpha}\delta & \bar{\beta}\delta \\ \gamma\bar{\gamma} & \gamma\bar{\delta} & \bar{\gamma}\delta & \delta\bar{\delta} \end{bmatrix} \begin{bmatrix} h_{11} \\ h_{12} \\ h_{21} \\ h_{22} \end{bmatrix}$$

which we will write more concisely as

$$\left[\hat{h}_{ij}\right] = R_A \left[h_{ij}\right].$$

Consequently, the map $[x^a] \to [\hat{x}^a]$ defined by (1.7.5) is given by

$$[x^a] \xrightarrow[G]{} [h_{ij}] \xrightarrow[R_A]{} [\hat{h}_{ij}] \xrightarrow[G^{-1}]{} [\hat{x}^a] \tag{1.7.6}$$

and the Lorentz transformation Λ_A determined via (1.7.5) [or (1.7.6)] by A is

$$\Lambda_A = G^{-1} R_A G.$$

Exercise 1.7.3. Calculate the product $G^{-1}R_AG$ explicitly to show that the entries $\Lambda^a{}_b$ of Λ_A are given by

$$\begin{aligned}
\Lambda^1{}_1 &= \tfrac{1}{2}(\alpha\bar{\delta} + \bar{\beta}\gamma + \beta\bar{\gamma} + \bar{\alpha}\delta), & \Lambda^1{}_2 &= \tfrac{i}{2}(\alpha\bar{\delta} + \bar{\beta}\gamma - \beta\bar{\gamma} - \bar{\alpha}\delta), \\
\Lambda^2{}_1 &= \tfrac{i}{2}(-\alpha\bar{\delta} + \bar{\beta}\gamma - \beta\bar{\gamma} + \bar{\alpha}\delta), & \Lambda^2{}_2 &= \tfrac{1}{2}(\alpha\bar{\delta} - \bar{\beta}\gamma - \beta\bar{\gamma} + \bar{\alpha}\delta), \\
\Lambda^3{}_1 &= \tfrac{1}{2}(\alpha\bar{\beta} - \gamma\bar{\delta} + \bar{\alpha}\beta - \bar{\gamma}\delta), & \Lambda^3{}_2 &= \tfrac{i}{2}(\alpha\bar{\beta} - \gamma\bar{\delta} - \bar{\alpha}\beta + \bar{\gamma}\delta), \\
\Lambda^4{}_1 &= \tfrac{1}{2}(\alpha\bar{\beta} + \gamma\bar{\delta} + \bar{\alpha}\beta + \bar{\gamma}\delta), & \Lambda^4{}_2 &= \tfrac{i}{2}(\alpha\bar{\beta} + \gamma\bar{\delta} - \bar{\alpha}\beta - \bar{\gamma}\delta), \\
\Lambda^1{}_3 &= \tfrac{1}{2}(\alpha\bar{\gamma} + \bar{\alpha}\gamma - \beta\bar{\delta} - \bar{\beta}\delta), & \Lambda^1{}_4 &= \tfrac{1}{2}(\alpha\bar{\gamma} + \bar{\alpha}\gamma + \beta\bar{\delta} + \bar{\beta}\delta), \\
\Lambda^2{}_3 &= \tfrac{i}{2}(-\alpha\bar{\gamma} + \bar{\alpha}\gamma + \beta\bar{\delta} - \bar{\beta}\delta), & \Lambda^2{}_4 &= \tfrac{i}{2}(-\alpha\bar{\gamma} + \bar{\alpha}\gamma - \beta\bar{\delta} + \bar{\beta}\delta), \\
\Lambda^3{}_3 &= \tfrac{1}{2}(\alpha\bar{\alpha} - \gamma\bar{\gamma} - \beta\bar{\beta} + \delta\bar{\delta}), & \Lambda^3{}_4 &= \tfrac{1}{2}(\alpha\bar{\alpha} - \gamma\bar{\gamma} + \beta\bar{\beta} - \delta\bar{\delta}), \\
\Lambda^4{}_3 &= \tfrac{1}{2}(\alpha\bar{\alpha} + \gamma\bar{\gamma} - \beta\bar{\beta} - \delta\bar{\delta}), & \Lambda^4{}_4 &= \tfrac{1}{2}(\alpha\bar{\alpha} + \beta\bar{\beta} + \gamma\bar{\gamma} + \delta\bar{\delta}).
\end{aligned} \tag{1.7.7}$$

Observe that the (4,4)-entry of Λ_A is positive so Λ_A is orthochronous. Moreover, $\det \Lambda_A = \det(G^{-1}R_AG) = (\det G^{-1})(\det R_A)(\det G) = \det R_A$ and one shows by direct calculation that $\det R_A = (\alpha\delta - \beta\gamma)^2(\bar{\alpha}\bar{\delta} - \bar{\beta}\bar{\gamma})^2 = 1$ so that Λ_A is proper. The map $A \to \Lambda_A$ of $SL(2,\mathbb{C})$ to $\mathcal{L}$ is called the *spinor map*. Note that if A and B are both in $SL(2,\mathbb{C})$, then

$$\Lambda_A \Lambda_B = (G^{-1}R_AG)(G^{-1}R_BG) = G^{-1}(R_A R_B)G. \tag{1.7.8}$$

But since $M_{AB}(H) = (AB)H(AB)^{CT} = ABHB^{CT}A^{CT} = A(BHB^{CT})A^{CT} = M_A(BHB^{CT}) = M_A(M_B(H)) = M_A \circ M_B(H)$ we conclude that $M_{AB} =$

$M_A \circ M_B$ and so $R_{AB} = R_A R_B$. Thus, (1.7.8) gives $\Lambda_A \Lambda_B = G^{-1} R_{AB} G$ and so

$$\Lambda_A \Lambda_B = \Lambda_{AB} . \tag{1.7.9}$$

Thus, the spinor map preserves matrix multiplication, i.e., is a group homomorphism of $SL(2, \mathbb{C})$ to $\mathcal{L}$. It is not one-to-one since it is clear from (1.7.7) that both A and $-A$ have the same image in $\mathcal{L}$. In fact, we claim that it is precisely two-to-one, i.e., that if A and B are in $SL(2, \mathbb{C})$ and $\Lambda_A = \Lambda_B$, then $A = \pm B$. To see this note that AB^{-1} is in $SL(2, \mathbb{C})$ and, since the spinor map is a homomorphism, $\Lambda_{AB^{-1}} = \Lambda_A \Lambda_{B^{-1}} = \Lambda_A (\Lambda_B)^{-1} = \Lambda_A (\Lambda_A)^{-1} =$ identity matrix.

Exercise 1.7.4. Let $AB^{-1} = \left[\begin{smallmatrix} \alpha & \beta \\ \gamma & \delta \end{smallmatrix}\right]$ and use (1.7.7) for $\Lambda_{AB^{-1}}$ (= identity) to show that $AB^{-1} = \pm \left[\begin{smallmatrix} 1 & 0 \\ 0 & 1 \end{smallmatrix}\right]$, i.e., that $A = \pm B$.

Exercise 1.7.5. For each real number θ define a 2×2 matrix $A(\theta)$ by

$$A(\theta) = \begin{bmatrix} \cosh \frac{\theta}{2} & -\sinh \frac{\theta}{2} \\ -\sinh \frac{\theta}{2} & \cosh \frac{\theta}{2} \end{bmatrix} .$$

Show that $A(\theta)$ is in $SL(2, \mathbb{C})$ and that

$$\Lambda_{A(\theta)} = L(\theta) = \begin{bmatrix} \cosh\theta & 0 & 0 & -\sinh\theta \\ 0 & 1 & 0 & 0 \\ 0 & 0 & 1 & 0 \\ -\sinh\theta & 0 & 0 & \cosh\theta \end{bmatrix} .$$

An element $A = \left[\begin{smallmatrix} \alpha & \beta \\ \gamma & \delta \end{smallmatrix}\right]$ of $SL(2, \mathbb{C})$ is said to be *unitary* if $A^{-1} = A^{CT}$, i.e., if

$$\begin{bmatrix} \alpha & \beta \\ \gamma & \delta \end{bmatrix} \begin{bmatrix} \bar\alpha & \bar\gamma \\ \bar\beta & \bar\delta \end{bmatrix} = \begin{bmatrix} \alpha\bar\alpha + \beta\bar\beta & \alpha\bar\gamma + \beta\bar\delta \\ \bar\alpha\gamma + \bar\beta\delta & \gamma\bar\gamma + \delta\bar\delta \end{bmatrix} = \begin{bmatrix} 1 & 0 \\ 0 & 1 \end{bmatrix} . \tag{1.7.10}$$

The set of all such matrices is denoted SU_2 and is a subgroup of $SL(2, \mathbb{C})$, i.e., SU_2 is also closed under the formation of products and inverses.

Exercise 1.7.6. Verify this.

Notice that if A is in SU_2, then, by (1.7.10), the (4, 4)-entry of Λ_A is $\frac{1}{2}(\alpha\bar\alpha + \beta\bar\beta + \gamma\bar\gamma + \delta\bar\delta) = \frac{1}{2}(1+1) = 1$ and so Λ_A is a rotation in $\mathcal{L}$ by Lemma 1.3.4. Thus, the spinor map carries SU_2 into the rotation subgroup $\mathcal{R}$ of $\mathcal{L}$. We show that, in fact, it maps SU_2 onto $\mathcal{R}$. To do this we borrow a result from linear algebra (or mechanics, depending on one's field) which asserts that any 3×3 rotation matrix $[R^i{}_j]_{i,j=1,2,3}$ can be represented in terms of its "Euler angles" ϕ_1, θ and ϕ_2 as

$$[R^i{}_j] = \begin{bmatrix} \cos\phi_2\cos\phi_1 & -\cos\phi_2\sin\phi_1 & \sin\phi_2\sin\theta \\ -\cos\theta\sin\phi_1\sin\phi_2 & -\cos\theta\cos\phi_1\sin\phi_2 & \\ & & \\ \sin\phi_2\cos\phi_1 & -\sin\phi_2\sin\phi_1 & -\cos\phi_2\sin\theta \\ +\cos\theta\sin\phi_1\cos\phi_2 & +\cos\theta\cos\phi_1\cos\phi_2 & \\ & & \\ \sin\theta\sin\phi_1 & \sin\theta\cos\phi_1 & \cos\theta \end{bmatrix}$$

(this is proved, for example, in [**GMS**]).

Exercise 1.7.7. Show that

$$A = \begin{bmatrix} \cos\frac{\theta}{2}e^{\frac{1}{2}i(\phi_1+\phi_2)} & i\sin\frac{\theta}{2}e^{-\frac{1}{2}i(\phi_2-\phi_1)} \\ i\sin\frac{\theta}{2}e^{\frac{1}{2}i(\phi_2-\phi_1)} & \cos\frac{\theta}{2}e^{-\frac{1}{2}i(\phi_1+\phi_2)} \end{bmatrix}$$

is in SU_2 and maps onto $\begin{bmatrix} & & & 0 \\ & [R^i{}_j] & & 0 \\ & & & 0 \\ 0 & 0 & 0 & 1 \end{bmatrix}$ under the spinor map.

With this we can now show that the spinor map is surjective, i.e., that every proper, orthochronous Lorentz transformation Λ is $\Lambda_{\pm A}$ for some A in $SL(2,\mathbb{C})$. By Theorem 1.3.5, there exists a real number θ and two rotations R_1 and R_2 in $\mathcal{L}$ such that $\Lambda = R_1 L(\theta) R_2$. There exist elements A_1 and A_2 of $SU_2 \subseteq SL(2,\mathbb{C})$ which the spinor map carries onto R_1 and R_2 respectively. Moreover, $A(\theta)$ (as defined in Exercise 1.7.6) maps onto $L(\theta)$. Since the spinor map is a homomorphism, $A_1 A(\theta) A_2$ maps onto $R_1 L(\theta) R_2 = \Lambda$ and the proof is complete.

And so the elements of $SL(2,\mathbb{C})$ generate Lorentz transformations. But they do other things as well, perhaps more familiar. Specificially, each 2×2 complex unimodular matrix defines a (normalized) fractional linear transformation of the Riemann sphere (extended complex plane). There is, in fact, a rather surprising connection between these two activities which we intend to explore since it sheds much light on both the mathematics and the kinematics of the Lorentz group. First though, a few preliminaries.

Thus far we have thought of a Lorentz transformation Λ exclusively as a coordinate transformation matrix; what some call a *passive* transformation (leaving points fixed, but changing coordinate systems). It will be useful now, however, to realize that Λ admits an equally natural interpretation as an *active* transformation (leaving the coordinate system fixed, but moving points about). More precisely, let us consider an orthogonal transformation $L\colon \mathcal{M} \to \mathcal{M}$ and fix a basis $\{e_a\}$. Then $\{\hat{e}_a\} = \{L\, e_a\}$ is the image basis and, if we write $e_b = \Lambda^a{}_b \, \hat{e}_a$, then the corresponding Lorentz transformation Λ is defined by

$$\Lambda = \begin{bmatrix} \Lambda^1{}_1 & \Lambda^1{}_2 & \Lambda^1{}_3 & \Lambda^1{}_4 \\ \Lambda^2{}_1 & \Lambda^2{}_2 & \Lambda^2{}_3 & \Lambda^2{}_4 \\ \Lambda^3{}_1 & \Lambda^3{}_2 & \Lambda^3{}_3 & \Lambda^3{}_4 \\ \Lambda^4{}_1 & \Lambda^4{}_2 & \Lambda^4{}_3 & \Lambda^4{}_4 \end{bmatrix}.$$

We emphasize again that Λ is the matrix of L^{-1} relative to the basis $\{\hat{e}_a\}$. Now, for each x in $\mathcal{M}$ we may write $x = x^a e_a = \hat{x}^a \hat{e}_a$, where $[\hat{x}^a] = \Lambda[\, x^a \,]$. Thus, we think of Λ as acting on the coordinates of a fixed point to give the coordinates of *the same point in a new coordinate system.* However, observe that $L^{-1}x = L^{-1}(\hat{x}^a \hat{e}_a) = \hat{x}^a L^{-1} \hat{e}_a = \hat{x}^a e_a$ so we may equally well view Λ as acting on the coordinates $[x^a]$ of some point relative to $\{e_a\}$ and yielding the coordinates $[\hat{x}^a]$ of *a new point (namely, $L^{-1}x$) in the same coordinate system.* It will be crucial somewhat later to observe that, with this new interpretation of Λ, $L^{-1}x$ *has the same position and time in $\mathcal{S}$ that x has in $\hat{\mathcal{S}}$.*

We will be much concerned in the remainder of this section with "past null directions" and the effect had on them by Lorentz transformations. For each x in the past null cone $C_N^-(0)$ at 0 in $\mathcal{M}$ we define the *past null direction* R_x^- through x by

$$R_x^- = \{\alpha x \colon \alpha \geq 0\}.$$

Future null directions are defined analogously and all of our results will have obvious "future duals". The *null direction* through x is the set of all real multiples of x, i.e., $R_{0,x}$. Obviously, if y is any positive scalar multiple of x, then $R_y^- = R_x^-$. Observe that if $L : \mathcal{M} \to \mathcal{M}$ is an orthogonal transformation corresponding to any orthochronous Lorentz transformation Λ, then $x \in C_N^-(0)$ implies $Lx \in C_N^-(0)$ so R_{Lx}^- is defined. Moreover, $L(R_x^-) = L(\{\alpha x \colon \alpha \geq 0\}) = \{L(\alpha x) \colon \alpha \geq 0\} = \{\alpha Lx \colon \alpha \geq 0\} = R_{Lx}^-$, i.e.,

$$L(R_x^-) = R_{Lx}^-. \tag{1.7.11}$$

Consequently, L (and therefore L^{-1} and so Λ also) can be regarded as a map on past null directions.

In order to unearth the connection between Lorentz and fractional linear transformations we observe that there is a natural one-to-one correspondence between past null directions and the points on a copy of the Riemann sphere. Specifically, we fix an admissible basis $\{e_a\}$ for $\mathcal{M}$ and denote by S^- the intersection of the past null cone $C_N^-(0)$ at 0 with the hyperplane $x^4 = -1$:

$$S^- = \{x = x^a e_a \colon x \in C_N^-(0), \quad x^4 = -1\}.$$

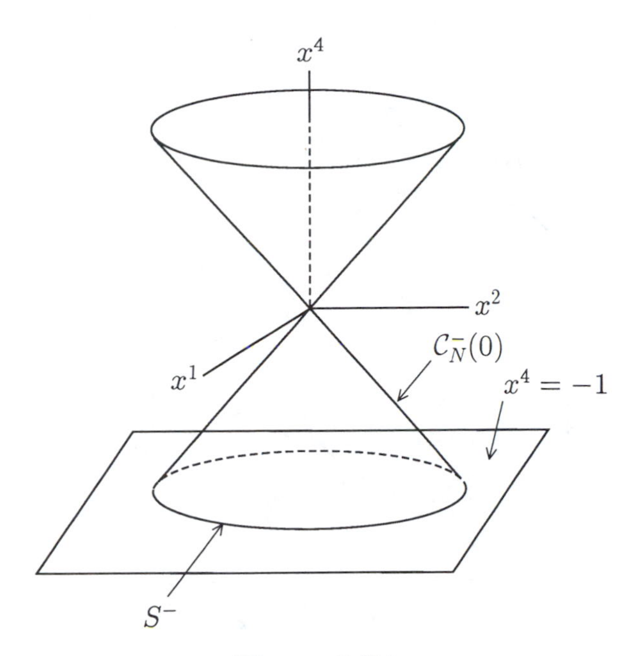

Figure 1.7.1

Observe that, since $x \in \mathcal{C}_N^-(0)$ if and only if $(x^1)^2 + (x^2)^2 + (x^3)^2 = (x^4)^2$, $S^- = \{x = x^a e_a : (x^1)^2 + (x^2)^2 + (x^3)^2 = 1\}$ and so is a copy of the ordinary 2-sphere S^2 in the instantaneous 3-space $x^4 = -1$ (see Figure 1.7.1).

Exercise 1.7.8. Show that any past null direction intersects S^- in a single point.

Conversely, every point on S^- determines a unique past null direction in $\mathcal{M}$. To obtain an explicit representation for this past null direction we wish to regard S^- as the Riemann sphere, that is, we wish to identify the points of S^- with extended complex numbers via stereographic projection (see, for example, [**A**]). To this end we take $N = (0, 0, 1, -1)$ in S^- as the north pole and project onto the 2-dimensional plane C in $x^4 = -1$ given by $x^3 = 0$ (see Figure 1.7.2). The relationship between a point $P(x^1, x^2, x^3, -1)$ other than N on S^- and its image ζ in the complex plane C under stereographic projection from N is easily calculated and is summarized in (1.7.12) and (1.7.13):

$$\zeta = \frac{x^1 + ix^2}{1 - x^3}, \tag{1.7.12}$$

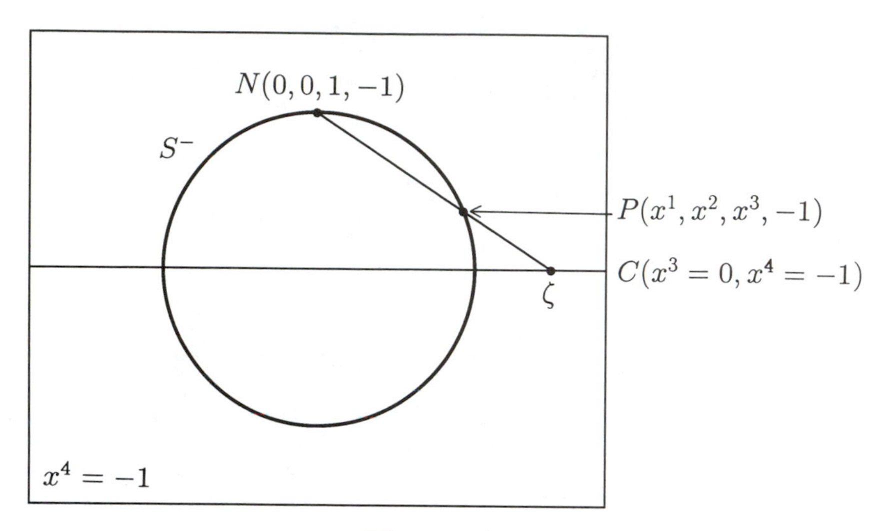

Figure 1.7.2

$$\begin{aligned}
x^1 &= \frac{\zeta+\bar{\zeta}}{\zeta\bar{\zeta}+1}, \\
x^2 &= \frac{\zeta-\bar{\zeta}}{i(\zeta\bar{\zeta}+1)}, \\
x^3 &= \frac{\zeta\bar{\zeta}-1}{\zeta\bar{\zeta}+1}, \\
x^4 &= -1 .
\end{aligned} \tag{1.7.13}$$

Of course, the north pole $N(0,0,1,-1)$ on S^- corresponds to the point at infinity in the extended complex plane $\bar{C}$. In order to avoid the need to deal with the point at infinity we prefer to represent extended complex numbers ζ in so-called "projective homogeneous coordinates", that is, by a pair $\left[\begin{smallmatrix}\xi\\ \eta\end{smallmatrix}\right]$ of complex numbers, not both zero, which satisfy

$$\zeta = \frac{\xi}{\eta}$$

(any pair $\left[\begin{smallmatrix}\xi\\ 0\end{smallmatrix}\right]$ with $\xi \neq 0$ gives the point at infinity).

Exercise 1.7.9. Show that if $\zeta = \frac{\xi'}{\eta'}$ also, then $\xi' = \lambda\xi$ and $\eta' = \lambda\eta$ for some nonzero complex number λ.

In terms of $\left[\begin{smallmatrix}\xi\\ \eta\end{smallmatrix}\right]$, (1.7.13) becomes

$$\begin{aligned} x^1 &= \frac{\xi\bar{\eta}+\bar{\xi}\eta}{\xi\bar{\xi}+\eta\bar{\eta}}, \\ x^2 &= \frac{\xi\bar{\eta}-\bar{\xi}\eta}{i(\xi\bar{\xi}+\eta\bar{\eta})}, \\ x^3 &= \frac{\xi\bar{\xi}-\eta\bar{\eta}}{\xi\bar{\xi}+\eta\bar{\eta}}, \\ x^4 &= -1\,. \end{aligned} \tag{1.7.14}$$

Reversing our point of view we find that any pair $\left[\begin{smallmatrix}\xi\\ \eta\end{smallmatrix}\right]$ of complex numbers, not both zero, gives rise to a point $P(x^1,x^2,x^3,-1)$ on S^- given by (1.7.14). Being on S^- [and therefore on $C_N^-(0)$] this point determines a past null direction R_P^- which, for emphasis, we prefer to denote $R^-_{\left[\begin{smallmatrix}\xi\\ \eta\end{smallmatrix}\right]}$. Multiplying P by the positive real number $\xi\bar{\xi}+\eta\bar{\eta}$ gives rise to another point X on $C_N^-(0)$: $X = X^a e_a$, where

$$\begin{aligned} X^1 &= \xi\bar{\eta}+\bar{\xi}\eta, & X^3 &= \xi\bar{\xi}-\eta\bar{\eta}, \\ X^2 &= \tfrac{1}{i}(\xi\bar{\eta}-\bar{\xi}\eta), & X^4 &= -(\xi\bar{\xi}+\eta\bar{\eta}). \end{aligned} \tag{1.7.15}$$

X, of course, also determines a past null direction R_X^- and, indeed,

$$R_X^- = R^-_{\left[\begin{smallmatrix}\xi\\ \eta\end{smallmatrix}\right]}\,. \tag{1.7.16}$$

Finally, we are in a position to tie all of these loose ends together. We begin with an element $A = \left[\begin{smallmatrix}\alpha & \beta\\ \gamma & \delta\end{smallmatrix}\right]$ of $SL(2,\mathbb{C})$. Then A defines a map which carries any pair $\left[\begin{smallmatrix}\xi\\ \eta\end{smallmatrix}\right]$, not both zero, onto another such pair which we denote

$$\begin{bmatrix}\hat{\xi}\\ \hat{\eta}\end{bmatrix} = A\begin{bmatrix}\xi\\ \eta\end{bmatrix} = \begin{bmatrix}\alpha & \beta\\ \gamma & \delta\end{bmatrix}\begin{bmatrix}\xi\\ \eta\end{bmatrix} = \begin{bmatrix}\alpha\xi+\beta\eta\\ \gamma\xi+\delta\eta\end{bmatrix}. \tag{1.7.17}$$

Observe that, thought of as a mapping on S^- (or $\bar{C}$), (1.7.17) defines a *fractional linear transformation.* Indeed, in terms of the extended complex number $\zeta = \xi/\eta$, (1.7.17) is equivalent to

$$\hat{\zeta} = \frac{\alpha\zeta+\beta}{\gamma\zeta+\delta}\,.$$

Now, $\left[\begin{smallmatrix}\hat{\xi}\\ \hat{\eta}\end{smallmatrix}\right]$ determines an $\hat{X}$ in $C_N^-(0)$ by (1.7.15) (with hats) and this, in turn, determines a past null direction $R_{\hat{X}}^- = R^-_{\left[\begin{smallmatrix}\hat{\xi}\\ \hat{\eta}\end{smallmatrix}\right]}$. On the other hand,

A also gives rise, via the spinor map, to a proper, orthochronous Lorentz transformation Λ_A which, regarded as an active transformation, carries X onto a point $\Lambda_A X$ on $\mathcal{C}_N^-(0)$. Our objective is to prove that $\hat{X}$ and $\Lambda_A X$ are, in fact, the same point so that, in particular, *the effect of the fractional linear transformation* (1.7.17) *determined by* A *on past null directions is the same as the effect of the Lorentz transformation* Λ_A *determined by* A, i.e.,

$$R^-_{\begin{bmatrix}\hat{\xi}\\ \hat{\eta}\end{bmatrix}} = R^-_{\Lambda_A X}\,. \tag{1.7.18}$$

To prove all of this we proceed as follows: Begin by solving (1.7.15) for the four products $\xi\bar{\eta}$, $\bar{\xi}\eta$, $\xi\bar{\xi}$ and $\eta\bar{\eta}$ to obtain

$$\xi\bar{\xi} = \tfrac{1}{2}(X^3 + X^4), \qquad \xi\bar{\eta} = \tfrac{1}{2}(X^1 + iX^2),$$
$$\bar{\xi}\eta = \tfrac{1}{2}(X^1 - iX^2), \qquad \eta\bar{\eta} = \tfrac{1}{2}(-X^3 + X^4),$$

so that

$$\frac{1}{2}\begin{bmatrix} X^3 + X^4 & X^1 + iX^2 \\ X^1 - iX^2 & -X^3 + X^4 \end{bmatrix} = \begin{bmatrix} \xi\bar{\xi} & \xi\bar{\eta} \\ \bar{\xi}\eta & \eta\bar{\eta} \end{bmatrix} = \begin{bmatrix} \xi \\ \eta \end{bmatrix}\begin{bmatrix} \bar{\xi} & \bar{\eta} \end{bmatrix}. \tag{1.7.19}$$

Now perform the unimodular transformation (1.7.17) to obtain $\begin{bmatrix}\hat{\xi}\\ \hat{\eta}\end{bmatrix}$. The corresponding point $\hat{X} = \hat{X}^a e_a$ given by (1.7.15) with hats must satisfy (1.7.19) with hats, i.e.,

$$\begin{aligned}
\frac{1}{2}\begin{bmatrix} \hat{X}^3 + \hat{X}^4 & \hat{X}^1 + i\hat{X}^2 \\ \hat{X}^1 - i\hat{X}^2 & -\hat{X}^3 + \hat{X}^4 \end{bmatrix} &= \frac{1}{2}\begin{bmatrix} \hat{X}^3 + \hat{X}^4 & \hat{X}^1 + i\hat{X}^2 \\ \hat{X}^1 - i\hat{X}^2 & -\hat{X}^3 + \hat{X}^4 \end{bmatrix}^{CT} \\
&= \left[\begin{bmatrix} \hat{\xi} \\ \hat{\eta} \end{bmatrix}\begin{bmatrix} \bar{\hat{\xi}} & \bar{\hat{\eta}} \end{bmatrix}\right]^{CT} \\
&= \begin{bmatrix} \bar{\hat{\xi}} & \bar{\hat{\eta}} \end{bmatrix}^{CT}\begin{bmatrix} \hat{\xi} \\ \hat{\eta} \end{bmatrix}^{CT} \\
&= \begin{bmatrix} \hat{\xi} \\ \hat{\eta} \end{bmatrix}\left[A\begin{bmatrix} \xi \\ \eta \end{bmatrix}\right]^{CT} \\
&= A\begin{bmatrix} \xi \\ \eta \end{bmatrix}\begin{bmatrix} \xi \\ \eta \end{bmatrix}^{CT} A^{CT} \\
&= A\left[\begin{bmatrix} \xi \\ \eta \end{bmatrix}\begin{bmatrix} \bar{\xi} & \bar{\eta} \end{bmatrix}\right] A^{CT} \\
&= \frac{1}{2}A\begin{bmatrix} X^3 + X^4 & X^1 + iX^2 \\ X^1 - iX^2 & -X^3 + X^4 \end{bmatrix} A^{CT}.
\end{aligned}$$

Thus,

$$\begin{bmatrix} \hat{X}^3+\hat{X}^4 & \hat{X}^1+i\hat{X}^2 \\ \hat{X}^1-i\hat{X}^2 & -\hat{X}^3+\hat{X}^4 \end{bmatrix} = A \begin{bmatrix} X^3+X^4 & X^1+iX^2 \\ X^1-iX^2 & -X^3+X^4 \end{bmatrix} A^{CT}. \quad (1.7.20)$$

Comparing (1.7.20) and (1.7.5) and the definition of Λ_A we find that, indeed,

$$\hat{X} = \Lambda_A X,$$

so that (1.7.18) is proved.

Since the spinor map is surjective, every element of $\mathcal{L}$ is Λ_A for some A in $SL(2, \mathbb{C})$ and so every element of $\mathcal{L}$ determines a fractional linear transformation of S^- which has the same effect on past null directions ($\pm A$ give rise to the same fractional linear transformation). Conversely, since the past null vectors span $\mathcal{M}$ (reconsider Exercise 1.2.1 and select only past-directed vectors), a Lorentz transformation is completely determined by its effect on past null directions. Some consequences of this correspondence between elements of $\mathcal{L}$ and fractional linear transformations of S^- are immediate.

Theorem 1.7.1. *A proper, orthochronous Lorentz transformation, if not the identity, leaves invariant at least one and at most two past null directions.*

This follows at once from the familiar fact that any fractional linear transformation of the Riemann sphere, if not the identity, has two (possibly coincident) fixed points (see [**A**]). Another well-known property of fractional linear transformations is that they are completely determined by their values on any three distinct points in the extended complex plane (see [**A**]). Hence:

Theorem 1.7.2. *A proper, orthochronous Lorentz transformation is completely determined by its effect on any three distinct past null directions. More precisely, given two sets of three distinct past null directions there is one and only one element of $\mathcal{L}$ which carries the first set (one-to-one) onto the second set.*

As our final application we will derive a remarkable result of Penrose [**Pen**] related to what has been called the "invisibility of the Lorentz contraction". An admissible observer $\mathcal{O}$ "observes" in a quite specific and well-defined way. One pictures the observer's frame of reference as a spatial coordinate grid with clocks located at the lattice points of the grid and either recording devices or assistants stationed with the clocks to take all of the required local readings. $\mathcal{O}$ then "observes", say, a moving sphere by either turning on the devices or alerting the assistants to record the arrival times at their locations of various points on the sphere. When things have calmed down again $\mathcal{O}$ will collect all of this data for analysis. He may then, for example, construct a "picture" of the sphere by selecting (arbitrarily) some instant of his time, collecting together all of the locations in his frame which recorded the passage of a point on the boundary of the sphere at that instant and "plotting" these points in his frame. In this way he will find himself constructing, not a sphere, but an ellipsoid due to length contraction in the direction of motion.

What our observer $\mathcal{O}$ actually "sees" (through his eye or a camera lens), however, is not so straightforward. We wish to construct an (admittedly idealized) geometrical representation in $\mathcal{M}$ of this "field of vision".

It is a clear evening and, as you stroll outside, you glance up and see the Big Dipper. More precisely, you direct the surface of your eye toward a group of incoming photons (idealize and assume one from each star in the constellation). Regardless of when they left their sources these photons arrive at this surface simultaneously (in your reference frame) and thereby create a pattern (image) which is recorded by your brain. This pattern is what you "see". Where can we find it in $\mathcal{M}$? Each of the photons you see has a worldline in $\mathcal{M}$ which lies along the past null cone $\mathcal{C}_N^-(0)$ (you are located at the origin of your coordinate system and the image is registered in your brain at $x^4 = 0$). Just slightly before $x^4 = 0$ the photons impacted the surface of your eye and formed their image. At $x^4 = -1$ the photons were all on a sphere of radius 1 about the origin of your coordinate system and formed on this sphere the same pattern that your eye registered a bit later. Projecting this image down to the plane $x^4 = -1$ in $\mathcal{M}$ we find the worldlines of these photons intersecting S^- in the very image that you "see". As a geometrical representation of what you see (at the event $x^1 = x^2 = x^3 = x^4 = 0$) we therefore take the intersections with S^- of the worldlines of all the photons that trigger your brain to record an image at $x^4 = 0$.

Now we ask the following question. Suppose that what you see is not the Big Dipper, but something with a *circular* outline, e.g., a sphere at rest in your reference frame. What is seen by another admissible observer, moving relative to your frame, but momentarily coincident with you at the origin? According to the new observer the sphere is *moving* and so certainly must "appear" contracted in the direction of motion. Surely, he must "see" an elliptical, not a circular image.

But he does not! We propose to argue that, despite the Lorentz contraction in the direction of motion, the sphere will still present a circular outline to $\hat{\mathcal{O}}$ (although, in a degenerate case, the circle may "appear" straight). Indeed, this is merely a reflection of yet another familiar property of fractional linear transformations of the Riemann sphere: they carry circles onto circles. Thus, if Λ is the Lorentz transformation relating $\mathcal{S}$ and $\hat{\mathcal{S}}$, then, regarded as an active transformation on past null directions, it carries any family of such null directions which intersect S^- in a circle onto another such family. In somewhat more detail we recall (page 80) that, for each x

in $\mathcal{M}$, $\Lambda(x)$ $[= L^{-1}(x)]$ has the same position and time in $\mathcal{S}$ that x has in $\hat{\mathcal{S}}$. In particular, $\Lambda(x) \in S^-$ if and only if $x \in \hat{S}^-$. Thus, $\Lambda(R_x^-) = R_{\Lambda(x)}^-$ "looks the same" to $\mathcal{O}$ at $x^4 = 0$ as R_x^- "looks" to $\hat{\mathcal{O}}$ at $\hat{x}^4 = 0$ (same relative position in the sky). Now, if we have a family $\mathcal{N}$ of past null directions (forming a certain "image" for $\mathcal{O}$ at $x^4 = 0$) it follows that the appearance of this image for $\hat{\mathcal{O}}$ at $\hat{x}^4 = 0$ will be the same as the appearance of $\Lambda(\mathcal{N})$ to $\mathcal{O}$ at $x^4 = 0$. If the rays in $\mathcal{N}$ present a circular outline to $\mathcal{O}$ at $x^4 = 0$, so will $\Lambda(\mathcal{N})$ and therefore $\hat{\mathcal{O}}$ will also see a circular outline at $\hat{x}^4 = 0$. $\mathcal{O}$ and $\hat{\mathcal{O}}$ both "see" a circular outline.

Exercise 1.7.10. Describe the "degenerate case" in which the circle "appears" straight.

Exercise 1.7.11. Offer a plausible physical explanation for this "invisibility of the Lorentz contraction". *Hint:* For $\mathcal{O}$ the photons which arrive simultaneously at the surface of his eye to form their image also left the sphere simultaneously. Is this true for $\hat{\mathcal{O}}$?

1.8 Particles and Interactions

A billiard ball rolling with constant speed in a straight line collides with another billiard ball, initially at rest, and the two balls rebound from the impact. The actual physical mechanisms involved in such an interaction are quite complicated, having to do with the electrical repulsion between electrons in the atoms at the surfaces of the two balls. Nevertheless, much can be said about the motion which results from such a collision even without detailed information about this electromagnetic interaction. What makes this possible is the idea (one of the most profound and powerful in all of physics) that such situations are often governed by *conservation laws.* Specifically, the conservation of Newtonian momentum has immediate implications for the motion of our billiard balls (for example, that, assuming the collision is glancing rather than head-on, they will separate along paths that form a right angle) and these predictions were well borne out by observation, at least until this century. However, Newtonian physics would make precisely the same predictions if the billiard balls were replaced by protons travelling at speeds comparable to that of light and here the observational evidence does not support these conclusions (e.g., the protons generally separate along paths which form an angle *less* than $90°$). In this section we shall investigate the relativistic alternative to the classical principles of the conservation of momentum and energy and draw some elementary consequences from it. First, though, some definitions.

A *material particle* in $\mathcal{M}$ is a pair (α, m), where $\alpha\colon I \to \mathcal{M}$ is a timelike worldline parametrized by proper time τ and m is a positive real number called the particle's *proper mass* (and is to be identified intuitively with the "inertial mass" of the particle from Newtonian mechanics). (α, m) is called a *free material particle* if α is of the form $\alpha(\tau) = x_0 + \tau U$ for some fixed event x_0 and unit timelike vector U. Recall that, for any timelike worldline $\alpha(\tau)$ the proper time derivative $\alpha'(\tau)$ is called the world velocity of α and denoted $U = U(\tau)$. The *world momentum* (or *energy-momentum*) of (α, m) is denoted P and defined by

$$P = P(\tau) = mU(\tau).$$

Notice that, since $U \cdot U = -1$ (Exercise 1.4.10), we have

$$P \cdot P = -m^2 . \tag{1.8.1}$$

Now fix an arbitrary admissible basis $\{e_a\}$. Writing $P = P^a e_a$ and using notation analogous to that established in (1.4.9) and (1.4.10) we have

$$P = (P^1, P^2, P^3, P^4) = m\gamma(\vec{u}, 1) = (\vec{p}, m\gamma),$$

where $\vec{p} = (P^1, P^2, P^3)$ is called the *relative 3-momentum of* (α, m) *in* $\{e_a\}$. Notice that if $\gamma = (1 - \beta^2)^{-\frac{1}{2}}$ is near 1, i.e., if the speed of (α, m) relative to $\{e_a\}$ is small, then $\vec{p}$ is approximately equal to $m\vec{u}$, the classical Newtonian momentum of (α, m) in $\{e_a\}$. The quantity $m\gamma = \frac{m}{\sqrt{1-\beta^2}}$ is sometimes referred to as the "relativistic mass" of (α, m) relative to $\{e_a\}$ since it permits one to retain a formal similarity between the Newtonian and relativistic definitions of momentum ("mass times velocity"). Inertial mass was regarded in classical physics as a measure of the particle's resistance to acceleration. From the relativistic point of view this resistance must become unbounded as $\beta \to 1$ and $m\gamma$ certainly has this property. We prefer, however, to avoid the quite misleading attitude that "mass increases with velocity" and simply abandon the Newtonian view that momentum is a linear function of velocity.

We shall denote by $|\vec{p}|$ the usual Euclidean magnitude of the relative 3-momentum in $\{e_a\}$, i.e., $|\vec{p}|^2 = (P^1)^2 + (P^2)^2 + (P^3)^2$. To see more clearly the relationship between P and more familiar Newtonian concepts we use the binomial expansion

$$\gamma = (1 - \beta^2)^{-\frac{1}{2}} = 1 + \tfrac{1}{2}\beta^2 + \tfrac{3}{8}\beta^4 + \cdots \tag{1.8.2}$$

of γ (valid since $|\beta| < 1$) to write

$$P^i = m\gamma u^i = mu^i + \tfrac{1}{2} mu^i \beta^2 + \cdots, \quad i = 1, 2, 3, \text{ and} \tag{1.8.3}$$

$$P^4 = m\gamma = m + \tfrac{1}{2} m\beta^2 + \cdots . \tag{1.8.4}$$

The nonlinear terms in (1.8.3) are absent from the Newtonian definition, but are crucial to the relativistic theory since they force $|\vec{p}|$ to become unbounded as $\beta \to 1$, i.e., they impose the "speed limit" on material particles relative to admissible frames of reference.

The physical interpretation of (1.8.4) is much more interesting. Notice, in particular, the appearance of the term $\frac{1}{2}m\beta^2$ corresponding to the classical kinetic energy. The presence of this term leads us to call P^4 the *total relativistic energy of* (α, m) *in* $\{e_a\}$ and denote it E.

$$E = -P \cdot e_4 = P^4 = m\gamma = m + \tfrac{1}{2}m\beta^2 + \dots \quad (1.8.5)$$

Exercise 1.8.1. Show that, relative to any admissible basis $\{e_a\}$,

$$m^2 = E^2 - |\,\vec{p}\,|^2 . \quad (1.8.6)$$

A few words of caution are in order here. The concept of "energy" in classical physics is quite a subtle one. Many different types of energy are defined in different situations, but each is in one way or another intuitively related to a system's "ability to do work". Now, simply calling P^4 the total relativistic energy of our particle does not ensure that this intuitive interpretation is still valid. Whether or not the name is appropriate can only be determined experimentally. In particular, one should determine whether or not the presence of the term m in (1.8.5) is consistent with this interpretation. Observe that when $\beta = 0$ (i.e., in the instantaneous rest frame of the particle) $P^4 = E = m$ ($= mc^2$ in traditional units) so that even when the particle is at rest relative to an admissible frame it still has "energy" in this frame, the amount being numerically equal to m. If this is really "energy" in the classical sense, it should be capable of doing work, i.e., it should be possible to "liberate" (and use) it. That this is indeed possible is demonstrated daily in particle physics laboratories and, fortunately not so often, in the explosion of atomic and nuclear bombs.

It is remarkable that the classically distinct concepts of momentum, energy and mass find themselves so naturally integrated into the single relativistic notion of world momentum (energy-momentum). We ask the reader to show that the process was indeed natural in the sense that if one believes that relativistic momentum should be represented by a vector in $\mathcal{M}$ and that the first three components of $P = mU$ are "right", then one has no choice about the fourth component.

Exercise 1.8.2. Show that two vectors v and w in $\mathcal{M}$ with the same spatial components relative to every admissible basis (i.e., $v^1 = w^1$, $v^2 = w^2$ and $v^3 = w^3$ for *every* $\{e_a\}$) must, in fact, be equal. *Hint:* It will be enough to show that a vector whose first three components are zero in every admissible coordinate system must be the zero vector.

Special relativity is of little interest to those who study colliding billiard balls (the relative speeds are so small that any "relativistic effects" are negligible). On the other hand, when the colliding objects are elementary particles (protons, neutrons, electrons, mesons, etc.) these relativistic effects are the dominant features. Such interactions between elementary particles, however, very often involve not only material particles, but photons as well and we wish to include these in our study. Now, a photon is, in many ways, analogous to a free material particle. Relative to any admissible frame of reference it travels along a straight line with constant speed, i.e., it has a linear worldline. Since this worldline is null, however, it has no proper time parametrization and so no world velocity. Nevertheless, photons do possess "momentum" and "energy" and so should have a "world momentum" (witness, for example, the photoelectric effect in which photons collide with and eject electrons from their orbits in an atom). Unlike a material particle, however, the photon's characteristic feature is not mass, but energy (frequency, wavelength) and this is highly observer-dependent [e.g., wavelengths of photons emitted from the atoms of a star are "red-shifted" (lengthened) relative to those measured on earth for the same atoms because the stars are receding from us due to the expansion of the universe]. A hint as to how these features can be modelled in $\mathcal{M}$ is provided by:

Exercise 1.8.3. Let N be a future-directed null vector in $\mathcal{M}$ and $\{e_a\}$ an admissible basis with $N = N^a e_a$. Show that

$$N = \epsilon(\vec{e} + e_4), \tag{1.8.7}$$

where $\epsilon = -N \cdot e_4 = N^4$ and $\vec{e}$ is the *direction 3-vector of N relative to $\{e_a\}$*, i.e.,

$$\vec{e} = \left((N^1)^2 + (N^2)^2 + (N^3)^2\right)^{-\frac{1}{2}} (N^1 e_1 + N^2 e_2 + N^3 e_3).$$

Now, we define a *photon*[2] in $\mathcal{M}$ to be a pair (α, N), where N is a future-directed null vector called the photon's *world momentum* and $\alpha : I \to \mathcal{M}$ (I an interval in $\mathbb{R}$ containing 0) is given by $\alpha(t) = x_0 + tN$ for some fixed event x_0 in $\mathcal{M}$ and all t in I. Relative to any admissible basis $\{e_a\}$ the positive real number

$$\epsilon = -N \cdot e_4 = N^4$$

is called the *energy of* (α, N) *in* $\{e_a\}$ (see Figure 1.8.1). The *frequency* ν

[2]No quantum mechanical subtleties are to be inferred from our use of the term "photon". Our definition is intended to model any "massless" particle travelling at the speed of light.

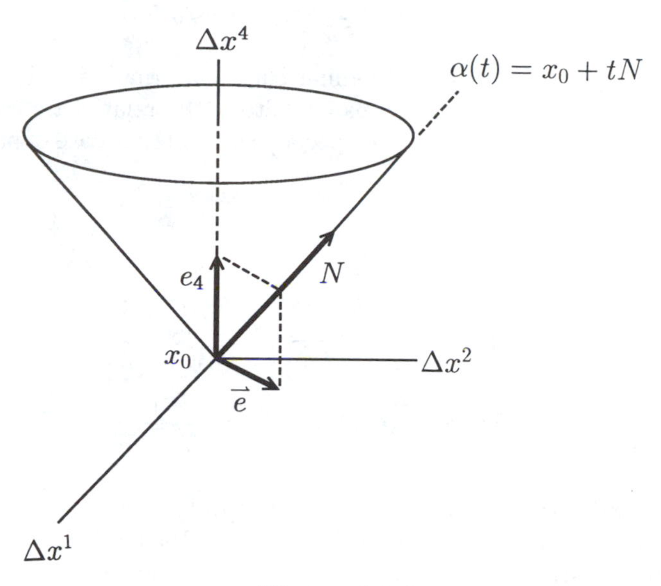

Figure 1.8.1

and *wavelength* λ of (α, N) in $\{e_a\}$ are defined by $\nu = \epsilon/h$ and $\lambda = 1/\nu$, where h is a constant (called *Planck's constant*).

It is interesting to compare the energies of a photon (α, N) in two different frames of reference. Thus, we let $\{e_a\}$ and $\{\hat{e}_a\}$ be two admissible bases and write $N = \epsilon(\vec{e} + e_4) = \hat{\epsilon}(\vec{\hat{e}} + \hat{e}_4)$, where $\epsilon = -N \cdot e_4$ and $\hat{\epsilon} = -N \cdot \hat{e}_4$.

Exercise 1.8.4. Show that $\hat{\epsilon} = \gamma\epsilon(1 - \beta(\vec{e} \cdot \vec{d}))$. *Hint:* Use Exercise 1.3.10.

But $\vec{e}$ and $\vec{d}$ both lie in the subspace spanned by e_1, e_2 and e_3 and the restriction of the Lorentz inner product to this subspace is just the usual positive definite inner product on $\mathbb{R}^3$. Thus, $\vec{e} \cdot \vec{d} = \cos\theta$, where θ is the angle in $\sum$ (the spatial coordinate system of the frame corresponding to $\{e_a\}$) between the direction of the photon and the direction of $\hat{\sum}$. We therefore obtain

$$\frac{\hat{\epsilon}}{\epsilon} = \frac{\hat{\nu}}{\nu} = \gamma(1 - \beta\cos\theta) = \frac{1 - \beta\cos\theta}{\sqrt{1 - \beta^2}} \tag{1.8.8}$$

which is the relativistic formula for the *Doppler effect.* Using the binomial expansion (1.8.2) for γ gives

$$\frac{\hat{\epsilon}}{\epsilon} = \frac{\hat{\nu}}{\nu} = (1-\beta\cos\theta)+\frac{1}{2}\beta^2(1-\beta\cos\theta)+\cdots . \tag{1.8.9}$$

The first term $1-\beta\cos\theta$ is the familiar classical formula for the Doppler effect, whereas the remaining terms constitute the relativistic correction contributed by time dilation. Three special cases of (1.8.8) are of particular interest.

$$\theta = 0(\text{so } \vec{d}=\vec{e}) \implies \frac{\hat{\nu}}{\nu} = \sqrt{\frac{1-\beta}{1+\beta}}, \tag{1.8.10}$$

$$\theta = \pi(\text{so } \vec{d}=-\vec{e}) \implies \frac{\hat{\nu}}{\nu} = \sqrt{\frac{1+\beta}{1-\beta}}, \tag{1.8.11}$$

$$\theta = \frac{\pi}{2}(\text{so } \vec{e}\cdot\vec{d}=0) \implies \frac{\hat{\nu}}{\nu} = \frac{1}{\sqrt{1-\beta^2}}. \tag{1.8.12}$$

The classical theory predicts no Doppler shift in the case $\theta = \pi/2$ so that the formula (1.8.12) for the so-called *transverse Doppler effect* represents a purely relativistic phenomenon. Experimental verification of (1.8.12) was first accomplished by Ives and Stilwell [**IS**] and is regarded as direct confirmation of the reality of time dilation.

Next we wish to compare the angles θ and $\hat{\theta}$ defined by $\cos\theta = \vec{e}\cdot\vec{d}$ and $\cos\hat{\theta} = \vec{\hat{e}}\cdot\vec{\hat{d}}$.

Exercise 1.8.5. Let $\vec{\hat{u}}$ denote the velocity 3-vector of $\mathcal{S}$ relative to $\hat{\mathcal{S}}$ [(1.3.12) and (1.3.15)] and show that

$$\vec{\hat{u}} = -\gamma\beta(\vec{d}+\beta e_4). \tag{1.8.13}$$

From (1.8.13) we conclude that

$$\vec{\hat{d}} = -\gamma(\vec{d}+\beta e_4). \tag{1.8.14}$$

Since $N = \epsilon(\vec{e}+e_4) = \hat{\epsilon}(\vec{\hat{e}}+\hat{e}_4)$ we obtain from the definitions of θ and $\hat{\theta}$

$$\vec{d}\cdot N = \epsilon\cos\theta \text{ and } \vec{\hat{d}}\cdot N = \hat{\epsilon}\cos\hat{\theta}. \tag{1.8.15}$$

Now, $\hat{\epsilon}\cos\hat{\theta} = \vec{\hat{d}}\cdot N = -\gamma(\vec{d}\cdot N+\beta e_4\cdot N) = -\gamma(\epsilon\cos\theta-\beta\epsilon) = -\gamma\epsilon\cos\theta+\gamma\beta\epsilon$. Thus,

$$\frac{\hat{\epsilon}}{\epsilon}\cos\hat{\theta} = \gamma(\beta-\cos\theta)$$

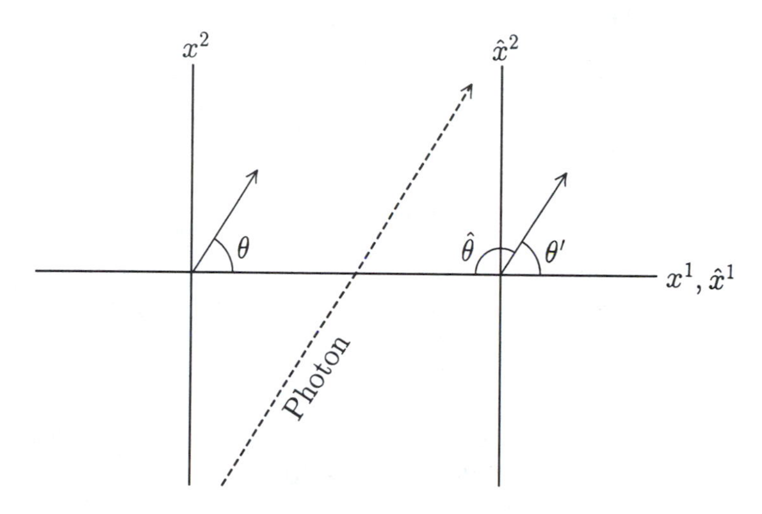

Figure 1.8.2

which, by (1.8.8), we may write as

$$\gamma(1 - \beta \cos\theta) \cos\hat{\theta} = \gamma(\beta - \cos\theta),$$

or

$$\cos\hat{\theta} = \frac{\beta - \cos\theta}{1 - \beta\cos\theta}. \tag{1.8.16}$$

Generally, however, one would be more interested in comparing the angles θ and $\theta' = \pi - \hat{\theta}$, e.g., when the spatial axes are in standard orientation as in Figure 1.8.2. Since $\cos\theta' = -\cos\hat{\theta}$, (1.8.16) becomes the standard *relativistic aberration formula*

$$\cos\theta' = \frac{\cos\theta - \beta}{1 - \beta\cos\theta}. \tag{1.8.17}$$

At this point we have assembled enough machinery to study some of the physical interactions to which the special theory of relativity is routinely applied. Henceforth, we shall use the term *free particle* to refer to either a free material particle or a photon. If $\mathcal{A}$ is a finite set of free particles, then each element of $\mathcal{A}$ has a unique world momentum vector. The sum of these vectors is called the *total world momentum of* $\mathcal{A}$. A *contact interaction* in $\mathcal{M}$ is a triple $(\mathcal{A}, x, \tilde{\mathcal{A}})$, where $\mathcal{A}$ and $\tilde{\mathcal{A}}$ are two finite sets of free particles, neither of which contains a pair of particles with linearly dependent world momenta, and x is an event such that

(a) x is the terminal point of all the particles in $\mathcal{A}$ [i.e., for each (α, m) in $\mathcal{A}$ with $\alpha : [a, b] \to \mathcal{M}$, we have $\alpha(b) = x$],
(b) x is the initial point of all the particles in $\tilde{\mathcal{A}}$, and
(c) the total world momentum of $\mathcal{A}$ equals the total world momentum of $\tilde{\mathcal{A}}$.

Intuitively, the event x should be regarded as the collision of all the particles in $\mathcal{A}$, from which emerge all the particles in $\tilde{\mathcal{A}}$ (which may be physically quite different than those in $\mathcal{A}$, e.g., it has been observed that the collision of two electrons can result in three electrons and a positron). The prohibition on pairs of particles with linearly dependent world momenta in the same set is based on the presumption that two such particles would be physically indistinguishable. Property (c) is called the *conservation of world momentum* and contains the appropriate relativistic generalizations of two classical conservation principles: the conservation of momentum and the conservation of energy.

Several conclusions concerning contact interactions can be drawn directly from the results we have available. Consider, for example, an interaction $(\mathcal{A}, x, \tilde{\mathcal{A}})$ in which $\tilde{\mathcal{A}}$ consists of a single photon. Then the total world momentum of $\tilde{\mathcal{A}}$ is null so the same must be true of $\mathcal{A}$. Since the world momenta of the individual particles in $\mathcal{A}$ are all either timelike or null and all are future-directed, Lemma 1.4.3 implies that all of these world momenta must be null and parallel. Since $\mathcal{A}$ cannot contain two distinct photons with parallel world momenta, $\mathcal{A}$ must also consist of a single photon which, by (c), must have the same world momentum as the photon in $\tilde{\mathcal{A}}$. In essence, "nothing happened at x". We conclude that *no nontrivial interaction of the type modelled by our definition can result in a single photon and nothing else.*

A contact interaction $(\mathcal{A}, x, \tilde{\mathcal{A}})$ is called a *disintegration* or *decay* if $\mathcal{A}$ consists of a single free particle.

Exercise 1.8.6. Analyze a disintegration $(\mathcal{A}, x, \tilde{\mathcal{A}})$ in which $\mathcal{A}$ consists of a single photon.

Suppose that $\mathcal{A}$ consists of a single free material particle of proper mass m_0 and $\tilde{\mathcal{A}}$ consists of two material particles with proper masses m_1 and m_2 (such disintegrations do, in fact, occur in nature, e.g., in α-emission). Let P_0, P_1 and P_2 be the world momenta of the particles with masses m_0, m_1 and m_2 respectively. Appealing to (1.8.1), the Reversed Triangle Inequality (Theorem 1.4.2) and the fact that P_1 and P_2 are linearly independent we find that

$$m_0 > m_1 + m_2. \tag{1.8.18}$$

The excess mass $m_0-(m_1+m_2)$ of the initial particle is regarded as a measure of the amount of energy required to split m_0 into two pieces. Stated somewhat differently, when the two particles in $\tilde{\mathcal{A}}$ were held together to form the single particle in $\mathcal{A}$ the "binding energy" contributed to the mass of this latter particle, while, after the decay, the difference in mass appears in the form of kinetic energy of the generated particles.

Exercise 1.8.7. Show that a free electron cannot emit or absorb a photon. *Hint:* The contradiction arises from the constancy of the proper mass m_e of an electron. A more complicated system such as an atom or molecule whose proper mass can vary with its energy state (these being determined by the principles of quantum mechanics) is not prohibited from absorbing or emitting photons.

Next we consider two examples of more detailed calculations for specific interactions, each of which models an important reaction in particle physics. We should emphasize at the outset, however, that the conservation of world momentum alone is almost never sufficient to determine all of the details of the resulting motion. Additional conservation laws (e.g., of "spin") can reduce the degree of indeterminacy, but quantum mechanics imposes a positive lower bound on the extent to which this is possible. As final preparation for our examples we will need to record the conservation of world momentum in component form relative to an arbitrary admissible basis $\{e_a\}$. Thus we write

$$\sum_{\mathcal{A}} m\gamma u^i + \sum_{\mathcal{A}} h\nu e^i = \sum_{\tilde{\mathcal{A}}} \tilde{m}\tilde{\gamma}\tilde{u}^i + \sum_{\tilde{\mathcal{A}}} h\tilde{\nu}\tilde{e}^i, \quad i=1,2,3, \tag{1.8.19}$$

$$\sum_{\mathcal{A}} m\gamma + \sum_{\mathcal{A}} h\nu = \sum_{\tilde{\mathcal{A}}} \tilde{m}\tilde{\gamma} + \sum_{\tilde{\mathcal{A}}} h\tilde{\nu}, \tag{1.8.20}$$

where the first and third sums in each are over all the material particles in $\mathcal{A}$ and $\tilde{\mathcal{A}}$ respectively, whereas the second and fourth sums are over all of the photons in $\mathcal{A}$ and $\tilde{\mathcal{A}}$ respectively.

In our first example we describe the so-called *Compton effect.* The physical situation we propose to model is the following: A photon collides with an electron and rebounds from it (generally with a different frequency), while the electron recoils from the collision. Thus, we consider a contact interaction $(\mathcal{A}, x, \tilde{\mathcal{A}})$, where $\mathcal{A}$ consists of a photon with world momentum N and a material particle with proper mass m_e and world velocity U and $\tilde{\mathcal{A}}$ consists of a photon with world momentum $\tilde{N}$ and a material particle with proper mass m_e and world velocity $\tilde{U}$. We analyze the interaction in a frame of reference in which the material particle in $\mathcal{A}$ is at rest (time axis parallel to the worldline of the particle). In this frame the conservation of world momentum equations (1.8.19) and (1.8.20) become

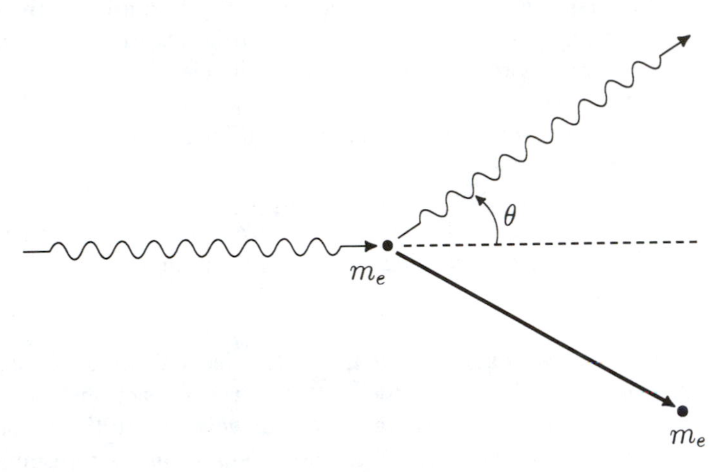

Figure 1.8.3

(since $u^i = 0,\ \gamma = 1$)

$$m_e\tilde{\gamma}\tilde{u}^i + h\tilde{\nu}\tilde{e}^i = h\nu e^i, \quad i = 1, 2, 3, \tag{1.8.21}$$

$$m_e\tilde{\gamma} + h\tilde{\nu} = m_e + h\nu. \tag{1.8.22}$$

Let $\xi = \tilde{\nu}/\nu$ and $k = h\nu/m_e$. We denote by θ the angle between the direction vectors of the two photons in the given frame of reference, i.e., $\cos\theta = e^1\tilde{e}^1 + e^2\tilde{e}^2 + e^3\tilde{e}^3$ (see Figure 1.8.3). With this notation (1.8.21) and (1.8.22) can be written

$$\tilde{\gamma}\tilde{u}^i = ke^i - \xi k\tilde{e}^i, \qquad i = 1, 2, 3, \tag{1.8.23}$$

$$\tilde{\gamma} - 1 = k(1 - \xi). \tag{1.8.24}$$

Since $\tilde{\beta}^2 = (\tilde{u}^1)^2 + (\tilde{u}^2)^2 + (\tilde{u}^3)^2 = 1 - \tilde{\gamma}^{-2}$, when we (Euclidean) dot each side of (1.8.23) with itself we obtain

$$\tilde{\gamma}^2\tilde{\beta}^2 = k^2(1 - 2\xi\cos\theta + \xi^2) = \tilde{\gamma}^2 - 1.$$

Thus,

$$\tilde{\gamma} + 1 = \frac{k^2(1 - 2\xi\cos\theta + \xi^2)}{\tilde{\gamma} - 1} = \frac{k(1 - 2\xi\cos\theta + \xi^2)}{1 - \xi} \tag{1.8.25}$$

by (1.8.24). Subtracting (1.8.24) from (1.8.25) we next obtain

$$2 = \frac{k(1-2\xi\cos\theta+\xi^2)-k(1-\xi)^2}{1-\xi} = \frac{k(2\xi-2\xi\cos\theta)}{1-\xi}$$
$$= \frac{2k\xi(1-\cos\theta)}{1-\xi} = \frac{4k\xi\sin^2(\frac{\theta}{2})}{1-\xi}.$$

Thus, $2k\xi\sin^2(\frac{\theta}{2})=1-\xi$ and therefore $\xi=\frac{1}{1+2k\sin^2(\theta/2)}$ so $\tilde{\nu}=\frac{\nu}{1+2k\sin^2(\theta/2)}$. From this we compute

$$\tilde{\lambda}-\lambda=\frac{1}{\tilde{\nu}}-\frac{1}{\nu}=\frac{1+2k\sin^2(\frac{\theta}{2})}{\nu}-\frac{1}{\nu}=\frac{2k\sin^2(\frac{\theta}{2})}{\nu}.$$

We conclude that

$$\tilde{\lambda}-\lambda = \frac{2h}{m_e}\sin^2(\tfrac{\theta}{2}) \tag{1.8.26}$$

which gives the change in wavelength of the photon as a function of the angle θ through which it is deflected (in the frame in which the electron is initially at rest). Observe that this change in wavelength does not depend on the wavelength λ of the incident photon, but only on the angle through which it is deflected. Moreover, this difference ranges from a minimum of 0 when $\theta = 0$ (the photon and electron do not interact physically) to a maximum of

$$\Delta\lambda_{max} = \frac{2h}{m_e} \tag{1.8.27}$$

when $\theta = \pi$ (the photon is thrown straight back). This maximum change in wavelength is a characteristic feature of the electron; the quantity h/m_e is called the *Compton wavelength* of the electron.

Next we consider an *inelastic collision* between two material particles. The situation we have in mind is as follows: two free material particles with masses m_1 and m_2 collide and coalesce to form a third material particle of mass m_3. Classically it is assumed that $m_3 = m_1 + m_2$ and on the basis of this assumption (and the conservation of Newtonian momentum) one finds that kinetic energy is lost during the collision. In Newtonian mechanics this lost kinetic energy disappears entirely from the mechanical picture in the sense that it is viewed as having taken the form of heat in the combined particle and therefore cannot be discussed further by the methods of mechanics. We shall see that this rather unsatisfactory feature of Newtonian mechanics is avoided in relativistic mechanics by observing that conservation of world momentum (which includes the conservation of energy) requires that the "hot" combined particle have a proper mass which is greater than the sum of the two masses from which it is formed, the difference $m_3 - (m_1 + m_2)$ being a measure of the energy required to bind the two particles together; this energy "acts like mass" in the combined particle.

We shall therefore consider a contact interaction $(\mathcal{A}, x, \tilde{\mathcal{A}})$, where $\mathcal{A}$ consists of two free material particles with proper masses m_1 and m_2 and world velocities U_1 and U_2 respectively and $\tilde{\mathcal{A}}$ consists of one free material particle with proper mass m_3 and world velocity U_3. Conservation of world momentum requires that

$$m_3 U_3 = m_1 U_1 + m_2 U_2. \tag{1.8.28}$$

Again observe that the Reversed Triangle Inequality (Theorem 1.4.2) gives $m_3 > m_1 + m_2$. Moreover, since $U_1 \cdot U_1 = U_2 \cdot U_2 = U_3 \cdot U_3 = -1$ we obtain [by dotting both sides of (1.8.28) with itself and using any admissible frame of reference]

$$\begin{aligned} m_3^2 &= m_1^2 + m_2^2 - 2m_1 m_2 U_1 \cdot U_2, \\ m_3^2 &= m_1^2 + m_2^2 - 2m_1 m_2 \gamma_1 \gamma_2 (\vec{u}_1, 1) \cdot (\vec{u}_2, 1), \\ m_3^2 &= m_1^2 + m_2^2 + 2m_1 m_2 \gamma_1 \gamma_2 (1 - \vec{u}_1 \cdot \vec{u}_2), \end{aligned} \tag{1.8.29}$$

which yields the resultant mass m_3 in terms of m_1, m_2 and the quantities u_1^i and u_2^i, $i = 1, 2, 3$, which can be measured in the given frame of reference. From (1.8.28) one can then compute U_3.

We wish to obtain an approximate formula for m_3 which can be compared with the Newtonian expression for the loss in kinetic energy. Assume that β_1 and β_2 are small so that γ_1 and γ_2 are approximately 1 (the frame of reference is then no longer arbitrary, of course). We will eventually take $\gamma_1 \gamma_2 \approx 1$, but first we consider the somewhat better approximations

$$\gamma_j \approx 1 + \tfrac{1}{2}\beta_j^2, \quad j = 1, 2,$$

obtained from the binomial expansion (1.8.2). Then

$$\gamma_1 \gamma_2 \approx (1 + \tfrac{1}{2}\beta_1^2)(1 + \tfrac{1}{2}\beta_2^2) = 1 + \tfrac{1}{2}\beta_1^2 + \tfrac{1}{2}\beta_2^2 + \tfrac{1}{4}\beta_1^2 \beta_2^2,$$

$$\gamma_1 \gamma_2 \approx 1 + \tfrac{1}{2}\beta_1^2 + \tfrac{1}{2}\beta_2^2. \tag{1.8.30}$$

Exercise 1.8.8. Show that (1.8.29) and (1.8.30) yield

$$m_3^2 \approx (m_1 + m_2)^2 + m_1 m_2 (\beta_1^2 + \beta_2^2 - 2\gamma_1 \gamma_2 (\vec{u}_1 \cdot \vec{u}_2)). \tag{1.8.31}$$

Now taking $\gamma_1 \gamma_2 \approx 1$ in (1.8.31) we obtain

$$m_3^2 \approx (m_1 + m_2)^2 + m_1 m_2 |\vec{v}|^2, \tag{1.8.32}$$

where $|\vec{v}|^2$ is the squared magnitude of the relative velocity $\vec{v} = \vec{u}_1 - \vec{u}_2$ of the two particles in $\mathcal{A}$ as measured in the given frame. From (1.8.32) we obtain

$$m_3 \approx m_1 + m_2 + \frac{m_1 m_2}{m_1 + m_2 + m_3} |\vec{v}|^2 .$$

Assuming that $m_3 \approx m_1 + m_2$ in the denominator we arrive at

$$m_3 \approx m_1 + m_2 + \frac{1}{2}\frac{m_1 m_2}{m_1 + m_2} |\vec{v}|^2, \tag{1.8.33}$$

where the last term represents the approximate gain in proper mass as a result of the collision.

Now, in Newtonian mechanics it is assumed that $m_3 = m_1 + m_2$ so that conservation of Newtonian momentum requires that

$$(m_1 + m_2)\vec{u}_3 = m_1 \vec{u}_1 + m_2 \vec{u}_2 . \tag{1.8.34}$$

Taking the Euclidean dot product of each side of (1.8.34) with itself then yields

$$(m_1 + m_2)^2 |\vec{u}_3|^2 = m_1^2 |\vec{u}_1|^2 + m_2^2 |\vec{u}_2|^2 + 2m_1 m_2 (\vec{u}_1 \cdot \vec{u}_2). \tag{1.8.35}$$

Exercise 1.8.9. Use (1.8.35) to show that the classical loss in kinetic energy due to the collision is given by

$$\frac{1}{2}m_1 |\vec{u}_1|^2 + \frac{1}{2}m_2 |\vec{u}_2|^2 - \frac{1}{2}(m_1 + m_2)|\vec{u}_3|^2 = \frac{1}{2}\frac{m_1 m_2}{m_1 + m_2}|\vec{v}|^2,$$

where $|\vec{v}|^2 = |\vec{u}_1 - \vec{u}_2|^2$.

Consequently, the Newtonian expression for the lost kinetic energy coincides with the relativistic formula (1.8.33) for the approximate gain in proper mass of the combined particle.

2

Skew-Symmetric Linear Transformations and Electromagnetic Fields

2.1 Motivation via the Lorentz Law

A *charged particle* in $\mathcal{M}$ is a triple (α, m, e), where (α, m) is a material particle and e is a nonzero real number called the *charge* of the particle. A *free charged particle* is a charged particle (α, m, e), where $(\alpha,\ m)$ is a free material particle. Charged particles do two things of interest to us. By their very presence they create electromagnetic fields and they also respond to the fields created by other charges. Our objective in this chapter is to isolate the appropriate mathematical object with which to model an electromagnetic field in $\mathcal{M}$, derive many of its basic properties and then investigate these two activities.

Charged particles "respond" to the presence of an electromagnetic field by experiencing changes in world momentum. The quantitative nature of this response is expressed by a differential equation relating the proper time derivative of the particle's world momentum to the field. This *equation of motion* is generally taken to be the so-called *Lorentz World Force Law* which expresses the rate at which the particle's world momentum changes at each point on the worldline as a *linear* function of the particle's world velocity:

$$\frac{dP}{d\tau} = eFU, \tag{2.1.1}$$

where $U = U(\tau)$ is the particle's world velocity, $P = mU$ its world momentum and, at each point, $F\colon \mathcal{M} \to \mathcal{M}$ is a linear transformation defined in terms of the classical "electric and magnetic 3-vectors $\vec{E}$ and $\vec{B}$" at that point [(2.1.1) is an abbreviated version of the somewhat more accurate and considerably more cumbersome $\frac{dP(\tau)}{d\tau} = eF_{\alpha(\tau)}(U(\tau))$, where $F_{\alpha(\tau)}$ is the appropriate linear transformation at $\alpha(\tau) \in \mathcal{M}$]. We should point out that (2.1.1) can be regarded as an appropriate equation of motion for charged particles in an electromagnetic field only if the charges whose motion is to be governed by it have negligible contribution to the ambient field. It must be possible to regard the field as "given" and the charged particles as "test charges". The much more difficult question of the interactions between the given field and the fields created by the moving charges will not be considered here (see [**Par**]).

We argue now that the form of the Lorentz Law (2.1.1) suggests that the linear transformations F must be of a particular type ("skew-symmetric"). Indeed, rewriting (2.1.1) at each fixed point of $\mathcal{M}$ as

$$FU = \frac{m}{e}\frac{dU}{d\tau}$$

and dotting both sides with U gives

$$FU \cdot U = \frac{m}{e}\frac{dU}{d\tau} \cdot U = \frac{m}{e}A \cdot U = 0$$

since a material particle's world velocity and world acceleration are always orthogonal (Exercise 1.4.12). Since any unit timelike vector $u \in \mathcal{M}$ is the world velocity of some charged particle (construct one!) we find that, for any such u, $Fu \cdot u = 0$. Linearity therefore implies that $Fv \cdot v = 0$ for all timelike v. Now, if u and v are timelike and future-directed, then $u + v$ is also timelike and so $0 = F(u + v) \cdot (u + v) = (Fu + Fv) \cdot (u + v) = Fu \cdot v + Fv \cdot u = Fu \cdot v + u \cdot Fv$. Thus, $Fu \cdot v = -u \cdot Fv$. But $\mathcal{M}$ has a basis of future-directed timelike vectors so it follows that F must satisfy

$$Fx \cdot y = -x \cdot Fy \tag{2.1.2}$$

for all x and y in $\mathcal{M}$. A linear transformation $F : \mathcal{M} \to \mathcal{M}$ which satisfies (2.1.2) for all x and y in $\mathcal{M}$ is said to be *skew-symmetric* (with respect to the Lorentz inner product on $\mathcal{M}$).

At each fixed point in $\mathcal{M}$ we therefore elect to model an electromagnetic field by a skew-symmetric linear transformation F whose job it is to assign to the world velocity U of a charged particle passing through that point the change in world momentum $\frac{dP}{d\tau} = eFU$ that the particle should expect to experience due to the field. One would picture the electromagnetic field *in toto* therefore as a smooth assignment of such a linear transformation to each point in (some region of) $\mathcal{M}$ (although we shall find that nature imposes a condition — Maxwell's Equations — on the manner in which such an assignment can be made). In the next four sections we carry out a general investigation of skew-symmetric linear transformations on $\mathcal{M}$ and then turn to some physical applications in the last two sections.

2.2 Elementary Properties

Throughout this section F will represent a nonzero, skew-symmetric linear transformation on $\mathcal{M}$. The most obvious consequence of the definition

(2.1.2) of skew-symmetry is that

$$Fx \cdot x = x \cdot Fx = 0 \tag{2.2.1}$$

for all x in $\mathcal{M}$. If $\{e_a\}_{a=1}^4$ is an arbitrary admissible basis for $\mathcal{M}$ and we write $Fe_b = F^a{}_b e_a = F^1{}_b e_1 + F^2{}_b e_2 + F^3{}_b e_3 + F^4{}_b e_4$, then (2.2.1) implies $F^a{}_a = 0$ for $a = 1,2,3,4$, i.e., the diagonal entries in the matrix of F are all zero. In addition, for $i,j = 1,2,3$, $F^j{}_i = -F^i{}_j$, whereas $F^4{}_i = F^i{}_4$. Thus, the matrix of F relative to any admissible basis has the form

$$[F^a{}_b] = \begin{bmatrix} 0 & F^1{}_2 & F^1{}_3 & F^1{}_4 \\ -F^1{}_2 & 0 & F^2{}_3 & F^2{}_4 \\ -F^1{}_3 & -F^2{}_3 & 0 & F^3{}_4 \\ F^1{}_4 & F^2{}_4 & F^3{}_4 & 0 \end{bmatrix}. \tag{2.2.2}$$

Observe that, due to the fact that the inner product on $\mathcal{M}$ is indefinite, the matrix of a skew-symmetric linear transformation on $\mathcal{M}$ is not a skew-symmetric matrix (in the "time" part).

In order to establish contact with the notation usually used in physics we introduce, in each admissible basis $\{e_a\}$, two 3-vectors $\vec{E} = E^1 e_1 + E^2 e_2 + E^3 e_3$ and $\vec{B} = B^1 e_1 + B^2 e_2 + B^3 e_3$, where $E^1 = F^1{}_4$, $E^2 = F^2{}_4$, $E^3 = F^3{}_4$, $B^1 = F^2{}_3$, $B^2 = -F^1{}_3$ and $B^3 = F^1{}_2$. Thus, (2.2.2) can be written

$$[F^a{}_b] = \begin{bmatrix} 0 & B^3 & -B^2 & E^1 \\ -B^3 & 0 & B^1 & E^2 \\ B^2 & -B^1 & 0 & E^3 \\ E^1 & E^2 & E^3 & 0 \end{bmatrix}. \tag{2.2.3}$$

If F is thought of as describing an electromagnetic field at some point of $\mathcal{M}$, then $\vec{E}$ and $\vec{B}$ are regarded as the classical electric and magnetic field 3-vectors at that point as measured in $\{e_a\}$.

We consider two simple examples which, in Section 2.4, we will show to be fully and uniquely representative in the sense that for any skew-symmetric $F : \mathcal{M} \to \mathcal{M}$ there exists a basis relative to which the matrix of F has one of these forms, but no basis in which it has the other. First fix an admissible basis $\{e_a\}$ and a positive real number α and define a linear transformation F_N on $\mathcal{M}$ whose matrix relative to this basis is

$$\begin{bmatrix} 0 & 0 & 0 & 0 \\ 0 & 0 & \alpha & 0 \\ 0 & -\alpha & 0 & \alpha \\ 0 & 0 & \alpha & 0 \end{bmatrix}.$$

Then F_N is clearly skew-symmetric, $\vec{E} = \alpha e_3$ and $\vec{B} = \alpha e_1$, so an observer in this frame measures electric and magnetic 3-vectors that are perpendicular and have the same magnitude.

Next, fix an admissible basis $\{e_a\}$ and two non-negative real numbers δ and ϵ and let $F_R\colon \mathcal{M} \to \mathcal{M}$ be the linear transformation whose matrix relative to $\{e_a\}$ is

$$\begin{bmatrix} 0 & \delta & 0 & 0 \\ -\delta & 0 & 0 & 0 \\ 0 & 0 & 0 & \epsilon \\ 0 & 0 & \epsilon & 0 \end{bmatrix}.$$

Again, F_R is skew-symmetric. Moreover, $\vec{E} = \epsilon e_3$ and $\vec{B} = \delta e_3$ so an observer in this frame will measure electric and magnetic 3-vectors in the same direction and of magnitude ϵ and δ respectively.

Relative to another admissible basis $\{\hat{e}_a\}$ all of the $\hat{F}^a{}_b$, $\hat{E}^i$ and $\hat{B}^i$ are defined in the same way. Thus, if Λ is the Lorentz transformation associated with the orthogonal transformation L that carries $\{e_a\}$ onto $\{\hat{e}_a\}$, i.e., the matrix of L^{-1} relative to $\{\hat{e}_a\}$, then the matrix of F relative to $\{\hat{e}_a\}$ is $\Lambda[F^a{}_b]\Lambda^{-1}$, i.e., $\hat{F}^a{}_b = \Lambda^a{}_\alpha \Lambda_b{}^\beta F^\alpha{}_\beta$.

Exercise 2.2.1. With $[F^a{}_b]$ as in (2.2.3) and $\Lambda = \Lambda(\beta)$ for some β in $(-1, 1)$, show that

$$\begin{aligned} \hat{E}^1 &= E^1, & \hat{E}^2 &= \gamma(E^2 - \beta B^3), & \hat{E}^3 &= \gamma(E^3 + \beta B^2), \\ \hat{B}^1 &= B^1, & \hat{B}^2 &= \gamma(\beta E^3 + B^2), & \hat{B}^3 &= -\gamma(\beta E^2 - B^3)\,. \end{aligned} \tag{2.2.4}$$

Exercise 2.2.2. Show that, for F_N, any other admissible observer measures electric and magnetic 3-vectors that are perpendicular and have the same magnitude. *Hint*: This is clear if the Lorentz transformation Λ relating the two frames is a rotation. Verify the statement for $\Lambda(\beta)$ and appeal to Theorem 1.3.5.

Exercise 2.2.3. Show that, for F_R, another admissible observer will, in general, *not* measure $\vec{E}$ and $\vec{B}$ in the same direction.

Of particular interest is the special case of (2.2.4) when either $\vec{B}$ or $\vec{E}$ is zero (so that $\mathcal{O}$ observes either a purely electric or a purely magnetic field): If $\vec{B} = \vec{O}$, then

$$\begin{aligned} \hat{E}^1 &= E^1, & \hat{E}^2 &= \gamma E^2, & \hat{E}^3 &= \gamma E^3, \\ \hat{B}^1 &= 0, & \hat{B}^2 &= \beta\gamma E^3, & \hat{B}^3 &= -\beta\gamma E^2. \end{aligned} \tag{2.2.5}$$

If $\vec{E} = \vec{O}$ we have

$$\begin{aligned} \hat{E}^1 &= 0, & \hat{E}^2 &= -\beta\gamma B^3, & \hat{E}^3 &= \beta\gamma B^2, \\ \hat{B}^1 &= B^1, & \hat{B}^2 &= \gamma B^2, & \hat{B}^3 &= \gamma B^3. \end{aligned} \tag{2.2.6}$$

The essential feature of (2.2.5) and (2.2.6) is that "purely electric" and "purely magnetic" are not relativistically meaningful notions since they are, in general, not invariant under Lorentz transformations. How much of an electromagnetic field is "electric" and how much "magnetic" depends on the frame of reference from which it is being observed. This is the familiar phenomemon of electromagnetic induction. For example, a charge deemed "at rest" in one frame will give rise to a purely electric field in that frame, but, viewed from another frame, will be "moving" and so will induce a nonzero magnetic field as well.

Since $\vec{E}$ and $\vec{B}$ are spacelike one can, beginning with any admissible basis $\{e_1, e_2, e_3, e_4\}$, choose a right-handed orthonormal basis $\{\hat{e}_1, \hat{e}_2, \hat{e}_3\}$ for $\text{Span}\{e_1, e_2, e_3\}$ such that $\vec{E}$ and $\vec{B}$ both lie in $\text{Span}\{\hat{e}_1, \hat{e}_2\}$ (so that $\hat{E}^3 = \hat{B}^3 = 0$). Choosing a rotation $[R^i{}_j]_{i,j=1,2,3}$ in this 3-dimensional Euclidean space that accomplishes the change of coordinates $\hat{x}^i = R^i{}_j x^j, \ i = 1, 2, 3,$ the corresponding rotation $[R^a{}_b]_{a,b=1,2,3,4}$ in $\mathcal{L}$ yields a new admissible coordinate system in which the third components of $\vec{E}$ and $\vec{B}$ are zero. The gist of all this is that one can, with little extra effort, work in a basis relative to which the matrix of F has the form

$$\begin{bmatrix} 0 & 0 & -B^2 & E^1 \\ 0 & 0 & B^1 & E^2 \\ B^2 & -B^1 & 0 & 0 \\ E^1 & E^2 & 0 & 0 \end{bmatrix}. \tag{2.2.7}$$

Next we collect a few facts that will be of use in the remainder of the chapter. We define the *range* and *kernel* (*null space*) of a linear transformation $T : \mathcal{M} \to \mathcal{M}$ by

$$\text{rng}\ T = \{ y \in \mathcal{M} : y = Tx \ \text{ for some } \ x \in \mathcal{M} \}$$

and

$$\ker\ T = \{ x \in \mathcal{M} : Tx = 0 \}.$$

Both rng T and ker T are obviously subspaces of $\mathcal{M}$ and, consequently, so are their orthogonal complements $(\text{rng}\ T)^\perp$ and $(\ker\ T)^\perp$ (Exercise 1.1.2).

Proposition 2.2.1. *If $F : \mathcal{M} \to \mathcal{M}$ is any nonzero, skew-symmetric linear transformation on $\mathcal{M}$, then*

(a) $\ker\ F = (\text{rng}\ F)^\perp$,

(b) $\text{rng}\ F = (\ker\ F)^\perp$,

(c) $\dim(\ker\ F)$ *is either* 0 *or* 2 *so* $\dim(\text{rng}\ F)$ *is either* 4 *or* 2, *respectively).*

Proof: (a) First let $x \in (\text{rng } F)^{\perp}$. Then $x \cdot Fy = 0$ for all y in $\mathcal{M}$. Thus, $Fx \cdot y = 0$ for all y in $\mathcal{M}$. But the inner product on $\mathcal{M}$ is nondegenerate so we must have $Fx = 0$, i.e., $x \in \ker F$. Next suppose $x \in \ker F$. Then $Fx = 0$ implies $Fx \cdot y = 0$ for all y in $\mathcal{M}$ so $x \cdot Fy = 0$ for all y in $\mathcal{M}$, i.e., $x \in (\text{rng } F)^{\perp}$.

Exercise 2.2.4. Show that, for any subspace $\mathcal{U}$ of $\mathcal{M}$, $(\mathcal{U}^{\perp})^{\perp} = \mathcal{U}$ and conclude from (a) that (b) is true. *Hint*: Use the fact that $\mathcal{U}$ and $\mathcal{U}^{\perp}$ have complementary dimensions, i.e., $\dim \mathcal{U} + \dim \mathcal{U}^{\perp} = \dim \mathcal{M}$ (see Theorem 16, Chapter 4 of [**La**]).

(c) Without loss of generality select a basis $\{e_a\}$ relative to which the matrix of F has the form (2.2.7). Then, for any $v \in \mathcal{M}$, Fv has components given by

$$\begin{bmatrix} 0 & 0 & -B^2 & E^1 \\ 0 & 0 & B^1 & E^2 \\ B^2 & -B^1 & 0 & 0 \\ E^1 & E^2 & 0 & 0 \end{bmatrix} \begin{bmatrix} v^1 \\ v^2 \\ v^3 \\ v^4 \end{bmatrix} = \begin{bmatrix} -B^2v^3 + E^1v^4 \\ B^1v^3 + E^2v^4 \\ B^2v^1 - B^1v^2 \\ E^1v^1 + E^2v^2 \end{bmatrix}$$

which is zero if and only if

$$\begin{cases} -B^2v^3 + E^1v^4 = 0, \\ \quad B^1v^3 + E^2v^4 = 0 \end{cases} \quad \text{and} \quad \begin{cases} B^2v^1 - B^1v^2 = 0, \\ E^1v^1 + E^2v^2 = 0. \end{cases}$$

Notice that the determinant of the coefficient matrix in the first system is $-\vec{E} \cdot \vec{B}$ and, in the second, $\vec{E} \cdot \vec{B}$. If $\vec{E} \cdot \vec{B} \neq 0$ both systems have only the trivial solution so the kernel of F consists of 0 alone and $\dim(\ker F) = 0$. Since $4 = \dim \mathcal{M} = \dim(\ker F) + \dim(\text{rng } F)$, $\dim(\text{rng } F) = 4$. If $\vec{E} \cdot \vec{B} = 0$, each system has nontrivial solutions, say, $(v^3, v^4) = (v_0^3, v_0^4)$ and $(v^1, v^2) = (v_0^1, v_0^2)$. Since $F \neq 0$, all of the nontrivial solutions to the first system are of the form $b(v_0^3, v_0^4)$, $b \in \mathbb{R}$, and, for the second, $a(v_0^1, v_0^2)$, $a \in \mathbb{R}$. Thus, the kernel of F is the set of

$$\begin{bmatrix} v^1 \\ v^2 \\ v^3 \\ v^4 \end{bmatrix} = a \begin{bmatrix} v_0^1 \\ v_0^2 \\ 0 \\ 0 \end{bmatrix} + b \begin{bmatrix} 0 \\ 0 \\ v_0^3 \\ v_0^4 \end{bmatrix},$$

so $\dim(\ker F) = 2$ and therefore $\dim(\text{rng } F) = 2$. ■

Recall that a real number λ is an *eigenvalue* of F if there exists a nonzero $x \in \mathcal{M}$ such that $Fx = \lambda x$ and that any such x is an *eigenvector* of F corresponding to λ. The *eigenspace* of F corresponding to λ is $\{x \in \mathcal{M} : Fx = \lambda x\}$, i.e., the set of eigenvectors for λ together with $0 \in \mathcal{M}$, and it is indeed a subspace of $\mathcal{M}$. A subspace $\mathcal{U}$ of $\mathcal{M}$ is said to be *invariant under* F if F maps $\mathcal{U}$ into $\mathcal{U}$, i.e., if $F\mathcal{U} \subseteq \mathcal{U}$. Any eigenspace of F is invariant under F since $Fx = \lambda x$ implies $F(Fx) = F(\lambda x) = \lambda Fx$.

Proposition 2.2.2. *If* $F : \mathcal{M} \to \mathcal{M}$ *is any nonzero, skew-symmetric linear transformation on* $\mathcal{M}$, *then*

(a) $Fx = \lambda x$ *implies that either* $\lambda = 0$ *or* x *is null (or both),*

(b) $F\mathcal{U} \subseteq \mathcal{U}$ *implies* $F(\mathcal{U}^\perp) \subseteq \mathcal{U}^\perp$.

Proof: (a) $Fx = \lambda x$ implies $Fx \cdot x = \lambda(x \cdot x)$ so $\lambda(x \cdot x) = 0$ and either $\lambda = 0$ or $x \cdot x = 0$.

(b) Suppose $F\mathcal{U} \subseteq \mathcal{U}$ and let $v \in \mathcal{U}^\perp$. Then, for every $u \in \mathcal{U}$, $Fv \cdot u = -v \cdot Fu = 0$ because $Fu \in \mathcal{U}$ and $v \in \mathcal{U}^\perp$. Thus, $Fv \in \mathcal{U}^\perp$ as required. ∎

The eigenvalues of a linear transformation are found by solving its characteristic equation. For a skew-symmetric linear transformation on $\mathcal{M}$ and with the notation established in (2.2.3) this equation is easy to write down and quite informative.

Theorem 2.2.3. *Let* $F\colon \mathcal{M} \to \mathcal{M}$ *be a skew-symmetric linear transformation and* $\{e_a\}_{a=1}^4$ *an arbitrary admissible basis for* $\mathcal{M}$. *With the matrix* $[F^a{}_b]$ *written in the form (2.2.3) and* I *the* 4×4 *identity matrix we have*

$$\det\left([F^a{}_b] - \lambda I\right) = \lambda^4 + \left(|\vec{B}|^2 - |\vec{E}|^2\right)\lambda^2 - \left(\vec{E} \cdot \vec{B}\right)^2, \qquad (2.2.8)$$

where $|\vec{E}|^2 = (E^1)^2 + (E^2)^2 + (E^3)^2$, $|\vec{B}|^2 = (B^1)^2 + (B^2)^2 + (B^3)^2$ *and* $\vec{E} \cdot \vec{B} = E^1B^1 + E^2B^2 + E^3B^3$.

Exercise 2.2.5. Prove Theorem 2.2.3. ∎

Consequently, the eigenvalues of F are the real solutions to

$$\lambda^4 + \left(|\vec{B}|^2 - |\vec{E}|^2\right)\lambda^2 - \left(\vec{E} \cdot \vec{B}\right)^2 = 0. \qquad (2.2.9)$$

Since the roots of the characteristic polynomial are independent of the choice of basis and since the leading coefficient on the left-hand side of (2.2.9) is one it follows that, while $\vec{E}$ and $\vec{B}$ will, in general, be different in different admissible bases, the algebraic combinations $|\vec{B}|^2 - |\vec{E}|^2$ and $\vec{E} \cdot \vec{B}$ are Lorentz invariants, i.e., the same in all admissible frames. In particular, if both are zero (i.e., if $\vec{E}$ and $\vec{B}$ are perpendicular and have the same magnitude) in one frame, the same will be true in any other frame. We shall say that F is *null* if $|\vec{B}|^2 - |\vec{E}|^2 = \vec{E} \cdot \vec{B} = 0$ in any (and therefore every) admissible basis; otherwise, F is *regular*. As defined earlier in this section, F_N is null and F_R is regular.

Exercise 2.2.6. Show that F is invertible iff $\vec{E} \cdot \vec{B} \neq 0$.

2.3 Invariant Subspaces

Our objective in this section is to obtain an intrinsic characterization of "null" and "regular" skew-symmetric linear transformations on $\mathcal{M}$ that will be used in the next section to derive their "canonical forms". Specifically, we will show that every skew-symmetric $F : \mathcal{M} \to \mathcal{M}$ has a 2-dimensional invariant subspace and that F is regular if and only if $\mathcal{U} \cap \mathcal{U}^\perp = \{0\}$ for some such subspace $\mathcal{U}$ (so F is null if and only if $\mathcal{U} \cap \mathcal{U} \neq \{0\}$ for every 2-dimensional invariant subspace $\mathcal{U}$).

We begin with a few observations on invariant subspaces in general. Let V be a real vector space of dimension at least 2 and $T : V \to V$ a nonzero linear transformation. Observe that if the characteristic equation $\det(T - \lambda I) = 0$ has a real root λ, then T has an eigenvector, i.e., there is a nonzero $v \in V$ such that $Tv = \lambda v$. Consequently, $\text{Span}\{v\}$ is a nontrivial invariant subspace for T. If there are no real roots the situation is less simple.

Lemma 2.3.1. *Let V be a real vector space of dimension greater than or equal to 2 and $T : V \to V$ a nonzero linear transformation. If $\det(T - \lambda I) = 0$ has a complex solution $\lambda = \alpha + \beta i$, $\beta \neq 0$, then there exist nonzero vectors x and y in V such that*

$$Tx = \alpha x - \beta y \quad \textit{and} \quad Ty = \alpha y + \beta x. \tag{2.3.1}$$

In particular, $\text{Span}\{x, y\}$ *is a nontrivial invariant subspace for T.*

Proof: Select a basis for V and let $[a_{ij}]_{i,j=1,\ldots,n}$ be the matrix of T in this basis. Since $\det(T - \lambda I) = 0$ when $\lambda = \alpha + \beta i$, the system

$$\begin{cases} a_{11}z_1 + \cdots + a_{1n}z_n = (\alpha + \beta i)z_1 \\ \vdots \qquad\qquad \vdots \qquad\qquad \vdots \\ a_{n1}z_1 + \cdots + a_{nn}z_n = (\alpha + \beta i)z_n \end{cases}$$

as a nontrivial *complex* solution $z_1 = x_1 + iy_1, \ldots, z_n = x_n + iy_n$. Thus,

$$\begin{cases} a_{11}(x_1 + iy_1) + \cdots + a_{1n}(x_n + iy_n) = (\alpha + \beta i)(x_1 + iy_1) \\ \vdots \qquad\qquad\qquad \vdots \qquad\qquad\qquad \vdots \\ a_{n1}(x_1 + iy_1) + \cdots + a_{nn}(x_n + iy_n) = (\alpha + \beta i)(x_n + iy_n). \end{cases}$$

Separating into real and imaginary parts gives

$$\begin{cases} a_{11}x_1 + \cdots + a_{1n}x_n = \alpha x_1 - \beta y_1 \\ \quad\vdots \qquad\qquad \vdots \qquad\qquad \vdots \\ a_{n1}x_1 + \cdots + a_{nn}x_n = \alpha x_n - \beta y_n \end{cases} \tag{2.3.2}$$

and

$$\begin{cases} a_{11}y_1 + \cdots + a_{1n}y_n = \alpha y_1 + \beta x_1 \\ \quad\vdots \qquad\qquad \vdots \qquad\qquad \vdots \\ a_{n1}y_1 + \cdots + a_{nn}y_n = \alpha y_n + \beta x_n. \end{cases} \tag{2.3.3}$$

Let x and y be the vectors in V whose components relative to our basis are $x_1, \ldots, x_n$ and $y_1, \ldots, y_n$ respectively. Then $Tx = \alpha x - \beta y$ and $Ty = \alpha y + \beta x$ as required. Notice that neither x nor y is zero since if $x = 0$, (2.3.2) and the fact that $\beta \neq 0$ imply $y = 0$ so $z_1 = \cdots = z_n = 0$, which is a contradiction. Similarly, $y = 0$ implies $x = 0$ so, in fact, neither can be zero. ∎

In order to apply this result to the case of interest to us we require two final preliminary results.

Lemma 2.3.2. *Let A and B be real numbers with $B \neq 0$. Then the equation $\lambda^4 + A\lambda^2 - B^2 = 0$ has a complex solution.*

Proof: Regard $\lambda^4 + A\lambda^2 - B^2 = 0$ as a quadratic in λ^2 to obtain $\lambda^2 = \frac{1}{2}(-A \pm \sqrt{A^2 + 4B^2})$. Choosing the minus sign gives a negative λ^2 and therefore complex λ. ∎

Lemma 2.3.3. *Let $F : \mathcal{M} \to \mathcal{M}$ be a nonzero, skew-symmetric linear transformation. If the characteristic equation*

$$\lambda^4 + (|\vec{B}|^2 - |\vec{E}|^2)\lambda^2 - (\vec{E} \cdot \vec{B})^2 = 0$$

has two distinct nonzero, real solutions, then there exists a 2-dimensional subspace $\mathcal{U}$ of $\mathcal{M}$ which is invariant under F and satisfies $\mathcal{U} \cap \mathcal{U}^\perp = \{0\}$.

Proof: Let λ_1 and λ_2 be the two distinct nonzero real eigenvalues. Then there exist nonzero vectors x and y such that $Fx = \lambda_1 x$ and $Fy = \lambda_2 y$. By Proposition 2.2.2(a), x and y are null. Observe next that x and y are linearly independent. Indeed, $ax + by = 0$ implies

$$\begin{aligned} aFx + bFy &= 0, \\ a(\lambda_1 x) + b(\lambda_2 y) &= 0, \\ \lambda_1(ax) + \lambda_2(by) &= 0, \\ \lambda_1(ax) + \lambda_2(-ax) &= 0, \\ (\lambda_1 - \lambda_2)ax &= 0. \end{aligned}$$

Since $\lambda_1 - \lambda_2 \neq 0$, $ax = 0$, but x is nonzero so $a = 0$. Similarly, $b = 0$ so x and y are independent. Thus, $\mathcal{U} = \text{Span}\{x, y\}$ is 2-dimensional; it is clearly invariant under F. Now suppose $ax + by \in \mathcal{U} \cap \mathcal{U}^\perp$. Then, in particular,

$$\begin{aligned} (ax + by) \cdot x &= 0, \\ a(x \cdot x) + b(x \cdot y) &= 0, \\ b(x \cdot y) &= 0. \end{aligned}$$

But x and y are null and nonparallel so $x \cdot y \neq 0$ and therefore $b = 0$. Similarly, $a = 0$ so $\mathcal{U} \cap \mathcal{U}^\perp = \{0\}$. ∎

Theorem 2.3.4. *Let $F \colon \mathcal{M} \to \mathcal{M}$ be a nonzero, skew-symmetric linear transformation on $\mathcal{M}$. If F is regular, then there exists a 2-dimensional subspace $\mathcal{U}$ of $\mathcal{M}$ which is invariant under F and satisfies $\mathcal{U} \cap \mathcal{U}^\perp = \{0\}$.*

Proof: Relative to any admissible basis, at least one of $\vec{E} \cdot \vec{B}$ or $|\vec{B}|^2 - |\vec{E}|^2$ must be nonzero and F's characteristic equation is

$$\lambda^4 + (|\vec{B}|^2 - |\vec{E}|^2)\lambda^2 - (\vec{E} \cdot \vec{B})^2 = 0. \tag{2.3.4}$$

We consider four cases:

1. $\vec{E} \cdot \vec{B} = 0$ and $|\vec{B}|^2 - |\vec{E}|^2 < 0$.

In this case (2.3.4) becomes $\lambda^2(\lambda^2 + (|\vec{B}|^2 - |\vec{E}|^2)) = 0$ and the solutions are $\lambda = 0$ and $\lambda = \pm\sqrt{|\vec{E}|^2 - |\vec{B}|^2}$. The latter are two distinct, nonzero real solutions so Lemma 2.3.3 yields the result.

2. $\vec{E} \cdot \vec{B} = 0$ and $|\vec{B}|^2 - |\vec{E}|^2 > 0$.

The solutions of (2.3.4) are now $\lambda = 0$ and $\lambda = \pm\beta i$, where $\beta = \sqrt{|\vec{B}|^2 - |\vec{E}|^2}$. Lemma 2.3.1 implies that there exist nonzero vectors x and y in $\mathcal{M}$ such that

$$Fx = -\beta y \quad \text{and} \quad Fy = \beta x. \tag{2.3.5}$$

We claim that x and y are linearly independent. Indeed, suppose $ax+by = 0$ with, say, $b \neq 0$. Then $y = kx$, where $k = -a/b$. Then $Fx = -\beta y$ implies $Fx = (-\beta k)x$. But F's only real eigenvalue is 0 and $\beta \neq 0$ so $k = 0$ and therefore $y = 0$, which is a contradiction. Thus, $b = 0$. Since $x \neq 0$, $ax = 0$ implies $a = 0$ and the proof is complete.

Thus, $\mathcal{U} = \text{Span}\{x, y\}$ is a 2-dimensional subspace of $\mathcal{M}$ that is invariant under F. We claim that $\mathcal{U} \cap \mathcal{U}^\perp = \{0\}$. Suppose $ax + by \in \mathcal{U} \cap \mathcal{U}^\perp$.

Then $ax + by$ is null so $(ax + by) \cdot (ax + by) = 0$, i.e.,

$$a^2(x \cdot x) + 2ab(x \cdot y) + b^2(y \cdot y) = 0\,.$$

But $x \cdot y = x \cdot (-\frac{1}{\beta}Fx) = -\frac{1}{\beta}(x \cdot Fx) = 0$ so

$$\begin{aligned}
a^2(x \cdot x) + b^2(y \cdot y) &= 0, \\
a^2 x \cdot \left(\frac{1}{\beta}Fy\right) + b^2 y \cdot \left(-\frac{1}{\beta}Fx\right) &= 0, \\
\left(\frac{a^2}{\beta}\right) x \cdot Fy - \left(\frac{b^2}{\beta}\right) y \cdot Fx &= 0, \\
\left(\frac{a^2}{\beta}\right) x \cdot Fy + \left(\frac{b^2}{\beta}\right) x \cdot Fy &= 0, \\
\left(\frac{a^2 + b^2}{\beta}\right) x \cdot Fy &= 0\,.
\end{aligned}$$

Now, if $a^2+b^2 \neq 0$, then $x \cdot Fy = 0$ so $x \cdot (\beta x) = 0$ and $x \cdot x = 0$. Similarly, $y \cdot y = 0$ so x and y are orthogonal null vectors and consequently parallel. But this is a contradiction since x and y are independent. Thus, $a^2 + b^2 = 0$ so $a = b = 0$ and $\mathcal{U} \cap \mathcal{U}^\perp = \{0\}$.

3. $\vec{E} \cdot \vec{B} \neq 0$ and $|\vec{B}|^2 - |\vec{E}|^2 = 0$.

In this case (2.3.4) becomes $\lambda^4 = (\vec{E} \cdot \vec{B})^2$ so $\lambda^2 = \pm\, |\vec{E} \cdot \vec{B}|$. $\lambda^2 = |\vec{E} \cdot \vec{B}|$ gives two distinct, nonzero real solutions so the conclusion follows from Lemma 2.3.3.

4. $\vec{E} \cdot \vec{B} \neq 0$ and $|\vec{B}|^2 - |\vec{E}|^2 \neq 0$.

Lemma 2.3.2 implies that (2.3.4) has a complex root $\alpha + \beta i$ $(\beta \neq 0)$. Thus, Lemma 2.3.1 yields nonzero vectors x and y in $\mathcal{M}$ with

$$Fx = \alpha x - \beta y \quad \text{and} \quad Fy = \alpha y + \beta x\,.$$

There are two possibilities:

i. x and y are linearly dependent. Then, since neither is zero, $y = kx$ for some $k \in \mathbb{R}$ with $k \neq 0$. Thus, $Fx = \alpha x - \beta y = \alpha x - k\beta x = (\alpha - k\beta)x$ and $Fy = \alpha y + \beta x = \alpha y + \frac{\beta}{k} y = (\alpha + \frac{\beta}{k})y$. Since $\alpha + \frac{\beta}{k} \neq \alpha - k\beta$ and since 0 is not a solution to (2.3.4) in this case we find that F has two distinct, nonzero real eigenvalues and again appeal to Lemma 2.3.3.

ii. x and y are linearly independent. Then $\mathcal{U} = \text{Span}\{x, y\}$ is a 2-dimensional subspace of $\mathcal{M}$ that is invariant under F.

Exercise 2.3.1. Complete the proof by showing that $\mathcal{U} \cap \mathcal{U}^{\perp} = \{0\}$. ■

To complete our work in this section we must show that a nonzero null skew-symmetric $F : \mathcal{M} \to \mathcal{M}$ has 2-dimensional invariant subspaces and that all of these intersect their orthogonal complements nontrivially. We address the question of existence first.

Proposition 2.3.5. *Let $F : \mathcal{M} \to \mathcal{M}$ be a nonzero, null, skew-symmetric linear transformation on $\mathcal{M}$. Then both* ker F *and* rng $F = (\text{ker } F)^{\perp}$ *are 2-dimensional invariant subspaces of $\mathcal{M}$ and their intersection is a 1-dimensional subspace of $\mathcal{M}$ spanned by a null vector.*

Proof: ker F and rng F are obviously invariant under F. Since F is null, $\vec{E} \cdot \vec{B} = 0$ so, by Exercise 2.2.6, F is not invertible. Thus, $\dim(\text{ker } F) \neq 0$. Proposition 2.2.1(c) then implies that $\dim(\text{ker } F) = 2$ and, consequently, $\dim(\text{rng } F) = 2$.

Now, since $\text{rng } F \cap \text{ker } F = \text{rng } F \cap (\text{rng } F)^{\perp}$ by Proposition 2.2.1(a), if this intersection is not $\{0\}$, it can contain only null vectors. Being a subspace of $\mathcal{M}$ it must therefore be 1-dimensional. We show that this intersection is, indeed, nontrivial as follows: For a null F the characteristic polynomial (2.3.4) reduces to $\lambda^4 = 0$. The Cayley-Hamilton Theorem (see [**H**]) therefore implies that

$$F^4 = 0 \qquad (F \text{ null}). \tag{2.3.6}$$

Next we claim that $\text{ker } F \subsetneq \text{ker } F^2$. $\text{ker } F \subseteq \text{ker } F^2$ is obvious. Now, suppose $\text{ker } F = \text{ker } F^2$, i.e., $Fx = 0 \ iff \ F^2x = 0$. Then

$$\begin{aligned} F^3x = F^2(Fx) = 0 &\implies Fx \in \text{ker } F^2 = \text{ker } F \\ &\implies F(Fx) = 0 \\ &\implies F^2x = 0 \\ &\implies Fx = 0 \quad \text{by assumption,} \end{aligned}$$

so $F^3x = 0 \implies Fx = 0$ and we conclude that $\text{ker } F^3 = \text{ker } F$. Repeating the argument gives $\text{ker } F^4 = \text{ker } F$. But by (2.3.6), $\text{ker } F^4 = \mathcal{M}$ so $\text{ker } F = \mathcal{M}$ and F is identically zero, contrary to hypothesis. Thus, $\text{ker } F \subsetneq \text{ker } F^2$ and we may select a nonzero $v \in \mathcal{M}$ such that $F^2v = 0$, but $Fv \neq 0$. Thus, $Fv \in \text{rng } F \cap \text{ker } F$ as required. ■

Exercise 2.3.2. Show that if $F : \mathcal{M} \to \mathcal{M}$ is a nonzero, null, skew-symmetric linear transformation on $\mathcal{M}$, then F^2v is null (perhaps 0) for every $v \in \mathcal{M}$. *Hint:* Begin with (2.3.6).

All that remains is to show that if F is null, then *every* 2-dimensional invariant subspace $\mathcal{U}$ satisfies $\mathcal{U} \cap \mathcal{U}^{\perp} \neq \{0\}$.

Lemma 2.3.6. *Let $F : \mathcal{M} \to \mathcal{M}$ be a nonzero, skew-symmetric linear transformation on $\mathcal{M}$. If there exists a 2-dimensional invariant subspace $\mathcal{U}$ for F with $\mathcal{U} \cap \mathcal{U}^\perp = \{0\}$, then $\mathcal{U}^\perp$ is also a 2-dimensional invariant subspace for F and there exists a real number α such that $F^2u = \alpha u$ for every $u \in \mathcal{U}$.*

Proof: $\mathcal{U}^\perp$ is a subspace of $\mathcal{M}$ (Exercise 1.1.2) and is invariant under F [Proposition 2.2.2(b)]. Notice that the restriction of the Lorentz inner product to $\mathcal{U}$ cannot be degenerate since this would contradict $\mathcal{U} \cap \mathcal{U}^\perp = \{0\}$. Thus, by Theorem 1.1.1, we may select an orthonormal basis $\{u_1, u_2\}$ for $\mathcal{U}$. Now, let x be an arbitrary element of $\mathcal{M}$. If u_1 and u_2 are both spacelike, then $v = x - [(x \cdot u_1)u_1 + (x \cdot u_2)u_2] \in \mathcal{U}^\perp$ and $x = v + [(x \cdot u_1)u_1 + (x \cdot u_2)u_2]$ so $x \in \mathcal{U} + \mathcal{U}^\perp$.

Exercise 2.3.3. Argue similarly that if $\{u_1, u_2\}$ contains one spacelike and one timelike vector, then any $x \in \mathcal{M}$ is in $\mathcal{U} + \mathcal{U}^\perp$ and explain why this is the only remaining possibility for the basis $\{u_1, u_2\}$.

Since $\mathcal{U} \cap \mathcal{U}^\perp = \{0\}$ we conclude that $\mathcal{M} = \mathcal{U} \bigoplus \mathcal{U}^\perp$ so dim $\mathcal{U}^\perp = 2$.

Now we let $\{u_1, u_2\}$ be an orthonormal basis for $\mathcal{U}$ and write $Fu_1 = au_1 + bu_2$ and $Fu_2 = cu_1 + du_2$. Then, since neither u_1 nor u_2 is null, we have $0 = Fu_1 \cdot u_1 = (au_1 + bu_2) \cdot u_1 = \pm a$ so $a = 0$ and, similarly, $d = 0$. Thus, $Fu_1 = bu_2$ and $Fu_2 = cu_1$, so $F^2u_1 = F(bu_2) = bFu_2 = bcu_1$ and $F^2u_2 = bcu_2$. Let $\alpha = bc$. Then, for any $u = \beta u_1 + \gamma u_2 \in \mathcal{U}$ we have $F^2u = \beta F^2u_1 + \gamma F^2u_2 = \beta(\alpha u_1) + \gamma(\alpha u_2) = \alpha(\beta u_1 + \gamma u_2) = \alpha u$ as required. ■

With this we can show that if F is null and nonzero and $\mathcal{U}$ is a 2-dimensional invariant subspace for F, then $\mathcal{U} \cap \mathcal{U}^\perp \neq \{0\}$. Suppose, to the contrary, that $\mathcal{U} \cap \mathcal{U}^\perp = \{0\}$. Lemma 2.3.6 implies the existence of an $\alpha \in \mathbb{R}$ such that $F^2u = \alpha u$ for all u in $\mathcal{U}$. Thus, $F^4u = F^2(F^2u) = F^2(\alpha u) = \alpha F^2u = \alpha^2 u$ for all $u \in \mathcal{U}$. But, by (2.3.6), $F^4u = 0$ for all $u \in \mathcal{U}$ so $\alpha = 0$ and $F^2 = 0$ on $\mathcal{U}$. Again by Lemma 2.3.6 we may apply the same argument to $\mathcal{U}^\perp$ to obtain $F^2 = 0$ on $\mathcal{U}^\perp$. Since $\mathcal{U}$ and $\mathcal{U}^\perp$ are 2-dimensional and $\mathcal{U} \cap \mathcal{U}^\perp = \{0\}$, $\mathcal{M} = \mathcal{U} \bigoplus \mathcal{U}^\perp$ so $F^2 = 0$ on all of $\mathcal{M}$. But then, for every $u \in \mathcal{M}$, $F^2u \cdot u = 0$ so $Fu \cdot Fu = 0$, i.e., rng F contains only null vectors. But then dim(rng F) $= 1$ and this contradicts Proposition 2.2.1(c) and we have proved:

Theorem 2.3.7. *Let $F : \mathcal{M} \to \mathcal{M}$ be a nonzero, skew-symmetric linear transformation on $\mathcal{M}$. If F is null, then F has 2-dimensional invariant subspaces and every such subspace $\mathcal{U}$ satisfies $\mathcal{U} \cap \mathcal{U}^\perp \neq \{0\}$.*

Combining this with Theorem 2.3.4 gives:

Corollary 2.3.8. *Let $F\colon \mathcal{M} \to \mathcal{M}$ be a nonzero, skew-symmetric linear transformation on $\mathcal{M}$. Then F has 2-dimensional invariant subspaces and F is regular iff there exists such a subspace $\mathcal{U}$ such that $\mathcal{U} \cap \mathcal{U}^\perp = \{0\}$ (so F is null iff $\mathcal{U} \cap \mathcal{U}^\perp \neq \{0\}$ for every such subspace).*

2.4 Canonical Forms

We now propose to use the results of the preceding section to prove that, for any skew-symmetric linear transformation $F\colon \mathcal{M} \to \mathcal{M}$, there exists a basis for $\mathcal{M}$ relative to which the matrix of F has one of the two forms

$$\begin{bmatrix} 0 & \delta & 0 & 0 \\ -\delta & 0 & 0 & 0 \\ 0 & 0 & 0 & \epsilon \\ 0 & 0 & \epsilon & 0 \end{bmatrix} \quad \text{or} \quad \begin{bmatrix} 0 & 0 & 0 & 0 \\ 0 & 0 & \alpha & 0 \\ 0 & -\alpha & 0 & \alpha \\ 0 & 0 & \alpha & 0 \end{bmatrix},$$

depending on whether F is regular or null respectively. We begin with the regular case.

Thus, we suppose $F : \mathcal{M} \to \mathcal{M}$ is a nonzero, skew-symmetric linear transformation and that (Corollary 2.3.8) there exists a 2-dimensional subspace $\mathcal{U}$ of $\mathcal{M}$ which satisfies $F\mathcal{U} \subseteq \mathcal{U}$ and $\mathcal{U} \cap \mathcal{U}^\perp = \{0\}$. Then (Lemma 2.3.6) $\mathcal{U}^\perp$ is also a 2-dimensional invariant subspace for F and there exist real numbers α and β such that

$$F^2 u = \alpha u \qquad \text{for all } u \in \mathcal{U} \text{ and} \tag{2.4.1}$$

$$F^2 v = \beta v \qquad \text{for all } v \in \mathcal{U}^\perp . \tag{2.4.2}$$

Since $\mathcal{M} = \mathcal{U} \bigoplus \mathcal{U}^\perp$ and $\mathcal{U}^{\perp\perp} = \mathcal{U}$ we may assume, without loss of generality, that the restriction of the Lorentz inner product to $\mathcal{U}$ has index 1 and its restriction to $\mathcal{U}^\perp$ has index 0.

We claim now that $\alpha \geq 0$ and $\beta \leq 0$. Indeed, dotting both sides of (2.4.1) with itself gives $F^2 u \cdot u = \alpha(u \cdot u)$, or $-Fu \cdot Fu = \alpha(u \cdot u)$ for any u in $\mathcal{U}$. Now if $u \in \mathcal{U}$ is timelike, then Fu is spacelike or zero so $u \cdot u < 0$ and $Fu \cdot Fu \geq 0$ and this implies $\alpha \geq 0$. Thus, we may write $\alpha = \epsilon^2$ with $\epsilon \geq 0$ so (2.4.1) becomes

$$F^2 u = \epsilon^2 u \qquad \text{for all } u \in \mathcal{U}. \tag{2.4.3}$$

Exercise 2.4.1. Show that, for some $\delta \geq 0$,

$$F^2 v = -\delta^2 v \qquad \text{for all } v \in \mathcal{U}^\perp. \tag{2.4.4}$$

Now, select a future-directed unit timelike vector e_4 in $\mathcal{U}$. Then Fe_4 is spacelike or zero and in $\mathcal{U}$ so we may select a unit spacelike vector e_3 in $\mathcal{U}$ with $Fe_4 = ke_3$ for some $k \geq 0$. Observe that $\epsilon^2 e_4 = F^2 e_4 = F(Fe_4) = F(ke_3) = kFe_3$. Thus,

$$\begin{aligned} kFe_3 \cdot e_4 &= \epsilon^2 e_4 \cdot e_4, \\ k(-e_3 \cdot Fe_4) &= \epsilon^2(-1), \\ k(e_3 \cdot (ke_3)) &= \epsilon^2, \\ k^2 e_3 \cdot e_3 &= \epsilon^2, \end{aligned}$$

so $k^2 = \epsilon^2$ and $k = \epsilon$ (since $k \geq 0$ and $\epsilon \geq 0$). Thus, we have

$$Fe_4 = \epsilon e_3 \,. \tag{2.4.5}$$

Notice that $\{e_3, e_4\}$ is an orthonormal basis for $\mathcal{U}$.

Exercise 2.4.2. Show that, in addition,

$$Fe_3 = \epsilon e_4 \,. \tag{2.4.6}$$

Now, let e_2 be an arbitrary unit spacelike vector in $\mathcal{U}^\perp$. Then Fe_2 is spacelike or zero and in $\mathcal{U}^\perp$ so we may select another unit spacelike vector e_1 in $\mathcal{U}^\perp$ and orthogonal to e_2 with $Fe_2 = ke_1$ for some $k \geq 0$ (if $e_1 \times e_2 \cdot e_3$ is -1 rather than 1, then relabel e_1 and e_2).

Exercise 2.4.3. Show that $k = \delta$.

Thus,

$$Fe_2 = \delta e_1 \tag{2.4.7}$$

and, as for (2.4.6),

$$Fe_1 = -\delta e_2 \,. \tag{2.4.8}$$

Now, $\{e_1, e_2, e_3, e_4\}$ is an orthonormal basis for $\mathcal{M}$ and any basis constructed in this way is called a *canonical basis for* F. From (2.4.5)-(2.4.8) we find that the matrix of F relative to such a basis is

$$\begin{bmatrix} 0 & \delta & 0 & 0 \\ -\delta & 0 & 0 & 0 \\ 0 & 0 & 0 & \epsilon \\ 0 & 0 & \epsilon & 0 \end{bmatrix} . \tag{2.4.9}$$

This is just the matrix of the F_R defined in Section 2.2. In a canonical basis an observer measures electric and magnetic fields in the x^3-direction and of magnitudes ϵ and δ respectively. We have shown that such a frame exists for any regular F. Observe that $|\vec{B}|^2 - |\vec{E}|^2 = \delta^2 - \epsilon^2$ and $\vec{E} \cdot \vec{B} = \delta\epsilon$. Since these two quantities are invariants, δ and ϵ can be calculated from the electric and magnetic 3-vectors in *any* frame. The *canonical form* (2.4.9) of F is particularly convenient for calculations. For example, the fourth power of the matrix (2.4.9) is easily computed and found to be

$$\begin{bmatrix} \delta^4 & 0 & 0 & 0 \\ 0 & \delta^4 & 0 & 0 \\ 0 & 0 & \epsilon^4 & 0 \\ 0 & 0 & 0 & \epsilon^4 \end{bmatrix},$$

so that, unlike the null case, $F^4 \neq 0$. The eigenvalues of F are of some interest and are also easy to calculate since the characteristic equation (2.3.4) becomes $\lambda^4 + (\delta^2 - \epsilon^2)\lambda^2 - \delta^2\epsilon^2 = 0$ i.e., $(\lambda^2 - \epsilon^2)(\lambda^2 + \delta^2) = 0$ whose only real solutions are $\lambda = \pm\epsilon$. The eigenspace corresponding to $\lambda = \epsilon$ is obtained by solving

$$\begin{bmatrix} 0 & \delta & 0 & 0 \\ -\delta & 0 & 0 & 0 \\ 0 & 0 & 0 & \epsilon \\ 0 & 0 & \epsilon & 0 \end{bmatrix} \begin{bmatrix} v^1 \\ v^2 \\ v^3 \\ v^4 \end{bmatrix} = \begin{bmatrix} \epsilon v^1 \\ \epsilon v^2 \\ \epsilon v^3 \\ \epsilon v^4 \end{bmatrix},$$

i.e.,

$$\begin{bmatrix} \delta v^2 \\ -\delta v^1 \\ \epsilon v^4 \\ \epsilon v^3 \end{bmatrix} = \begin{bmatrix} \epsilon v^1 \\ \epsilon v^2 \\ \epsilon v^3 \\ \epsilon v^4 \end{bmatrix}.$$

If $\epsilon = 0$ and $\delta \neq 0$, then $v^1 = v^2 = 0$, whereas v^3 and v^4 are arbitrary. Thus, the eigenspace is $\text{Span}\{e_3, e_4\}$. Similarly, if $\epsilon \neq 0$ and $\delta = 0$, $v^1 = v^2 = 0$ and $v^3 = v^4$ so the eigenspace is $\text{Span}\{e_3 + e_4\}$. If $\epsilon\delta \neq 0$, $\delta v^2 = \epsilon v^1$ and $-\delta v^1 = \epsilon v^2$ again imply $v^1 = v^2 = 0$; in addition, $v^3 = v^4$ so the eigenspace is $\text{Span}\{e_3 + e_4\}$. In the first case the eigenspace contains two independent null directions (those of $e_3 + e_4$ and $e_3 - e_4$), whereas in the last two cases, there is only one $(e_3 + e_4)$. For $\lambda = -\epsilon$, the result is obviously the same in the first case, while in the second and third the eigenspace is spanned by $e_3 - e_4$. The null directions corresponding to $e_3 \pm e_4$ are called the *principal null directions* of F.

Now we turn to the case of a nonzero, null, skew-symmetric linear transformation $F : \mathcal{M} \to \mathcal{M}$ and construct an analogous "canonical basis". Begin with an arbitrary future-directed unit timelike vector e_4 in $\mathcal{M}$.

Exercise 2.4.4. Show that Fe_4 is spacelike. *Hint:* $Fe_4 = 0$ would imply $e_4 \in (\text{rng } F)^\perp$.

Thus, we may select a unit spacelike vector e_3 in $\mathcal{M}$ such that $e_3 \cdot e_4 = 0$ and

$$Fe_4 = \alpha e_3 \tag{2.4.10}$$

for some $\alpha > 0$. Observe that $e_3 = F(\frac{1}{\alpha}e_4) \in \text{rng } F$. Next we claim that Fe_3 is a nonzero vector in $\text{rng } F \cap \ker F$. $Fe_3 \neq 0$ is clear since $Fe_3 = 0 \Longrightarrow e_3 \in \text{rng } F \cap \ker F$, but e_3 is spacelike and this contradicts Proposition 2.3.5. $Fe_3 \in \text{rng } F$ is obvious. Now, by Exercise 2.3.2, F^2e_3 is either zero or null and nonzero. $F^2e_3 = 0$ implies $F(Fe_3) = 0$ so $Fe_3 \in \ker F$ as required. Suppose, on the other hand, that F^2e_3 is null and nonzero. $Fe_3 \cdot F^2e_3 = 0$ implies that Fe_3 is not timelike. $Fe_3 \cdot e_3 = 0$ implies that Fe_3 is not spacelike since then $\text{rng } F$ would contain a null and two orthogonal spacelike vectors, contradicting Proposition 2.3.5. Thus, Fe_3 is null and nonzero. But then $\{e_3, Fe_3\}$ is a basis for $\text{rng } F$ and Fe_3 is orthogonal to both so $Fe_3 \in (\text{rng } F)^\perp = \ker F$ as required.

Now we wish to choose a unit spacelike vector e_2 such that $e_2 \cdot e_4 = 0$, $e_2 \cdot e_3 = 0$ and $\text{Span}\{e_2 + e_4\} = \text{rng } F \cap \ker F$. To see how this is done select any null vector N spanning $\text{rng } F \cap \ker F$ such that $N \cdot e_4 = -1$. Then let $e_2 = N - e_4$. It follows that $e_2 \cdot e_2 = (N - e_4) \cdot (N - e_4) = N \cdot N - 2N \cdot e_4 + e_4 \cdot e_4 = 0 - 2(-1) - 1 = 1$ so e_2 is unit spacelike. Moreover, $e_2 + e_4 = N$ spans $\text{rng } F \cap \ker F$. Also, $e_2 \cdot e_4 = (N - e_4) \cdot e_4 = N \cdot e_4 - e_4 \cdot e_4 = -1 - (-1) = 0$. Finally, $e_2 + e_4 \in (\text{rng } F)^\perp$ implies $0 = (e_2 + e_4) \cdot e_3 = e_2 \cdot e_3 + e_4 \cdot e_3 = e_2 \cdot e_3$ and the construction is complete. Now, there exists an $\alpha' > 0$ such that $Fe_3 = \alpha'(e_2 + e_4)$. But $\alpha = e_3 \cdot (\alpha e_3) = e_3 \cdot Fe_4 = -e_4 \cdot [\alpha'(e_2 + e_4)] = -\alpha'[e_4 \cdot e_2 + e_4 \cdot e_4] = -\alpha'[0 - 1] = \alpha'$ so

$$Fe_3 = \alpha(e_2 + e_4). \tag{2.4.11}$$

Next we compute $Fe_2 = F(N - e_4) = FN - Fe_4 = 0 - \alpha e_3$ so

$$Fe_2 = -\alpha e_3. \tag{2.4.12}$$

Finally, we select a unit spacelike vector e_1 which is orthogonal to e_2, e_3 and e_4 and satisfies $e_1 \times e_2 \cdot e_3 = 1$ to obtain an admissible basis $\{e_a\}_{a=1}^4$.

Exercise 2.4.5. Show that

$$Fe_1 = 0. \tag{2.4.13}$$

Hint: Show that $Fe_1 \cdot e_a = 0$ for $a = 1, 2, 3, 4$.

A basis for $\mathcal{M}$ constructed in the manner just described is called a *canonical basis* for F. The matrix of F relative to such a basis [read off from (2.4.10)-(2.4.13)] is

$$\begin{bmatrix} 0 & 0 & 0 & 0 \\ 0 & 0 & \alpha & 0 \\ 0 & -\alpha & 0 & \alpha \\ 0 & 0 & \alpha & 0 \end{bmatrix} \tag{2.4.14}$$

and is called a *canonical form* for F. This is, of course, just the matrix of the transformation F_N introduced in Section 2.2 and we now know that every null F takes this form in some basis. An observer in the corresponding frame sees electric $\vec{E} = \alpha e_3$ and magnetic $\vec{B} = \alpha e_1$ 3-vectors that are perpendicular and have the same magnitude α.

Exercise 2.4.6. Calculate the *third* power of the matrix (2.4.14) and improve (2.3.5) by showing

$$F^3 = 0 \quad (F \text{ null}). \tag{2.4.15}$$

For any two vectors u and v in $\mathcal{M}$ define a linear transformation $u \wedge v$: $\mathcal{M} \to \mathcal{M}$ by $u \wedge v(x) = u(v \cdot x) - v(u \cdot x)$.

Exercise 2.4.7. Show that, if F is null, then, relative to a canonical basis $\{e_a\}_{a=1}^4$,

$$F = Fe_3 \wedge e_3 \,. \tag{2.4.16}$$

The only eigenvalue of a null F is, of course, $\lambda = 0$.

Exercise 2.4.8. Show that, relative to a canonical basis $\{e_a\}_{a=1}^4$, the eigenspace of F corresponding to $\lambda = 0$, i.e., ker F, is Span$\{e_1, e_2 + e_4\}$ and so contains precisely one null direction (which is called the *principal null direction* of F).

2.5 The Energy-Momentum Transformation

Let $F: \mathcal{M} \to \mathcal{M}$ be a nonzero, skew-symmetric linear transformation on $\mathcal{M}$. The linear transformation $T: \mathcal{M} \to \mathcal{M}$ defined by

$$T = \frac{1}{4\pi}\left[\frac{1}{4}\mathrm{tr}(F^2)I - F^2\right], \tag{2.5.1}$$

where $F^2 = F \circ F$, I is the identity transformation $I(x) = x$ for every x in $\mathcal{M}$ and $\operatorname{tr}(F^2)$ is the trace of F^2, i.e., the sum of the diagonal entries in the matrix of F^2 relative to any basis, is called the *energy-momentum transformation* associated with F. Observe that T is *symmetric* with respect to the Lorentz inner product, i.e.,

$$Tx \cdot y = x \cdot Ty \tag{2.5.2}$$

for all x and y in $\mathcal{M}$.

Exercise 2.5.1. Prove (2.5.2).

Moreover, since $\operatorname{tr}(I) = 4$, T is *trace-free*, i.e.,

$$\operatorname{tr} T = 0. \tag{2.5.3}$$

Relative to any admissible basis for $\mathcal{M}$ the matrix $[T^a{}_b]$ of T has entries given by

$$T^a{}_b = \frac{1}{4\pi}\left[\frac{1}{4}F^\alpha{}_\beta F^\beta{}_\alpha \delta^a_b - F^a{}_\alpha F^\alpha{}_b\right], \qquad a, b = 1, 2, 3, 4. \tag{2.5.4}$$

Although not immediately apparent from the definition, T contains all of the information relevant to describing the classical "energy" and "momentum" content of the electromagnetic field represented by F in each admissible frame. To see this we need the matrix of T in terms of the electric and magnetic 3-vectors $\vec{E}$ and $\vec{B}$.

Exercise 2.5.2. With the matrix of F relative to $\{e_a\}$ written in the form (2.2.3), calculate the matrix of F^2 relative to $\{e_a\}$ and show that it can be written as

$$\begin{bmatrix} (E^1)^2-(B^2)^2-(B^3)^2 & E^1E^2+B^1B^2 & E^1E^3+B^1B^3 & E^2B^3-E^3B^2 \\ E^1E^2+B^1B^2 & (E^2)^2-(B^1)^2-(B^3)^2 & E^2E^3+B^2B^3 & E^3B^1-E^1B^3 \\ E^1E^3+B^1B^3 & E^2E^3+B^2B^3 & (E^3)^2-(B^1)^2-(B^2)^2 & E^1B^2-E^2B^1 \\ E^3B^2-E^2B^3 & E^1B^3-E^3B^1 & E^2B^1-E^1B^2 & |\vec{E}|^2 \end{bmatrix} \tag{2.5.5}$$

Now, $\frac{1}{4\pi}$ times the off-diagonal entries in (2.5.5) are the off-diagonal entries in $[T^a{}_b]$. Adding the diagonal entries in (2.5.5) gives $\operatorname{tr}(F^2) = 2(|\vec{E}|^2 - |\vec{B}|^2)$ so $\frac{1}{4}\operatorname{tr}(F^2) = \frac{1}{2}((E^1)^2 + (E^2)^2 + (E^3)^2 - (B^1)^2 - (B^2)^2 - (B^3)^2)$. Subtracting the diagonal entries in (2.5.5) from the corresponding diagonal entries in $\frac{1}{4}\operatorname{tr}(F^2)I$ gives 4π times the diagonal entries in $[T^a{}_b]$. Thus,

$$\begin{aligned}
T^1{}_1 &= \tfrac{1}{8\pi}\left[-(E^1)^2+(E^2)^2+(E^3)^2-(B^1)^2+(B^2)^2+(B^3)^2\right], \\
T^2{}_2 &= \tfrac{1}{8\pi}\left[(E^1)^2-(E^2)^2+(E^3)^2+(B^1)^2-(B^2)^2+(B^3)^2\right], \\
T^3{}_3 &= \tfrac{1}{8\pi}\left[(E^1)^2+(E^2)^2-(E^3)^2+(B^1)^2+(B^2)^2-(B^3)^2\right], \\
T^4{}_4 &= -\tfrac{1}{8\pi}\left[|\vec{E}|^2+|\vec{B}|^2\right].
\end{aligned} \tag{2.5.6}$$

Notice once again that the nonzero index of the Lorentz inner product has the unfortunate consequence that the matrix of a symmetric linear transformation on $\mathcal{M}$ is not (quite) a symmetric matrix.

In classical electromagnetic theory the quantity $\frac{1}{8\pi}[\,|\vec{E}|^2+|\vec{B}|^2\,]$ $(= -T^4{}_4)$ is called the *energy density* measured in the given frame of reference for the electromagnetic field with electric and magnetic 3-vectors $\vec{E}$ and $\vec{B}$. The 3-vector $\frac{1}{4\pi}\,\vec{E}\times\vec{B} = (E^2B^3-E^3B^2)e_1+(E^3B^1-E^1B^3)e_2+(E^1B^2-E^2B^1)e_3 = T^1{}_4e_1+T^2{}_4e_2+T^3{}_4e_3 = -(T^4{}_1e_1+T^4{}_2e_2+T^4{}_3e_3)$ is called the *Poynting 3-vector* and describes the energy density flux of the field. Finally, the 3×3 matrix $[T^i{}_j]_{i,j=1,2,3}$ is known as the *Maxwell stress tensor* of the field in the given frame. Thus, the entries in the matrix of T relative to an admissible basis all have something to say about the energy content of the field F measured in the corresponding frame.

Notice that the $(4,4)$-entry in the matrix $[T^a{}_b]$ of T relative to $\{e_a\}$ is $T^4{}_4 = -Te_4\cdot e_4 = -\frac{1}{8\pi}[|\,\vec{E}\,|^2+|\,\vec{B}\,|^2]$. Thus, we define, for every future-directed unit timelike vector U, the *energy density* of F in any admissible basis with $e_4 = U$ to be $TU\cdot U$. In the sense of the following result, the energy density completely determines the energy-momentum transformation.

Theorem 2.5.1. *Let S and T be two nonzero linear transformations on $\mathcal{M}$ which are symmetric with respect to the Lorentz inner product, i.e., satisfy (2.5.2). If $SU\cdot U = TU\cdot U$ for every future-directed unit timelike vector U, then $S = T$.*

Proof: Observe first that the hypothesis, together with the linearity of S and T imply that $SV\cdot V = TV\cdot V$ for all timelike vectors V. Now select a basis $\{U_a\}_{a=1}^4$ for $\mathcal{M}$ consisting exclusively of future-directed unit timelike vectors (convince yourself that such things exist). Thus, $SU_a\cdot U_a = TU_a\cdot U_a$ for each $a = 1,2,3,4$. Next observe that, for all $a,b = 1,2,3,4$, Lemma 1.4.3 implies that U_a+U_b is timelike and future-directed so that

$$\begin{aligned}
S(U_a+U_b)\cdot(U_a+U_b) &= T(U_a+U_b)\cdot(U_a+U_b), \\
SU_a\cdot U_a+2SU_a\cdot U_b+SU_b\cdot U_b &= TU_a\cdot U_a+2TU_a\cdot U_b+TU_b\cdot U_b, \\
SU_a\cdot U_b &= TU_a\cdot U_b.
\end{aligned}$$

Exercise 2.5.3. Show that

$$Sx \cdot y = Tx \cdot y \tag{2.5.7}$$

for all x and y in $\mathcal{M}$.

Now, let $\{e_a\}_{a=1}^4$ be an orthonormal basis for $\mathcal{M}$. Then (2.5.7) gives

$$Se_a \cdot e_b = Te_a \cdot e_b \tag{2.5.8}$$

for all $a, b = 1, 2, 3, 4$. But (2.5.8) shows that the matrices of S and T relative to $\{e_a\}$ are identical so $S = T$. ∎

We investigate the eigenvalues and eigenvectors of T by working in a canonical basis for F. First suppose F is regular and $\{e_a\}$ is a canonical basis for F. Then the matrix $[F^a{}_b]$ of F relative to $\{e_a\}$ has the form (2.4.9) and a simple calculation gives

$$[F^a{}_b]^2 = \begin{bmatrix} -\delta^2 & 0 & 0 & 0 \\ 0 & -\delta^2 & 0 & 0 \\ 0 & 0 & \epsilon^2 & 0 \\ 0 & 0 & 0 & \epsilon^2 \end{bmatrix}$$

so $\operatorname{tr}(F^2) = 2(\epsilon^2 - \delta^2)$ and therefore $[T^a{}_b] = \frac{1}{4\pi}[\,\frac{1}{4}\operatorname{tr}(F^2)I - [F^a{}_b]^2\,]$ is given by

$$[T^a{}_b] = \frac{1}{8\pi}(\epsilon^2 + \delta^2) \begin{bmatrix} 1 & 0 & 0 & 0 \\ 0 & 1 & 0 & 0 \\ 0 & 0 & -1 & 0 \\ 0 & 0 & 0 & -1 \end{bmatrix}.$$

$\det(T - \lambda I) = 0$ therefore gives $(\lambda + \frac{\epsilon^2+\delta^2}{8\pi})^2(\lambda - \frac{\epsilon^2+\delta^2}{8\pi})^2 = 0$ so $\lambda = \pm\frac{\epsilon^2+\delta^2}{8\pi} = \mp T^4{}_4$ (the energy density). The eigenvectors corresponding to $\lambda = \frac{\epsilon^2+\delta^2}{8\pi}$ are obtained by solving

$$\frac{\epsilon^2 + \delta^2}{8\pi} \begin{bmatrix} 1 & 0 & 0 & 0 \\ 0 & 1 & 0 & 0 \\ 0 & 0 & -1 & 0 \\ 0 & 0 & 0 & -1 \end{bmatrix} \begin{bmatrix} v^1 \\ v^2 \\ v^3 \\ v^4 \end{bmatrix} = \frac{\epsilon^2 + \delta^2}{8\pi} \begin{bmatrix} v^1 \\ v^2 \\ v^3 \\ v^4 \end{bmatrix},$$

i.e.,

$$\begin{bmatrix} v^1 \\ v^2 \\ -v^3 \\ -v^4 \end{bmatrix} = \begin{bmatrix} v^1 \\ v^2 \\ v^3 \\ v^4 \end{bmatrix},$$

so $v^3 = v^4 = 0$, whereas v^1 and v^2 are arbitrary. Thus, the eigenspace is $\operatorname{Span}\{e_1, e_2\}$ which contains only spacelike vectors. Similarly, the eigenspace corresponding to $\lambda = -\frac{\epsilon^2+\delta^2}{8\pi}$ is $\operatorname{Span}\{e_3, e_4\}$ which contains two independent null directions $(e_3 \pm e_4)$ called the *principal null directions* of T.

If F is null and $\{e_a\}$ is a canonical basis, then $[F^a{}_b]$ has the form (2.4.14) so

$$[F^a{}_b]^2 = \begin{bmatrix} 0 & 0 & 0 & 0 \\ 0 & -\alpha^2 & 0 & \alpha^2 \\ 0 & 0 & 0 & 0 \\ 0 & -\alpha^2 & 0 & \alpha^2 \end{bmatrix}$$

and therefore $\operatorname{tr} F^2 = 0$ so

$$[T^a{}_b] = -\frac{1}{4\pi}[F^a{}_b]^2 = \frac{\alpha^2}{4\pi}\begin{bmatrix} 0 & 0 & 0 & 0 \\ 0 & 1 & 0 & -1 \\ 0 & 0 & 0 & 0 \\ 0 & 1 & 0 & -1 \end{bmatrix}.$$

Exercise 2.5.4. Show that $\lambda = 0$ is the only eigenvalue of T and that the corresponding eigenspace is $\operatorname{Span}\{e_1, e_3, e_2 + e_4\}$, which contains only one null direction (that of $e_2 + e_4$), again called the *principal null direction* of T.

Exercise 2.5.5. Show that every eigenvector of F is also an eigenvector of T (corresponding to a different eigenvalue, in general).

Exercise 2.5.6. Show that the energy-momentum transformation T satisfies the *dominant energy condition,* i.e., has the property that if u and v are timelike or null and both are future-directed, then

$$Tu \cdot v \geq 0. \tag{2.5.9}$$

Hint: Work in canonical coordinates for the corresponding F.

2.6 Motion in Constant Fields

Thus far we have concentrated our attention on the formal mathematical structure of the object we have chosen to model an electromagnetic field at a fixed point of $\mathcal{M}$, that is, a skew-symmetric linear transformation. In order to reestablish contact with the physics of relativistic electrodynamics we must address the issue of how a given collection of charged particles gives rise to these linear transformations at each point of $\mathcal{M}$ and then study how the worldline of another charge introduced into the system will respond to the presence of the field. The first problem we defer to Section 2.7. In this section we consider the motion of a charged particle in the simplest of all electromagnetic fields, i.e., those that are constant. Thus, we presume the existence of a system of particles that determines a *single* skew-symmetric linear transformation $F\colon \mathcal{M} \to \mathcal{M}$ with the property that any charged particle (α, m, e) introduced into the system will experience changes in world momentum at *every* point on its worldline described by (2.1.1). More particularly, we have in mind fields with the property that there exists a frame of reference in which the field is constant and either purely magnetic ($\vec{E} = \vec{0}$) or purely electric ($\vec{B} = \vec{0}$). To a reasonable degree of approximation such fields exist in nature and are of considerable practical importance. Such a field, however, can obviously not be null (without being identically zero) so we shall restrict our attention to the regular case and will work exclusively in a canonical basis.

Suppose then that $F: \mathcal{M} \to \mathcal{M}$ is nonzero, skew-symmetric and regular. Then there exists an admissible basis $\{e_a\}_{a=1}^4$ for $\mathcal{M}$ and two real numbers $\epsilon \geq 0$ and $\delta \geq 0$ so that the matrix of F in $\{e_a\}$ is

$$[F^a{}_b] = \begin{bmatrix} 0 & \delta & 0 & 0 \\ -\delta & 0 & 0 & 0 \\ 0 & 0 & 0 & \epsilon \\ 0 & 0 & \epsilon & 0 \end{bmatrix},$$

so that $\vec{E} = \epsilon e_3$ and $\vec{B} = \delta e_3$. Let (α, m, e) be a charged particle with world velocity $U = U(\tau) = U^a(\tau)e_a$ which satisfies

$$\frac{dU}{d\tau} = \frac{e}{m} FU \tag{2.6.1}$$

at each point of α. Thus,

$$\begin{bmatrix} dU^1/d\tau \\ dU^2/d\tau \\ dU^3/d\tau \\ dU^4/d\tau \end{bmatrix} = \frac{e}{m} \begin{bmatrix} 0 & \delta & 0 & 0 \\ -\delta & 0 & 0 & 0 \\ 0 & 0 & 0 & \epsilon \\ 0 & 0 & \epsilon & 0 \end{bmatrix} \begin{bmatrix} U^1 \\ U^2 \\ U^3 \\ U^4 \end{bmatrix} = \begin{bmatrix} \omega U^2 \\ -\omega U^1 \\ \nu U^4 \\ \nu U^3 \end{bmatrix},$$

where $\omega = \frac{\delta e}{m}$ and $\nu = \frac{\epsilon e}{m}$. Thus, we have

$$\begin{cases} \dfrac{dU^1}{d\tau} = \omega U^2 \\ \dfrac{dU^2}{d\tau} = -\omega U^1 \end{cases} \tag{2.6.2}$$

and

$$\begin{cases} \dfrac{dU^3}{d\tau} = \nu U^4 \\ \dfrac{dU^4}{d\tau} = \nu U^3 \,. \end{cases} \tag{2.6.3}$$

We (temporarily) assume that neither ϵ nor δ is zero so that $\omega\nu \neq 0$. Differentiating the first equation in (2.6.2) with respect to τ and using the second equation gives

$$\frac{d^2U^1}{d\tau^2} = -\omega^2 U^1 \tag{2.6.4}$$

and similarly for (2.6.3),

$$\frac{d^2U^3}{d\tau^2} = \nu^2 U^3\,. \tag{2.6.5}$$

The general solution to (2.6.4) is

$$U^1 = A\,\sin\omega\tau + B\,\cos\omega\tau \tag{2.6.6}$$

and, since $U^2 = \frac{1}{\omega}\frac{dU^1}{d\tau}$,

$$U^2 = A\,\cos\omega\tau - B\,\sin\omega\tau\,. \tag{2.6.7}$$

Similarly,

$$U^3 = C\,\sinh\nu\tau + D\,\cosh\nu\tau \tag{2.6.8}$$

and

$$U^4 = C\,\cosh\nu\tau + D\,\sinh\nu\tau\,. \tag{2.6.9}$$

Exercise 2.6.1. Integrate (2.6.6) and (2.6.7) and show that the result can be written in the form

$$x^1(\tau) = a\,\sin(\omega\tau + \phi) + x_0^1 \tag{2.6.10}$$

and

$$x^2(\tau) = a\,\cos(\omega\tau + \phi) + x_0^2\,, \tag{2.6.11}$$

where a, ϕ, x_0^1 and x_0^2 are constants and $a > 0$.

Integrating (2.6.8) and (2.6.9) gives

$$x^3(\tau) = \frac{C}{\nu}\,\cosh\nu\tau + \frac{D}{\nu}\,\sinh\nu\tau + x_0^3 \tag{2.6.12}$$

and

$$x^4(\tau) = \frac{C}{\nu}\sinh\nu\tau + \frac{D}{\nu}\cosh\nu\tau + x_0^4. \tag{2.6.13}$$

Observe now that if $\epsilon = 0$ and $\delta \neq 0$, (2.6.10) and (2.6.11) are unchanged, whereas $\frac{dU^3}{d\tau} = \frac{dU^4}{d\tau} = 0$ imply that (2.6.12) and (2.6.13) are replaced by

$$x^3(\tau) = C^3\tau + x_0^3 \qquad (\epsilon = 0) \tag{2.6.14}$$

and

$$x^4(\tau) = C^4\tau + x_0^4 \qquad (\epsilon = 0). \tag{2.6.15}$$

Similarly, if $\epsilon \neq 0$ and $\delta = 0$, then (2.6.12) and (2.6.13) are unchanged, but (2.6.10) and (2.6.11) become

$$x^1(\tau) = C^1\tau + x_0^1 \qquad (\delta = 0) \tag{2.6.16}$$

and

$$x^2(\tau) = C^2\tau + x_0^2 \qquad (\delta = 0). \tag{2.6.17}$$

Now we consider two special cases. First suppose that $\epsilon = 0$ and $\delta \neq 0$ (so that an observer in $\{e_a\}$ sees a constant and purely magnetic field in the e_3-direction). Then (2.6.10), (2.6.11), (2.6.14) and (2.6.15) give

$$\alpha(\tau) = (a\sin(\omega\tau+\phi) + x_0^1, a\cos(\omega\tau+\phi) + x_0^2, C^3\tau + x_0^3, C^4\tau + x_0^4)$$

so that

$$U(\tau) = \big(a\omega\,\cos(\omega\tau+\phi), -a\omega\sin(\omega\tau+\phi), C^3, C^4\big).$$

Now, $U \cdot U = -1$ implies $a^2\omega^2 + (C^3)^2 - (C^4)^2 = -1$. Since $C^4 = U^4 = \gamma > 0$, $C^4 = (1 + a^2\omega^2 + (C^3)^2)^{\frac{1}{2}}$ so

$$\alpha(\tau) = (x_0^1, x_0^2, x_0^3, x_0^4) + (a\sin(\omega\tau+\phi), a\cos(\omega\tau+\phi),$$

$$C^3\tau, (1 + a^2\omega^2 + (C^3)^2)^{\frac{1}{2}}\,\tau). \tag{2.6.18}$$

Note that $(x^1 - x_0^1)^2 + (x^2 - x_0^2)^2 = a^2$. Thus, if $C^3 \neq 0$, the trajectory in $\{e_1, e_2, e_3\}$-space is a spiral along the e_3-direction (i.e., along the magnetic field lines). If $C^3 = 0$, the trajectory is a circle. This latter case is of some practical significance since one can introduce constant magnetic fields in a bubble chamber in such a way as to induce a particle of interest to follow a circular path. We show now that by making relatively elementary measurements one can in this way determine the charge-to-mass ratio $\frac{e}{m}$ for the particle. Indeed, with $C^3 = 0$, (2.6.18) yields by differentiation

$$U(\tau) = \left(a\omega\cos(\omega\tau+\phi), -a\omega\sin(\omega\tau+\phi), 0, (1+a^2\omega^2)^{\frac{1}{2}}\right). \tag{2.6.19}$$

But $U = \gamma(\vec{u}, 1)$ by (1.4.10) so $\vec{u} = (\frac{a\omega}{\gamma}\cos(\omega\tau+\phi), -\frac{a\omega}{\gamma}\sin(\omega\tau+\phi), 0)$ and thus

$$\beta^2 = |\,\vec{u}\,|^2 = \frac{a^2\omega^2}{\gamma^2} = \frac{a^2\omega^2}{1+a^2\omega^2} = \frac{1}{\frac{m^2}{a^2e^2\delta^2}+1}.$$

Exercise 2.6.2. Assume $e > 0$ and $\beta > 0$ and solve for $\frac{e}{m}$ to obtain

$$\frac{e}{m} = \frac{1}{a\,|\,\delta\,|}\frac{\beta}{\sqrt{1-\beta^2}}.$$

Finally, we suppose that $\delta = 0$ and $\epsilon \neq 0$ (constant and purely electric field in the e_3-direction). Then (2.6.12), (2.6.13), (2.6.16) and (2.6.17) give

$$\begin{aligned}\alpha(\tau) = \; & \big(C^1\tau + x_0^1, C^2\tau + x_0^2, \tfrac{C}{\nu}\cosh\nu\tau + \tfrac{D}{\nu}\sinh\nu\tau + x_0^3, \\ & \tfrac{C}{\nu}\sinh\nu\tau + \tfrac{D}{\nu}\cosh\nu\tau + x_0^4\big).\end{aligned}$$

Consequently,

$$U(\tau) = \left(C^1, C^2, C\sinh\nu\tau + D\cosh\nu\tau, C\cosh\nu\tau + D\sinh\nu\tau\right).$$

We consider the case in which $\alpha(0) = 0$ so that $x_0^1 = x_0^2 = 0$, $x_0^3 = -\frac{C}{\nu}$ and $x_0^4 = -\frac{D}{\nu}$. Next we suppose that $\vec{u}\,(0) = e_1$ (the initial velocity of the particle relative to $\{e_a\}_{a=1}^4$ has magnitude 1 and direction perpendicular to that of the field $\vec{E} = \epsilon e_3$). Then $C^1 = 1$, $C^2 = 0$ and $D = 0$, i.e., $U(\tau) = (1, 0, C\sinh\nu\tau, C\cosh\nu\tau)$. Moreover, $U \cdot U = -1$ gives $-1 = 1^2 + 0^2 + C^2\sinh^2\nu\tau - C^2\cosh^2\nu\tau = 1 - C^2$ so $C^2 = 2$. Since $C = \gamma(0) > 0$, we have $C = \sqrt{2}$. Thus,

$$\alpha(\tau) = \left(\tau, 0, \frac{\sqrt{2}}{\nu}(\cosh\nu\tau - 1), \frac{\sqrt{2}}{\nu}\sinh\nu\tau\right).$$

The trajectory in $\{e_1, e_2, e_3\}$-space is the curve $\tau \to (\tau, 0, \frac{\sqrt{2}}{\nu}(\cosh\nu\tau - 1))$. Thus, $x^3 = \frac{\sqrt{2}}{\nu}(\cosh(\nu x^1) - 1)$, i.e.,

$$x^3 = \frac{m\sqrt{2}}{e\epsilon}\left(\cosh\left(\frac{e\epsilon}{m}x^1\right) - 1\right)$$

which is a catenary in the x^1x^3-plane (see Figure 2.6.1).

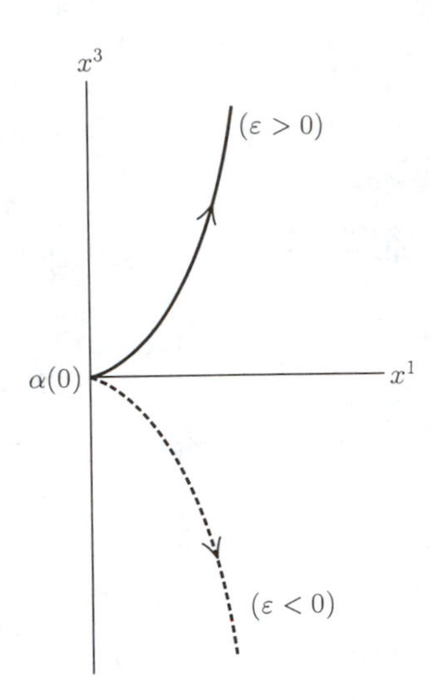

Figure 2.6.1

2.7 Variable Electromagnetic Fields

Most electromagnetic fields encountered in nature are not constant. That is, the linear transformations that tell a charged particle how to respond to the field generally vary from point to point along the particle's worldline. To discuss such phenomena we shall require a few preliminaries.

A subset R of $\mathcal{M}$ is said to be *open* in $\mathcal{M}$ if, for each $x_0 \in R$, there exists a positive real number ε such that the set $N_\varepsilon^E(x_0) = \{x \in \mathcal{M} : ((x^1 - x_0^1)^2 + (x^2 - x_0^2)^2 + (x^3 - x_0^3)^2 + (x^4 - x_0^4)^2)^{\frac{1}{2}} < \varepsilon\}$ is contained entirely in R (in Section A.1 of Appendix A we show that this definition does not depend on the particular admissible basis relative to which the coordinates are calculated). This is the usual Euclidean notion of an open set in $\mathbb{R}^4$ so, intuitively, one thinks of open sets in $\mathcal{M}$ as those sets which do not contain any of their "boundary points". Open sets in $\mathcal{M}$ will be called *regions* in $\mathcal{M}$. A real-valued function $f\colon R \to \mathbb{R}$ defined on some region R in $\mathcal{M}$ is said to be *smooth* if it has continuous partial derivatives of all orders and types with respect to x^1, x^2, x^3 and x^4 for any (and therefore all) admissible coordinate systems on $\mathcal{M}$. For convenience, we shall denote the partial derivative $\frac{\partial f}{\partial x^a}$ of such a function by $f_{,a}$. Now, suppose we have assigned to each point p in some region R of $\mathcal{M}$ a linear transformation $F(p)\colon \mathcal{M} \to \mathcal{M}$. Relative to an admissible basis each $F(p)$ will have a matrix $[F^a{}_b(p)]$. If the entries in this matrix are smooth on R we say that the assignment $p \xrightarrow{F} F(p)$ itself is *smooth*. If each of the linear transformations $F(p)$ is skew-symmetric, the smooth assignment $p \xrightarrow{F} F(p)$ is a reasonable first approximation to the definition of an "electromagnetic field on R". However, nature does not grant us so much freedom as to allow us to make such assignments arbitrarily. The rules by which we must play the game consist of a system of partial differential equations known as "Maxwell's equations". In regions that are free of charge and in terms of the electric and magnetic 3-vectors $\vec{E}$ and $\vec{B}$ these equations require that

$$\begin{aligned} \operatorname{div} \vec{E} &= 0, \qquad \operatorname{curl} \vec{B} - \frac{\partial \vec{E}}{\partial x^4} = \vec{0}, \\ \operatorname{div} \vec{B} &= 0, \qquad \operatorname{curl} \vec{E} + \frac{\partial \vec{B}}{\partial x^4} = \vec{0}, \end{aligned} \tag{2.7.1}$$

where div and curl are the familiar divergence and curl from vector analysis in $\mathbb{R}^3$. We now translate (2.7.1) into the language of Minkowski spacetime.

A mapping $V: R \to \mathcal{M}$ which assigns to each p in some region R of $\mathcal{M}$ a vector $V(p)$ in $\mathcal{M}$ is called a *vector field* on R. Relative to any admissible basis $\{e_a\}$ for $\mathcal{M}$ we write $V(p) = V^a(p)e_a$, where $V^a: R \to \mathbb{R}$, $a = 1, 2, 3, 4$, are the *component functions* of V relative to $\{e_a\}$. A vector field is said to be *smooth* if its component functions relative to any (and therefore every) admissible basis are smooth. Now consider a smooth assignment $p \xrightarrow{F} F(p)$ of a linear transformation to each $p \in R$. We define a vector field div F, called the *divergence* of F, by specifying that its component functions relative to any $\{e_a\}$ are given by

$$(\operatorname{div} F)^b = \eta^{b\beta} F^\alpha{}_{\beta,\alpha}, \qquad b = 1, 2, 3, 4. \tag{2.7.2}$$

Thus, $(\operatorname{div} F)^i = F^\alpha{}_{i,\alpha}$ for $i = 1, 2, 3$ and $(\operatorname{div} F)^4 = -F^\alpha{}_{4,\alpha}$.

Exercise 2.7.1. A vector v in $\mathcal{M}$ has components relative to two admissible bases that are related by $\hat{v}^a = \Lambda^a{}_b v^b$. Show that (2.7.2) does indeed define a vector in $\mathcal{M}$ by showing that it has the correct "transformation law":

$$\left(\widehat{\operatorname{div} F}\right)^a = \Lambda^a{}_b (\operatorname{div} F)^b, \qquad a = 1, 2, 3, 4, \tag{2.7.3}$$

where $(\widehat{\operatorname{div} F})^a = \eta^{a\gamma} \hat{F}^\alpha{}_{\gamma,\alpha}$ and $\hat{F}^a{}_{b,c} = \frac{\partial}{\partial \hat{x}^c} \hat{F}^a{}_b$. *Hint:* Use the change of basis formula

$$\hat{F}^a{}_b = \Lambda^a{}_\alpha \Lambda_b{}^\beta F^\alpha{}_\beta \tag{2.7.4}$$

and the chain rule to show first that

$$\hat{F}^a{}_{b,c} = \Lambda^a{}_\alpha \Lambda_b{}^\beta \Lambda_c{}^\gamma F^\alpha{}_{\beta,\gamma}\,. \tag{2.7.5}$$

Exercise 2.7.2. Show that if $p \xrightarrow{F} F(p)$ and $p \xrightarrow{G} G(p)$ are two smooth assignments of linear transformations to points in the region R and $F+G$ is defined at each $p \in R$ by $(F+G)(p) = F(p) + G(p)$, then

$$\operatorname{div}(F+G) = \operatorname{div} F + \operatorname{div} G. \tag{2.7.6}$$

Exercise 2.7.3. Show that, if each $F(p)$ is skew-symmetric, then, in terms of the 3-vectors $\vec{E}$ and $\vec{B}$,

$$(\operatorname{div} F)^i = \left[\frac{\partial \vec{E}}{\partial x^4} - \operatorname{curl} \vec{B}\right] \cdot e_i, \qquad i = 1, 2, 3, \tag{2.7.7}$$

$$(\operatorname{div} F)^4 = -\operatorname{div} \vec{E}. \tag{2.7.8}$$

We conclude from Exercise 2.7.3 that the first pair of equations in (2.7.1) is equivalent to the single equation

$$\operatorname{div} F = 0, \tag{2.7.9}$$

where 0 is, of course, the zero vector in $\mathcal{M}$.

The second pair of equations in (2.7.1) is most conveniently expressed in terms of a mathematical object closely related to F, but with a matrix that *is* skew-symmetric. Thus, we define for each skew-symmetric linear transformation $F\colon \mathcal{M} \to \mathcal{M}$ an associated bilinear form

$$\tilde{F}\colon \mathcal{M} \times \mathcal{M} \to \mathbb{R}$$

by

$$\tilde{F}(u, v) = u \cdot Fv \tag{2.7.10}$$

for all u and v in $\mathcal{M}$. Then $\tilde{F}$ is *skew-symmetric,* i.e., satisfies

$$\tilde{F}(v, u) = -\tilde{F}(u, v). \tag{2.7.11}$$

The *matrix* $[F_{ab}]$ *of* $\tilde{F}$ relative to an admissible basis $\{e_a\}$ has entries given by

$$F_{ab} = \tilde{F}(e_a, e_b) = e_a \cdot Fe_b = \eta_{ac} F^c{}_b \tag{2.7.12}$$

and is clearly a skew-symmetric matrix (notice how the position of the indices is used to distinguish the matrix of $\widetilde{F}$ from the matrix of F).

Exercise 2.7.4. Show that if $u = u^a e_a$ and $v = v^b e_b$, then $\widetilde{F}(u,v) = F_{ab} u^a v^b$.

The entries F_{ab} are often called the *components* of $\widetilde{F}$ in the basis $\{e_a\}$. If $\{\hat{e}_a\}$ is another admissible basis, related to $\{e_a\}$ by the Lorentz transformation $[\Lambda^a{}_b]$, then the components of $\widetilde{F}$ in the two bases are related by

$$\hat{F}_{ab} = \Lambda_a{}^\alpha \Lambda_b{}^\beta F_{\alpha\beta}, \qquad a,b = 1,2,3,4. \tag{2.7.13}$$

To prove this we observe that, by definition, $\hat{F}_{ab} = \eta_{ac} \hat{F}^c{}_b = \eta_{ac} \Lambda^c{}_\gamma \Lambda_b{}^\beta F^\gamma{}_\beta$. Now, (1.2.12) gives $\Lambda^c{}_\gamma = \eta^{c\rho} \eta_{\gamma\alpha} \Lambda_\rho{}^\alpha$ so $\hat{F}_{ab} = \eta_{ac} \eta^{c\rho} \eta_{\gamma\alpha} \Lambda_\rho{}^\alpha \Lambda_b{}^\beta F^\gamma{}_\beta = \eta_{ac} \eta^{c\rho} \Lambda_\rho{}^\alpha \Lambda_b{}^\beta (\eta_{\gamma\alpha} F^\gamma{}_\beta) = \delta_a^\rho \Lambda_\rho{}^\alpha \Lambda_b{}^\beta F_{\alpha\beta} = \Lambda_a{}^\alpha \Lambda_b{}^\beta F_{\alpha\beta}$ as required.

Computing the quantities $\eta_{ac} F^c{}_b$ in terms of $\vec{E}$ and $\vec{B}$ gives

$$[F_{ab}] = \begin{bmatrix} 0 & B^3 & -B^2 & E^1 \\ -B^3 & 0 & B^1 & E^2 \\ B^2 & -B^1 & 0 & E^3 \\ -E^1 & -E^2 & -E^3 & 0 \end{bmatrix}. \tag{2.7.14}$$

Every smooth assignment $p \xrightarrow{F} F(p)$ of a skew-symmetric linear transformation to each point in some region in $\mathcal{M}$ therefore gives rise to an assignment $p \xrightarrow{\widetilde{F}} \widetilde{F}(p)$ which is likewise smooth in the sense that the entries in the matrix (2.7.14) are smooth real-valued functions. As usual, we denote the derivatives $\partial F_{ab}/\partial x^c$ by $F_{ab,\,c}$.

Exercise 2.7.5. Show that the second pair of equations in (2.7.1) is equivalent to

$$F_{ab,\,c} + F_{bc,\,a} + F_{ca,\,b} = 0, \qquad a,b,c = 1,2,3,4. \tag{2.7.15}$$

Now we define an *electromagnetic field* on a region R in $\mathcal{M}$ to be a smooth assignment $p \xrightarrow{F} F(p)$ of a skew-symmetric linear transformation to each point p in R such that it and its associated assignment $p \xrightarrow{\tilde{F}} \tilde{F}(p)$ of skew-symmetric bilinear forms satisfy *Maxwell's equations* (2.7.9) and (2.7.15).

We remark in passing that a skew-symmetric bilinear form is often referred to as a *bivector* and a smooth assignment of one such to each p in

R is called a *2-form* on R. In the language of exterior calculus the left-hand side of (2.7.15) specifies what is called the *exterior derivative* of $\widetilde{F}$ (a 3-form) and denoted $d\widetilde{F}$. Then (2.7.15) becomes

$$d\widetilde{F} = 0\,.$$

Since most modern expositions of electromagnetic theory are phrased in terms of these differential forms and because it will be of interest to us in Chapter 3, we show next that the first pair of equations in (2.7.1) [or equivalently, (2.7.9)] can be written in a similar way. Indeed, the reader may have noticed a certain "duality" between the first and second pairs of equations in (2.7.1). Specifically, the first pair can be obtained from the second by formally changing the $\vec{B}$ to an $\vec{E}$ and the $\vec{E}$ to $-\vec{B}$ (and adjusting a sign). This suggests defining the "dual" of the 2-form $\widetilde{F}$ to be a 2-form ${}^*\widetilde{F}$ whose matrix at each point is obtained from (2.7.14) by formally making the substitutions $B^i \to E^i$ and $E^i \to -B^i$ so that the first pair of equations in (2.7.1) would be equivalent to $d{}^*\widetilde{F} = 0$. In order to carry out this program rigorously we will require a few preliminaries. First we introduce the *Levi-Civita symbol* ϵ_{abcd} defined by

$$\epsilon_{abcd} = \begin{cases} 1 & \text{if } abcd \text{ is an even permutation of } 1234 \\ -1 & \text{if } abcd \text{ is an odd permutation of } 1234 \\ 0 & \text{otherwise}. \end{cases}$$

Thus, for example, $\epsilon_{1234} = \epsilon_{3412} = \epsilon_{4321} = 1$, $\epsilon_{1324} = \epsilon_{3142} = -1$ and $\epsilon_{1224} = \epsilon_{1341} = 0$. The Levi-Civita symbol arises most naturally in the theory of determinants where it is shown that, for any 4×4 matrix $M = [M^a{}_b]_{a,b=1,2,3,4}$,

$$M^\alpha{}_a\, M^\beta{}_b\, M^\gamma{}_c\, M^\delta{}_d\, \epsilon_{\alpha\beta\gamma\delta} \;=\; \epsilon_{abcd}(\det M)\,. \tag{2.7.16}$$

Exercise 2.7.6. Let F be a skew-symmetric linear transformation on $\mathcal{M}$ and $\widetilde{F}$ its associated bilinear form. For $a, b = 1, 2, 3, 4$ define

$${}^*F_{ab} \;=\; -\tfrac{1}{2}\epsilon_{\alpha\beta ab}\, F^{\alpha\beta}\,, \tag{2.7.17}$$

where $F^{\alpha\beta} = \eta^{\alpha\mu}\,\eta^{\beta\nu}\,F_{\mu\nu}$. Show that, in terms of $\vec{E}$ and $\vec{B}$, the matrix $[{}^*F_{ab}]$ is just (2.7.14) after the substitutions $B^i \to E^i$ and $E^i \to -B^i$ have been made, e.g., ${}^*F_{12} = E^3$. *Hint:* Just calculate $-\frac{1}{2}\epsilon_{\alpha\beta ab}\,F^{\alpha\beta}$ in terms of $\vec{E}$ and $\vec{B}$ for various choices of a and b and use the skew-symmetry of ${}^*F_{ab}$ and F_{ab} to minimize the number of such choices you must make.

Exercise 2.7.7. Let $\{e_a\}$ and $\{\hat{e}_a\}$ be two admissible bases for $\mathcal{M}$, F a skew-symmetric linear transformation on $\mathcal{M}$ and $\tilde{F}$ its associated bilinear form. Define ${}^*F_{ab} = -\frac{1}{2}\epsilon_{\alpha\beta ab}\,F^{\alpha\beta}$ and ${}^*\hat{F}_{ab} = -\frac{1}{2}\epsilon_{\alpha\beta ab}\,\hat{F}^{\alpha\beta}$, where $\hat{F}^{\alpha\beta} = \eta^{\alpha\mu}\,\eta^{\beta\nu}\,\hat{F}_{\mu\nu}$ and $\hat{F}_{\mu\nu} = \eta_{\mu\sigma}\,\hat{F}^{\sigma}{}_{\nu}$. Show that for any two vectors $u = u^a\,e_a = \hat{u}^a\,\hat{e}_a$ and $v = v^b\,e_b = \hat{v}^b\,\hat{e}_b$ in $\mathcal{M}$,

$$ {}^*F_{ab}\,u^a\,v^b \;=\; {}^*\hat{F}_{ab}\,\hat{u}^a\,\hat{v}^b\,. \tag{2.7.18}$$

Hint: First show that (2.7.13) is equivalent to

$$ \hat{F}^{ab} \;=\; \Lambda^a{}_\alpha\,\Lambda^b{}_\beta\,F^{\alpha\beta} \tag{2.7.19}$$

and use (2.7.16).

The equality in (2.7.18) legitimizes the following definition: If F is a skew-symmetric linear transformation on $\mathcal{M}$ and $\tilde{F}$ is its associated bilinear form we define the *dual* of $\tilde{F}$ to be the bilinear form ${}^*\tilde{F}\colon \mathcal{M}\times\mathcal{M}\to\mathbb{R}$ whose value at $(u,v)\in\mathcal{M}\times\mathcal{M}$ is

$$ {}^*\tilde{F}(u,v) \;=\; {}^*F_{ab}\,u^a v^b\,. \tag{2.7.20}$$

Exercise 2.7.7 assures us that this definition is independent of the particular admissible basis in which the calculations are performed. Moreover, Exercise 2.7.6 and the above-mentioned duality between the first and second pairs of equations in (2.7.1) make it clear that the first of Maxwell's equations (2.7.9) is equivalent to

$$ {}^*F_{ab,\,c} + {}^*F_{bc,\,a} + {}^*F_{ca,\,b} \;=\; 0, \qquad a,b,c = 1,2,3,4, \tag{2.7.21}$$

or, more concisely,

$$ d\,{}^*\tilde{F} = 0\,. $$

We should point out that the linear transformation F, its associated bilinear form $\tilde{F}$ and the dual ${}^*\tilde{F}$ of $\tilde{F}$ all contain precisely the same information from both the mathematical and the physical points of view (examine their matrices in terms of $\vec{E}$ and $\vec{B}$). Some matters are more conveniently discussed in terms of F. For others, the appropriate choice is $\tilde{F}$ or ${}^*\tilde{F}$. Some calculations are simplest when carried out with the $F^a{}_b$, whereas for others one might prefer to work with F_{ab}, or F^{ab}, or ${}^*F_{ab}$. One must become comfortable with this sort of shifting perspective. In particular, one must develop a facility for the "index gymnastics" that, as we have seen already in this section, are necessitated by such a shift. To reenforce this point, to prepare gently for Chapter 3 and to derive a very important property of the energy-momentum transformation, we pause to provide a bit more practice.

Exercise 2.7.8. Show that, for any skew-symmetric linear transformation $F: \mathcal{M} \to \mathcal{M}$, $\frac{1}{2} F_{ab} F^{ab} = |\vec{B}|^2 - |\vec{E}|^2$ and $\frac{1}{4} {}^*F_{ab}F^{ab} = \vec{E} \cdot \vec{B}$.

Next we consider a skew-symmetric linear transformation $F: \mathcal{M} \to \mathcal{M}$ and its associated energy-momentum transformation $T: \mathcal{M} \to \mathcal{M}$ given by (2.5.1). Define a bilinear form $\widetilde{T}: \mathcal{M} \times \mathcal{M} \to \mathbb{R}$ by $\widetilde{T}(u,v) = u \cdot Tv$ for all $(u,v) \in \mathcal{M} \times \mathcal{M}$. Then $\widetilde{T}$ is symmetric, i.e., $\widetilde{T}(v,u) = \widetilde{T}(u,v)$ by (2.5.2). Now let $\{e_a\}$ be an admissible basis and $[T^a{}_b]$ the matrix of T relative to this basis [see (2.5.4)]. For all $a, b = 1,2,3,4$, we let $T_{ab} = T(e_a, e_b) = e_a \cdot Te_b = \eta_{a\gamma} T^\gamma{}_b$. Then, if $u = u^a e_a$ and $v = v^b e_b$, we have $T(u,v) = T_{ab}\, u^a v^b$ just as in Exercise 2.7.4. As an exercise in index manipulation and because we will need the result in Chapter 3 we show that T_{ab} can be written in the form

$$T_{ab} = \frac{1}{4\pi}\left[F_{a\alpha}\, F_b{}^\alpha - \frac{1}{4}\,\eta_{ab}\, F_{\alpha\beta}\, F^{\alpha\beta}\right], \tag{2.7.22}$$

where $F_b{}^\alpha = \eta_{b\mu}\, F^{\mu\alpha}$. Begin with (2.5.4).

$$\begin{aligned}
4\pi\, T_{ab} &= 4\pi\,\eta_{a\gamma}\, T^\gamma{}_b = \eta_{a\gamma}\left[\tfrac{1}{4}\, F^\alpha{}_\beta\, F^\beta{}_\alpha\, \delta^\gamma_b - F^\gamma{}_\alpha\, F^\alpha{}_b\right] \\
&= \tfrac{1}{4}\, F^\alpha{}_\beta\, F^\beta{}_\alpha (\eta_{a\gamma}\,\delta^\gamma_b) - (\eta_{a\gamma}\, F^\gamma{}_\alpha)\, F^\alpha{}_b \\
&= \tfrac{1}{4}\, F^\alpha{}_\beta\, F^\beta{}_\alpha\, \eta_{ab} - F_{a\alpha}\, F^\alpha{}_b \\
&= \tfrac{1}{4}\,\eta_{ab}(\eta^{\alpha\gamma}\, F_{\gamma\beta})(\eta_{\alpha\sigma}\, F^{\beta\sigma}) - F_{a\alpha}\,\eta_{b\gamma}\, F^{\alpha\gamma} \\
&= \tfrac{1}{4}\,\eta_{ab}(\eta^{\alpha\gamma}\,\eta_{\alpha\sigma})\, F_{\gamma\beta}\, F^{\beta\sigma} + F_{a\alpha}\,\eta_{b\gamma}\, F^{\gamma\alpha} \\
&= \tfrac{1}{4}\,\eta_{ab}\,\delta^\gamma_\sigma\, F_{\gamma\beta}\, F^{\beta\sigma} + F_{a\alpha}\, F_b{}^\alpha = \tfrac{1}{4}\,\eta_{ab}\, F_{\gamma\beta}\, F^{\beta\gamma} + F_{a\alpha}\, F_b{}^\alpha \\
&= F_{a\alpha}\, F_b{}^\alpha - \tfrac{1}{4}\,\eta_{ab}\, F_{\gamma\beta}\, F^{\gamma\beta} = F_{a\alpha}\, F_b{}^\alpha - \tfrac{1}{4}\,\eta_{ab}\, F_{\alpha\beta}\, F^{\alpha\beta}
\end{aligned}$$

as required.

Exercise 2.7.9. Show that if $u = u^a e_a$ and $v = v^b e_b$ are timelike or null and both are future-directed, then the dominant energy condition (2.5.9) can be written

$$T_{ab}\, u^a\, v^b \geq 0\,.$$

Now let $p \xrightarrow{F} F(p)$ be an electromagnetic field on some region R in $\mathcal{M}$. Assign to each p in R a linear transformation $T(p)$ which is the energy-momentum transformation of $F(p)$.

Exercise 2.7.10. Show that the assignment $p \xrightarrow{T} T(p)$ is smooth and that

$$\operatorname{div} T = 0. \tag{2.7.23}$$

Hints: From (2.5.4) and the product rule show that $4\pi T^a{}_{b,c} = -F^a{}_\alpha F^\alpha{}_{b,c} - F^\alpha{}_b F^a{}_{\alpha,c} + \frac{1}{4}(F^\alpha{}_\beta F^\beta{}_{\alpha,c} + F^\beta{}_\alpha F^\alpha{}_{\beta,c})\delta^a_b$. Next show that $4\pi T^a{}_{b,a} = -F^a{}_{\alpha,a} F^\alpha{}_b - F^a{}_\alpha F^\alpha{}_{b,a} + \frac{1}{2} F^\alpha{}_{\beta,b} F^\beta{}_\alpha$. Finally, observe that $F^a{}_\alpha F^\alpha{}_{b,a} = F^{a\alpha} F_{\alpha b,a}(F^{a\alpha} - F^{\alpha a}) = -\frac{1}{2} F^{\alpha a}(F_{\alpha b,a} - F_{ab,\,\alpha})$ and $F^a{}_{\alpha,a} F^\alpha{}_b = (\eta^{c\gamma} F^a{}_{\gamma,a}) F_{cb}$.

With the definitions behind us we can now spend some time looking at examples and applications. Of course, we have already encountered several examples since any assignment of the *same* skew-symmetric linear transformation to each p in R is obviously smooth and satisfies Maxwell's equations and these *constant electromagnetic fields* were investigated in Section 2.6. As our first nontrivial example we examine the so-called *Coulomb field* of a single free charged particle.

We begin with a free charged particle $(\alpha,\ m,\ e)$. Since $\alpha : \mathbb{R} \to \mathcal{M}$ we may let $W = \alpha(\mathbb{R})$. Then W is a timelike straight line which we may assume, without loss of generality, to be a time axis with $\alpha(0) = 0$. Let $\{e_a\}_{a=1}^4$ be an admissible basis with $W = \operatorname{Span}\{e_4\}$, i.e., a rest frame for the particle. We define an electromagnetic field F on $\mathcal{M} - W$ by specifying, at each point, its matrix relative to $\{e_a\}$ and decreeing that its matrix in any other basis is obtained from the change of basis formula (2.7.4). Thus, at each point of $\mathcal{M} - W$ we define the matrix of the *Coulomb field* $F = F(x^1, x^2, x^3, x^4)$ of (α, m, e) relative to a rest frame for (α, m, e) to be

$$[F^a{}_b] = e \begin{bmatrix} 0 & 0 & 0 & x^1/r^3 \\ 0 & 0 & 0 & x^2/r^3 \\ 0 & 0 & 0 & x^3/r^3 \\ x^1/r^3 & x^2/r^3 & x^3/r^3 & 0 \end{bmatrix}, \tag{2.7.24}$$

where $r^3 = ((x^1)^2 + (x^2)^2 + (x^3)^2)^{3/2}$. Thus, $\vec{B} = \vec{0}$ and $\vec{E} = \frac{e}{r^3}\,\vec{r}$, where $\vec{r} = x^1 e_1 + x^2 e_2 + x^3 e_3$. Thus, $|\vec{E}|^2 = (\frac{e^2}{r^6})\,\vec{r}\cdot\vec{r} = \frac{e^2}{r^4}$ so $|\vec{E}| = \frac{|e|}{r^2}$. Any two bases $\{e_a\}$ and $\{\hat{e}_a\}$ with $W = \operatorname{Span}\{e_4\}$ are related by a rotation in $\mathcal{R}$ (by Lemma 1.3.4). We ask the reader to show that our definition of the Coulomb field is invariant under rotations and so the field is well-defined.

Exercise 2.7.11. Suppose $R = [R^a{}_b]_{a,b=1,2,3,4} \in \mathcal{R}$ is a rotation and $\hat{x}^a = R^a{}_b\, x^b$, $a = 1,2,3,4$. Show that $\hat{r}^2 = (\hat{x}^1)^2 + (\hat{x}^2)^2 + (\hat{x}^3)^2 = r^2$ and that the matrix $[\hat{F}^a{}_b] = R[F^a{}_b]R^{-1}$ of the Coulomb field (2.7.24) in the hatted coordinate system is

$$e\begin{bmatrix} 0 & 0 & 0 & \hat{x}^1/\hat{r}^3 \\ 0 & 0 & 0 & \hat{x}^2/\hat{r}^3 \\ 0 & 0 & 0 & \hat{x}^3/\hat{r}^3 \\ \hat{x}^1/\hat{r}^3 & \hat{x}^2/\hat{r}^3 & \hat{x}^3/\hat{r}^3 & 0 \end{bmatrix}.$$

To justify referring to the Coulomb field as an electromagnetic field we must, of course, observe that it is smooth on the region $\mathcal{M}-W$ and verify Maxwell's equations (2.7.9) and (2.7.15). Since $(\text{div } F)^b = \eta^{b\beta} F^\alpha{}_{\beta,\alpha}$ we obtain, from (2.7.24), $(\text{div } F)^i = \eta^{\beta i} F^\alpha{}_{\beta,\alpha} = F^\alpha{}_{i,\alpha} = F^1{}_{i,1} + F^2{}_{i,2} + F^3{}_{i,3} + F^4{}_{i,4} = 0+0+0+0 = 0$. Moreover,

$$\begin{aligned}
(\text{div } F)^4 &= \eta^{\beta 4} F^\alpha{}_{\beta,\alpha} = -F^\alpha{}_{4,\alpha} \\
&= -e\left[\tfrac{\partial}{\partial x^1}\left(\tfrac{x^1}{r^3}\right) + \tfrac{\partial}{\partial x^2}\left(\tfrac{x^2}{r^3}\right) + \tfrac{\partial}{\partial x^3}\left(\tfrac{x^3}{r^3}\right) + 0\right] \\
&= -\tfrac{e}{r^6}\left[r^3 - x^1\left(3r^2\tfrac{\partial r}{\partial x^1}\right) + r^3 - x^2\left(3r^2\tfrac{\partial r}{\partial x^2}\right) + r^3 - x^3\left(3r^2\tfrac{\partial r}{\partial x^3}\right)\right] \\
&= -\tfrac{e}{r^6}\left[3r^3 - x^1\left(3r^2\left(\tfrac{x^1}{r}\right)\right) - x^2\left(3r^2\left(\tfrac{x^2}{r}\right)\right) - x^3\left(3r^2\left(\tfrac{x^3}{r}\right)\right)\right] \\
&= -\tfrac{e}{r^6}\left[3r^3 - 3r\left((x^1)^2 + (x^2)^2 + (x^3)^2\right)\right] \\
&= -\tfrac{e}{r^6}\left[3r^3 - 3r^3\right] = 0\,.
\end{aligned}$$

Next observe that, from (2.7.24) and (2.7.14) we obtain

$$[F_{ab}] = e\begin{bmatrix} 0 & 0 & 0 & x^1/r^3 \\ 0 & 0 & 0 & x^2/r^3 \\ 0 & 0 & 0 & x^3/r^3 \\ -x^1/r^3 & -x^2/r^3 & -x^3/r^3 & 0 \end{bmatrix}.$$

Thus, (2.7.15) is automatically satisfied if all of a, b and c are in $\{1,2,3\}$. The remaining possibilities are all easily checked one-by-one, e.g., if $a = 1$, $b = 2$ and $c = 4$ we obtain

$$\begin{aligned}
F_{12,\,4} + F_{24,\,1} + F_{41,\,2} &= \tfrac{\partial}{\partial x^4}(0) + \tfrac{\partial}{\partial x^1}\left(\tfrac{x^2}{r^3}\right) + \tfrac{\partial}{\partial x^2}\left(-\tfrac{x^1}{r^3}\right) \\
&= 0 + x^2\left(-3r^{-4}\left(\tfrac{x^1}{r}\right)\right) + x^1\left(3r^{-4}\left(\tfrac{x^2}{r}\right)\right) \\
&= 0\,.
\end{aligned}$$

Exercise 2.7.12. Calculate the matrix of the energy-momentum transformation (2.5.1) for the Coulomb field (2.7.24) in its rest frames and show, in particular, that $T^4{}_4 = -\frac{e^2}{8\pi r^4}$.

Recalling that $-T^4{}_4$ is interpreted as the energy density of the electromagnetic field F as measured in the given frame of reference, we seem forced to conclude from Exercise 2.7.12 that the total energy contained in a sphere of radius $R > 0$ about a point charge (which would be obtained by integrating the energy density over the sphere) is

$$\int_0^{2\pi}\int_0^{\pi}\int_0^{R}\frac{e^2}{8\pi r^4}r^2\sin\phi\,dr\,d\phi\,d\theta = \frac{e^2}{2}\int_0^R\frac{1}{r^2}\,dr$$

and this is an improper integral which diverges. The energy contained in such a sphere would seem to be infinite. But then (1.8.6) would suggest an infinite mass for the charge in its rest frames. This is, of course, absurd since finite applied forces are found to produce nonzero accelerations of point charges. Although classical electromagnetic theory is quite beautiful and enormously successful in predicting the behavior of physical systems there are, as this calculation indicates, severe logical difficulties at the very foundations of the subject and, even today, these have not been resolved to everyone's satisfaction (see [**Par**] for more on this).

As an application we wish to calculate the field of a uniformly moving charge. Special relativity offers a particularly elegant solution to this problem since, according to the Relativity Principle, it matters not at all whether we view the charge as moving relative to a "fixed" frame of reference or the frame as moving relative to a "stationary" charge. Thus, in effect, we need only transform the Coulomb field to a new reference frame, moving relative to the rest frame of the charge. More specifically, we wish to calculate the field due to a charge moving uniformly in a straight line with speed β relative to some admissible frame $\hat{\mathcal{S}}$ at the instant the charge passes through that frame's spatial origin. We may clearly assume, without loss of generality, that the motion is along the negative $\hat{x}^1$-axis and that the charge passes through $(\hat{x}^1, \hat{x}^2, \hat{x}^3) = (0,0,0)$ at $\hat{x}^4 = 0$. If $\mathcal{S}$ is the frame in which the charge is at rest we need only transform the Coulomb field to $\hat{\mathcal{S}}$ with a boost $\Lambda(\beta)$ and evaluate at $x^4 = \hat{x}^4 = 0$. The Coulomb field in $\mathcal{S}$ has $E^i = e(x^i/r^3)$, $i = 1,2,3$, and $B^i = 0$, $i = 1,2,3$, so, from Exercise 2.2.1,

$$\begin{array}{lll} \hat{E}^1 = e\left(\frac{x^1}{r^3}\right), & \hat{E}^2 = e\gamma\left(\frac{x^2}{r^3}\right), & \hat{E}^3 = e\gamma\left(\frac{x^3}{r^3}\right), \\ \hat{B}^1 = 0, & \hat{B}^2 = e\beta\gamma\left(\frac{x^3}{r^3}\right), & \hat{B}^3 = -e\beta\gamma\left(\frac{x^2}{r^3}\right). \end{array}$$

We wish to express these in terms of measurements made in $\hat{\mathcal{S}}$. Setting $\hat{x}^4 = 0$ in (1.3.29) gives $x^1 = \gamma\hat{x}^1$, $x^2 = \hat{x}^2$ and $x^3 = \hat{x}^3$ so that $r^2 = (x^1)^2 + (x^2)^2 + (x^3)^2 = \gamma^2(\hat{x}^1)^2 + (\hat{x}^2)^2 + (\hat{x}^3)^2$, which we now denote $\tilde{r}^2$. Thus,

$$\begin{array}{lll} \hat{E}^1 = e\gamma(\hat{x}^1/\tilde{r}^3), & \hat{E}^2 = e\gamma(\hat{x}^2/\tilde{r}^3), & \hat{E}^3 = e\gamma(\hat{x}^3/\tilde{r}^3), \\ \hat{B}^1 = 0, & \hat{B}^2 = e\beta\gamma(\hat{x}^3/\tilde{r}^3), & \hat{B}^3 = -e\beta\gamma(\hat{x}^2/\tilde{r}^3), \end{array}$$

so

$$\vec{\hat{E}} = \frac{e\gamma}{\tilde{r}^3}\left(\hat{x}^1\hat{e}_1 + \hat{x}^2\hat{e}_2 + \hat{x}^3\hat{e}_3\right) = \frac{e\gamma}{\tilde{r}^3}\,\vec{\hat{r}}$$

and

$$\begin{aligned}
\vec{\hat{B}} &= \frac{e\gamma}{\tilde{r}^3}\left(0 \cdot \hat{e}_1 + \beta\hat{x}^3\hat{e}_2 - \beta\hat{x}^2\hat{e}_3\right) \\
&= \frac{e\gamma}{\tilde{r}^3}\left(\beta\hat{x}^3\hat{e}_2 - \beta\hat{x}^2\hat{e}_3\right) \\
&= \frac{e\gamma}{\tilde{r}^3}\begin{vmatrix} \hat{e}_1 & \hat{e}_2 & \hat{e}_3 \\ -\beta & 0 & 0 \\ \hat{x}^1 & \hat{x}^2 & \hat{x}^3 \end{vmatrix} \\
&= \frac{e\gamma}{\tilde{r}^3}\left(\beta(-\hat{e}_1) \times \vec{\hat{r}}\right) \\
&= \frac{e\gamma}{\tilde{r}^3}\left(\vec{\hat{u}} \times \vec{\hat{r}}\right).
\end{aligned}$$

Observe that, in the nonrelativistic limit $(\gamma \approx 1)$ we obtain

$$\vec{\hat{E}} \approx \frac{e}{\tilde{r}^3}\vec{\hat{r}} \qquad (\gamma \approx 1)$$

and

$$\vec{\hat{B}} \approx \frac{e}{\tilde{r}^3}\left(\vec{\hat{u}} \times \vec{\hat{r}}\right) \qquad (\gamma \approx 1).$$

The first of these equations asserts that the field of a slowly moving charge is approximately the Coulomb field, whereas the second is called the *Biot-Savart Law.*

Observe that the Coulomb field is certainly regular at each point of $\mathcal{M} - W$ since $|\vec{B}|^2 - |\vec{E}|^2 = 0 - \frac{|e|}{r^2} = -\frac{|e|}{r^2}$ which is nonzero. As a nontrivial example of an electromagnetic field that is null we consider next what are called "simple, plane electromagnetic waves".

Let $K : \mathcal{M} \to \mathcal{M}$ denote some fixed, nonzero, skew-symmetric linear transformation on $\mathcal{M}$ and $S : \mathcal{M} \to \mathbb{R}$ a smooth, nonconstant real-valued function on $\mathcal{M}$. Define, for each $x \in \mathcal{M}$, a linear transformation $F(x) : \mathcal{M} \to \mathcal{M}$ by $F(x) = S(x)K$. Then the assignment $x \xrightarrow{F} F(x)$ is obviously smooth and one could determine necessary and sufficient conditions on S and K to ensure that F satisfies Maxwell's equations and so represents an electromagnetic field. We limit our attention to a special case. For this we begin with a smooth, nonconstant function $P : \mathbb{R} \to \mathbb{R}$ and a fixed, nonzero vector $k \in \mathcal{M}$. Now take $S(x) = P(k \cdot x)$ so that

$$F(x) = P(k \cdot x)K. \tag{2.7.25}$$

Observe that F takes the same value for all $x \in \mathcal{M}$ for which $k \cdot x$ is a constant, i.e., F is constant on the 3-dimensional hyperplanes $\{x \in \mathcal{M} : k \cdot x = r_0\}$ for some real constant r_0. We now set about determining conditions on P, k and K which ensure that (2.7.25) defines an electromagnetic field on $\mathcal{M}$.

Fix an admissible basis $\{e_a\}_{a=1}^4$. Let $k = k^a e_a$ and $x = x^a e_a$ and suppose the matrix of K relative to this basis is $[K^a{}_b]$. Then $F^a{}_b = P(k\cdot x)K^a{}_b = P(\eta_{\alpha\beta}\, k^\alpha\, x^\beta)K^a{}_b$. First we consider the equation div $F = 0$. Now, $(\text{div } F)^i = F^\alpha{}_{i,\alpha}$, $i = 1,2,3$, and $(\text{div } F)^4 = -F^\alpha{}_{4,\alpha}$. But

$$\begin{aligned} F^a{}_{b,c} &= \frac{\partial}{\partial x^c}\left(P(k\cdot x)K^a{}_b\right) \\ &= P'(k\cdot x)\,\frac{\partial}{\partial x^c}\,(k\cdot x)K^a{}_b \end{aligned}$$

so

$$F^a{}_{b,i} = P'(k\cdot x)k^i\, K^a{}_b\,, \qquad i = 1,2,3,$$

and

$$F^a{}_{b,4} = -P'(k\cdot x)k^4\, K^a{}_b\,.$$

Now, for $i = 1,2,3$,

$$\begin{aligned} (\text{div } F)^i &= F^1{}_{i,1} + F^2{}_{i,2} + F^3{}_{i,3} + F^4{}_{i,4} \\ &= P'(k\cdot x)k^1K^1{}_i + P'(k\cdot x)k^2K^2{}_i + P'(k\cdot x)k^3K^3{}_i - P'(k\cdot x)k^4K^4{}_i \\ &= P'(k\cdot x)\left[k^1\,K^1{}_i + k^2\,K^2{}_i + k^3\,K^3{}_i - k^4\,K^4{}_i\right]. \end{aligned}$$

But $P'(k\cdot x)$ is not identically zero since P is not constant so $(\text{div } F)^i = 0$ implies

$$k^1\,K^1{}_i + k^2\,K^2{}_i + k^3\,K^3{}_i - k^4\,K^4{}_i = 0\,, \qquad i = 1,2,3,$$

that is,

$$\eta_{ab}\, k^a\, K^b{}_i = 0\,, \qquad i = 1,2,3.$$

Exercise 2.7.13. Show that $(\text{div } F)^4 = 0$ requires that $\eta_{ab}\, k^a\, K^b{}_4 = 0$.

Thus, div $F = 0$ for an F given by (2.7.25) becomes

$$\eta_{ab}\, k^a\, K^b{}_c = 0\,, \qquad c = 1,2,3,4. \tag{2.7.26}$$

Next we consider (2.7.15). For this we observe that $[F_{ab}] = [P(k\cdot x)K_{ab}]$ so $F_{ab,\,c} = \frac{\partial}{\partial x^c}(P(k\cdot x)K_{ab}) = P'(k\cdot x)\frac{\partial}{\partial x^c}(k\cdot x)K_{ab}$ and therefore

$$F_{ab,\,i} = P'(k\cdot x)k^i\, K_{ab}$$

and

$$F_{ab,\,4} = -P(k\cdot x)k^4\, K_{ab}\,.$$

Thus, $F_{ab,\,c} + F_{bc,\,a} + F_{ca,\,b} = 0$ implies

$$P'(k\cdot x)\left[K_{ab}\,\frac{\partial}{\partial x^c}\,(k\cdot x) + K_{bc}\,\frac{\partial}{\partial x^a}\,(k\cdot x) + K_{ca}\,\frac{\partial}{\partial x^b}\,(k\cdot x)\right] = 0\,.$$

Again, $P'(k\cdot x) \not\equiv 0$ so the expression in brackets must be zero, i.e.,

$$K_{ab}\,\frac{\partial}{\partial x^c}\,(k\cdot x) + K_{bc}\,\frac{\partial}{\partial x^a}\,(k\cdot x) + K_{ca}\,\frac{\partial}{\partial x^b}\,(k\cdot x) = 0\,.$$

If a, b and c are chosen from $\{1,2,3\}$ this becomes

$$K_{ab}\,k^c + K_{bc}\,k^a + K_{ca}\,k^b \;=\; 0\,, \qquad a,b,c = 1,2,3. \tag{2.7.27}$$

If any of a, b or c is 4, then the terms with a k^4 have a minus sign. This, and (2.7.26) also, become easier to write if we introduce the notation

$$k_b = \eta_{ab}\,k^a\,, \qquad b = 1,2,3,4.$$

Thus, $k_i = k^i$ for $i = 1,2,3$, but $k_4 = -k^4$. Now (2.7.26), (2.7.27) and the equation corresponding to (2.7.27) when a, b or c is 4 can be written

$$k_b\,K^b{}_c \;=\; 0\,, \qquad c = 1,2,3,4, \tag{2.7.28}$$

and

$$K_{ab}\,k_c + K_{bc}\,k_a + K_{ca}\,k_b \;=\; 0\,, \qquad a,b,c = 1,2,3,4, \tag{2.7.29}$$

and we have proved:

Theorem 2.7.1. *Let $K\colon \mathcal{M} \to \mathcal{M}$ be a nonzero, skew-symmetric linear transformation of $\mathcal{M}$, k a nonzero vector in $\mathcal{M}$ and $P : \mathbb{R} \to \mathbb{R}$ a smooth, nonconstant function. Then $F(x) = P(k \cdot x)K$ defines a smooth assignment of a skew-symmetric linear transformation to each $x \in \mathcal{M}$ and satisfies Maxwell's equations if and only if (2.7.28) and (2.7.29) are satisfied.*

Any $F(x)$ of the type described in Theorem 2.7.1 for which (2.7.28) and (2.7.29) are satisfied is therefore an electromagnetic field and is called a *simple plane electromagnetic wave.* We have already observed that such fields are constant on hyperplanes of the form

$$k^1\,x^1 + k^2\,x^2 + k^3\,x^3 - k^4\,x^4 \;=\; r_0 \tag{2.7.30}$$

and we now investigate some of their other characteristics. First observe that if x and x_0 are two points in the hyperplane, then the displacement vector $x - x_0$ between them is orthogonal to k since $(x - x_0)\cdot k = x \cdot k - x_0 \cdot k = r_0 - r_0 = 0$. Thus, k is the normal vector to these hyperplanes. We show next that k is necessarily null. Begin with (2.7.29). Multiply through by k^c and sum as indicated.

$$K_{ab}\,k_c\,k^c + K_{bc}\,k_a\,k^c + K_{ca}\,k_b\,k^c = 0, \qquad a,b = 1,2,3,4.$$

Thus,

$$K_{ab}(k \cdot k) + (K_{bc}\,k^c)k_a + (K_{ca}\,k^c)k_b = 0\,, \quad a,b = 1,2,3,4. \quad (2.7.31)$$

But now observe that, by (2.7.28),

$$\begin{aligned} 0 &= K^b{}_c\,k_b = \eta^{b\beta}\,K_{\beta c}\,\eta_{\alpha b}\,k^\alpha \\ &= (\eta^{\beta b}\,\eta_{\alpha b})K_{\beta c}\,k^\alpha = \delta^\beta_\alpha\,K_{\beta c}\,k^\alpha \\ &= K_{\alpha c}\,k^\alpha = K_{bc}\,k^b = -K_{bc}\,k^c\,. \end{aligned}$$

Thus, $K_{bc}\,k^c = 0 = K_{ca}\,k^c$, so (2.7.31) gives $K_{ab}(k \cdot k) = 0$, for all $a,b = 1,2,3,4$. But for some choice of a and b, $K_{ab} \neq 0$ so

$$k \cdot k = 0$$

and so k is null.

Next we show that a simple plane electromagnetic wave $F(x) = P(k \cdot x)K$ is null at each point x. Indeed, suppose $x_0 \in \mathcal{M}$ and $F(x_0) = P(k \cdot x_0)K$ is regular (and, in particular, nonzero). Then $P(k \cdot x_0) \neq 0$ so K must be regular (compute $\vec{E} \cdot \vec{B}$ and $|\vec{B}|^2 - |\vec{E}|^2$). Relative to a canonical basis for K we have

$$[K^a{}_b] = \begin{bmatrix} 0 & K^1{}_2 & 0 & 0 \\ K^2{}_1 & 0 & 0 & 0 \\ 0 & 0 & 0 & K^3{}_4 \\ 0 & 0 & K^4{}_3 & 0 \end{bmatrix} = \begin{bmatrix} 0 & \delta & 0 & 0 \\ -\delta & 0 & 0 & 0 \\ 0 & 0 & 0 & \epsilon \\ 0 & 0 & \epsilon & 0 \end{bmatrix}.$$

We write out (2.7.28) for $c = 1,2,3$ and 4:

$$\begin{aligned} c = 1: &\quad k_b\,K^b{}_1 = 0 = k_2\,K^2{}_1 = -\delta\,k_2, \\ c = 2: &\quad k_b\,K^b{}_2 = 0 = k_1\,K^1{}_2 = \delta\,k_1, \\ c = 3: &\quad k_b\,K^b{}_3 = 0 = k_4\,K^4{}_3 = \epsilon\,k_4, \\ c = 4: &\quad k_b\,K^b{}_4 = 0 = k_3\,K^3{}_4 = \epsilon\,k_3. \end{aligned}$$

Now, k is null so $k^4 \neq 0$ and therefore $\epsilon = 0$. Thus, $\delta \neq 0$ so $k_1 = k_2 = 0$. Next we write out (2.7.29) with $a = 1$, $b = 2$ and $c = 3$:

$$\begin{aligned} K_{12}\,k_3 + K_{23}\,k_1 + K_{31}\,k_2 &= 0\,, \\ K_{12}\,k_3 &= 0\,, \\ \delta\,k_3 &= 0\,. \end{aligned}$$

But $\delta = 0$ would imply $K = 0$ and $k_3 = 0$ would imply $k_4 = 0$ and so $k = 0$. Either is a contradiction so F must be null at each point.

Next we tie these last two bits of information together and show that the null vector k is actually in the principal null direction of the null transformation K. We select a canonical basis for K so that

$$[K^a{}_b] = \begin{bmatrix} 0 & 0 & 0 & 0 \\ 0 & 0 & \alpha & 0 \\ 0 & -\alpha & 0 & \alpha \\ 0 & 0 & \alpha & 0 \end{bmatrix} \qquad (\alpha \neq 0)\,.$$

Now we write out (2.7.28) for $c = 2$ and 3 ($c = 1$ contains no information and $c = 4$ is redundant):

$$\begin{aligned} c = 2: &\qquad k_b\, K^b{}_2 = 0 = -\alpha\, k_3 \Longrightarrow k_3 = 0\,, \\ c = 3: &\qquad k_b\, K^b{}_3 = 0 = \alpha\, k_2 + \alpha\, k_4 \Longrightarrow k_4 = -k_2\,. \end{aligned}$$

Thus, $k^3 = 0$ and $k^4 = k^2$ so k null implies $k^1 = 0$, i.e., $k = k^2(e_2 + e_4)$ in canonical coordinates. But $e_2 + e_4$ is in the principal null direction of K (Exercise 2.4.8) so we have proved half of the following theorem.

Theorem 2.7.2. *Let $K\colon \mathcal{M} \to \mathcal{M}$ be a nonzero, skew-symmetric linear transformation of $\mathcal{M}$, k a nonzero vector in $\mathcal{M}$ and $P : \mathbb{R} \to \mathbb{R}$ a smooth nonconstant function. Then $F(x) = P(k \cdot x)K$ defines a simple plane electromagnetic wave [i.e., satisfies Maxwell's equations (2.7.28) and (2.7.29)] if and only if K is null and k is in the principal null direction of K.*

Proof: We have already proved the necessity. For the sufficiency we assume K is null and k is in its principal null direction. Relative to canonical coordinates, the only nonzero entries in $[K^a{}_b]$ and $[K_{ab}]$ are $K^2{}_3 = K^3{}_4 = K^4{}_3 = -K^3{}_2 = \alpha$ and $K_{23} = K_{34} = -K_{43} = -K_{32} = \alpha$. Moreover, k is a multiple of $e_2 + e_4$, say, $k = m(e_2 + e_4)$ so $k^1 = k^3 = k_1 = k_3 = 0$ and $k^2 = k^4 = k_2 = -k_4 = m$.

Exercise 2.7.14. Verify (2.7.28) and (2.7.29). ∎

Thus, we can manufacture simple plane electromagnetic waves by beginning with a nonzero null $K : \mathcal{M} \to \mathcal{M}$, finding a nonzero null vector k in the principal null direction of K, selecting any smooth, nonconstant $P : \mathbb{R} \to \mathbb{R}$ and setting $F(x) = P(k \cdot x)K$. In fact, it is even easier than this for, as we now show, given an arbitrary nonzero null vector k we can produce a nonzero null $K : \mathcal{M} \to \mathcal{M}$ which has k as a principal null direction. To see this, select a nonzero vector l in $\mathrm{Span}\{k\}^\perp$ and set $K = k \wedge l$ (see Exercise 2.4.7). Thus, for every $v \in \mathcal{M}$, $Kv = (k \wedge l)v = k(l \cdot v) - l(k \cdot v)$.

Exercise 2.7.15. Show that, relative to an arbitrary admissible basis $\{e_a\}$, $K^a{}_b = k^a\, l_b - l^a\, k_b$ and $K_{ab} = k_a\, l_b - l_a\, k_b$.

Now one easily verifies (2.7.28) and (2.7.29). Indeed, $k_b K^b{}_c = k_b(k^b l_c - l^b k_c) = (k_b k^b)l_c - (k_b l^b)k_c = (k \cdot k)l_c - (k \cdot l)k_c = 0 \cdot l_c - 0 \cdot k_c = 0$ since k is null and $l \in \mathrm{Span}\{k\}^\perp$.

Exercise 2.7.16. Verify (2.7.29).

Since K is obviously skew-symmetric we may select an arbitrary smooth nonconstant $P\colon \mathbb{R} \to \mathbb{R}$ and be assured that $F(x) = P(k \cdot x)K$ represents a simple plane electromagnetic wave. Most choices of $P : \mathbb{R} \to \mathbb{R}$, of course, yield physically unrealizable solutions F. One particular choice that is important not only because it gives rise to an observable field, but also because, mathematically, many electromagnetic waves can be regarded (via Fourier analysis) as superpositions of such waves, is

$$P(t) = \sin nt,$$

where n is a positive integer. Thus, we begin with an arbitrary nonzero, null, skew-symmetric $K\colon \mathcal{M} \to \mathcal{M}$ and let $\{e_a\}$ be a canonical basis for K. Then $k = e_2 + e_4$ is along the principal null direction of K so

$$\begin{aligned} F(x) &= \sin(nk \cdot x)K \\ &= \sin(n(e_2 + e_4) \cdot x)K \\ &= \sin\big(n(x^2 - x^4)\big)K \end{aligned}$$

defines a simple plane electromagnetic wave. For some nonzero α in $\mathbb{R}$,

$$[F^a{}_b] = \begin{bmatrix} 0 & 0 & 0 & 0 \\ 0 & 0 & \alpha\sin\big(n(x^2 - x^4)\big) & 0 \\ 0 & -\alpha\sin\big(n(x^2 - x^4)\big) & 0 & \alpha\sin\big(n(x^2 - x^4)\big) \\ 0 & 0 & \alpha\sin\big(n(x^2 - x^4)\big) & 0 \end{bmatrix}.$$

Thus, $\vec{E} = \alpha\sin(n(x^2 - x^4))e_3$ and $\vec{B} = \alpha\sin(n(x^2 - x^4))e_1$. F is constant on the 3-dimensional hyperplanes $x^2 - x^4 = r_0$. At each fixed instant $x^4 = x_0^4$ an observer in the canonical reference frame sees his instantaneous 3-space layered with planes $x^2 = x_0^4 + r_0$ on which F is constant (see Figure 2.7.1). Next, fix not x^4, but $x^2 = x_0^2$ so that $\vec{E} = \alpha\sin\big(n(x_0^2 - x^4)\big)e_3$ and $\vec{B} = \alpha\sin(n(x_0^2 - x^4))e_1$. Thus, at a given location, $\vec{E}$ and $\vec{B}$ will always be in the same directions (except for reversals when sin changes sign), but the intensities vary periodically with time.

Exercise 2.7.17. Show that, for any electromagnetic field, each of the functions F_{ab} satisfies the *wave equation*

$$\frac{\partial^2 F_{ab}}{(\partial x^1)^2} + \frac{\partial^2 F_{ab}}{(\partial x^2)^2} + \frac{\partial^2 F_{ab}}{(\partial x^3)^2} = \frac{\partial^2 F_{ab}}{(\partial x^4)^2}. \tag{2.7.32}$$

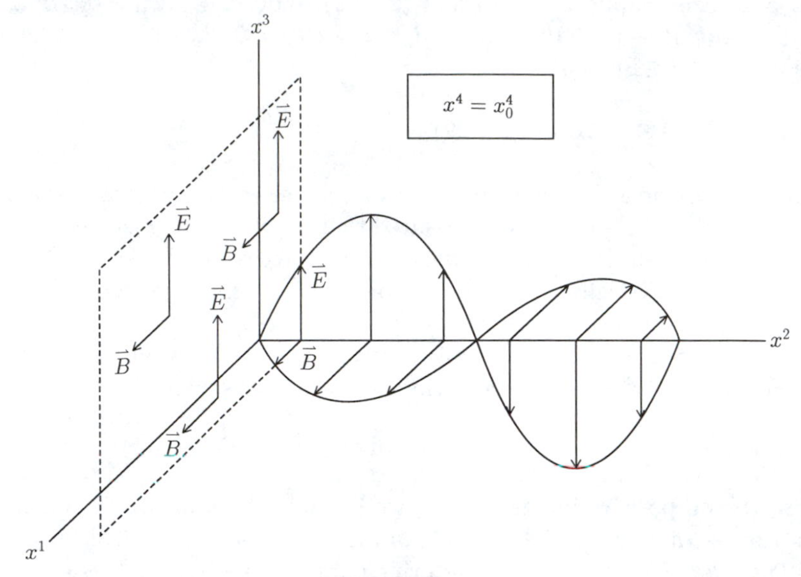

Figure 2.7.1

Hints: Differentiate (2.7.15) with respect to x^μ, multiply by $\eta^{\mu c}$ and sum as indicated. Then use (2.7.9) to show that two of the three terms must vanish.

Of course, not everything that satisfies a wave equation is "wavelike" [e.g., constant fields satisfy (2.7.32)]. However, historically the result of Exercise 2.7.17 first suggested to Maxwell that there might exist electromagnetic fields with wavelike characteristics (and which propagate with speed 1). Our last examples are obviously of this sort and the electromagnetic theory of light is based on the study of such solutions to Maxwell's equations.

3

The Theory of Spinors

3.1 Representations of the Lorentz Group

The concept of a "spinor" emerged from the work of E. Cartan on the representations of simple Lie algebras. However, it was not until Dirac employed a special case in the construction of his relativistically invariant equation for the electron with "spin" that the notion acquired its present name or its current stature in mathematical physics. In this chapter we present an elementary introduction to the algebraic theory of spinors in Minkowski spacetime and illustrate its utility in special relativity by recasting in spinor form much of what we have learned about the structure of the electromagnetic field in Chapter 2. We shall not stray into quantum mechanics and, in particular, will not discuss the Dirac equation (for this, see the encyclopedic monograph [**PR**] of Penrose and Rindler). Since it is our belief that an intuitive appreciation of the notion of a spinor is best acquired by approaching them by way of group representations, we have devoted this first section to an introduction to these ideas and how they arise in special relativity. Since this section is primarily motivational, we have not felt compelled to prove everything we say and have, at several points, contented ourselves with a reference to a proof in the literature.

A vector v in $\mathcal{M}$ (e.g., a world momentum) is an object that is decribed in each admissible frame of reference by four numbers (components) with the property that if $v = v^a e_a = \hat{v}^a \hat{e}_a$ and $[\Lambda^a{}_b]$ is the Lorentz transformation relating $\{e_a\}$ and $\{\hat{e}_a\}$ (i.e., $e_b = \Lambda^a{}_b \hat{e}_a$), then the components v^a and $\hat{v}^a$ are related by the "transformation law"

$$\hat{v}^a = \Lambda^a{}_b v^b, \quad a = 1, 2, 3, 4. \tag{3.1.1}$$

A linear transformation $L : \mathcal{M} \to \mathcal{M}$ (e.g., an electromagnetic field) is another type of object that is again described in each admissible basis by a set of numbers (the entries in its matrix relative to that basis) with the property that if $[L^a{}_b]$ and $[\hat{L}^a{}_b]$ are the matrices of L in $\{e_a\}$ and $\{\hat{e}_a\}$, then

$$\hat{L}^a{}_b = \Lambda^a{}_\alpha \Lambda_b{}^\beta L^\alpha{}_\beta, \quad a, b = 1, 2, 3, 4, \tag{3.1.2}$$

where $[\Lambda_a{}^b]$ is the inverse of $[\Lambda^a{}_b]$, i.e., $\Lambda^a{}_\alpha \Lambda_b{}^\alpha = \Lambda_\alpha{}^a \Lambda^\alpha{}_b = \delta^a_b$ [(3.1.2) is just the familiar change of basis formula]. As we found in Chapter 2, it is often convenient to associate with such a linear transformation a corresponding bilinear form $\tilde{L} : \mathcal{M} \times \mathcal{M} \to \mathbb{R}$ defined by $\tilde{L}(u,v) = u \cdot Lv$. Again, $\tilde{L}$ is described in each admissible basis by its set of components $L_{ab} = \tilde{L}(e_a, e_b)$ and components in different bases are related by a specific transformation law:

$$\hat{L}_{ab} = \Lambda_a{}^\alpha \Lambda_b{}^\beta L_{\alpha\beta}, \quad a,b = 1,2,3,4. \tag{3.1.3}$$

Such bilinear forms can, of course, arise naturally of their own accord without reference to any linear transformation. The Lorentz inner product is itself such an example. Indeed, if we define $g : \mathcal{M} \times \mathcal{M} \to \mathbb{R}$ by

$$g(u,v) = u \cdot v,$$

then, in all admissible bases, $g_{ab} = g(e_a, e_b) = e_a \cdot e_b = \eta_{ab} = g(\hat{e}_a, \hat{e}_b) = \hat{g}_{ab}$. In this very special case the components are the same in all admissible bases, but, nevertheless, (1.2.14) shows that the same transformation law is satisfied:

$$\hat{g}_{ab} = \Lambda_a{}^\alpha \Lambda_b{}^\beta g_{\alpha\beta}, \quad a,b = 1,2,3,4.$$

The point of all of this is that examples of this sort abound in geometry and physics. In each case one has under consideration an "object" of geometrical or physical significance (an inner product, a world momentum vector, an electromagnetic field transformation, etc.) which is described in each admissible basis by a set of numerical "components" and with the property that components in different bases are related by a specific linear transformation law that depends on the Lorentz transformation relating the two bases. Different "types" of objects are distinguished by their number of components in each basis and by the precise form of the transformation law. Classically, such objects were called "world tensors" or "4-tensors" (we give the precise definition shortly). World tensors are well suited to the task of expressing "Lorentz invariant" relationships since, for example, a statement which asserts the equality, in some basis, of the components of two world tensors of the same type necessarily implies that their components in any other basis must also be equal (since the "transformation law" to the new basis components is the same for both). This is entirely analogous to the use of 3-vectors in classical physics and Euclidean geometry to express relationships that are true in all Cartesian coordinate systems if they are true in any one. For many years it was tacitly assumed that *any* valid Lorentz invariant statement (in particular, any law of relativistic physics) should be expressible as a world tensor equation. Dirac put an end to this in 1928 when he proposed a law (equation) to describe the relativistic electron with spin that was manifestly Lorentz invariant, but not expressed in terms of world tensors. To understand precisely what world tensors are and why they did not suffice for Dirac's purposes we must take a more careful look at "transformation laws" in general.

Observe that if v is a vector with components v^a and $\hat{v}^a$ in two admissible bases and if we write these components as column vectors, then the transformation law (3.1.1) can be written as a matrix product:

$$\begin{bmatrix} \hat{v}^1 \\ \hat{v}^2 \\ \hat{v}^3 \\ \hat{v}^4 \end{bmatrix} = \begin{bmatrix} \Lambda^1{}_1 & \Lambda^1{}_2 & \Lambda^1{}_3 & \Lambda^1{}_4 \\ \Lambda^2{}_1 & \Lambda^2{}_2 & \Lambda^2{}_3 & \Lambda^2{}_4 \\ \Lambda^3{}_1 & \Lambda^3{}_2 & \Lambda^3{}_3 & \Lambda^3{}_4 \\ \Lambda^4{}_1 & \Lambda^4{}_2 & \Lambda^4{}_3 & \Lambda^4{}_4 \end{bmatrix} \begin{bmatrix} v^1 \\ v^2 \\ v^3 \\ v^4 \end{bmatrix}.$$

By virtue of their linearity the same is true of (3.1.2) and (3.1.3). For example, writing the $L^a{}_b$ and $\hat{L}^a{}_b$ as column matrices, (3.1.2) can be written in terms of the 16×16 matrix $[\Lambda^a{}_\alpha \Lambda_b{}^\beta]$ as

$$\begin{bmatrix} \hat{L}^1{}_1 \\ \hat{L}^1{}_2 \\ \vdots \\ \hat{L}^4{}_4 \end{bmatrix} = \begin{bmatrix} \Lambda^1{}_1 \Lambda_1{}^1 & \Lambda^1{}_1 \Lambda_1{}^2 & \cdots & \Lambda^1{}_4 \Lambda_1{}^4 \\ \Lambda^1{}_1 \Lambda_2{}^1 & \Lambda^1{}_1 \Lambda_2{}^2 & \cdots & \Lambda^1{}_4 \Lambda_2{}^4 \\ \vdots & \vdots & & \vdots \\ \Lambda^4{}_1 \Lambda_4{}^1 & \Lambda^4{}_1 \Lambda_4{}^2 & \cdots & \Lambda^4{}_4 \Lambda_4{}^4 \end{bmatrix} \begin{bmatrix} L^1{}_1 \\ L^1{}_2 \\ \vdots \\ L^4{}_4 \end{bmatrix}.$$

Exercise 3.1.1. Write (3.1.3) as a matrix product.

In this way one can think of a transformation law as a rule which assigns to each $\Lambda \in \mathcal{L}$ a certain matrix D_Λ which transforms components in one basis $\{e_a\}$ to those in another $\{\hat{e}_a\}$, related to $\{e_a\}$ by Λ. Observe that, for each of the examples we have considered thus far, these rules $\Lambda \to D_\Lambda$ carry the identity matrix in $\mathcal{L}$ onto the corresponding identity "transformation matrix" (as is only fair since, if the basis is not changed, the components of the "object" should not change). Moreover, if Λ_1 and Λ_2 are in $\mathcal{L}$ and $\Lambda_1 \Lambda_2$ is their product (still in $\mathcal{L}$), then $\Lambda_1 \Lambda_2 \to D_{\Lambda_1 \Lambda_2} = D_{\Lambda_1} D_{\Lambda_2}$ [this is obvious for (3.1.1) since $D_\Lambda = \Lambda$ and follows for (3.1.2) and (3.1.3) either from a rather dreary calculation or from standard facts about change of basis matrices]. This also makes sense, of course, since the components in any basis are uniquely determined so that changing components from basis #1 to basis #2 and then from basis #2 to basis #3 should give the same result as changing directly from basis #1 to basis #3. In order to say all of this more efficiently we introduce some terminology.

Let n be a positive integer. A *matrix group of order* n is a collection $\mathcal{G}$ of $n \times n$ invertible matrices that is closed under the formation of products and inverses (i.e., if G, G_1, and G_2 are in $\mathcal{G}$, then G^{-1} and $G_1 G_2$ are also in $\mathcal{G}$). We have seen numerous examples, e.g., the Lorentz group $\mathcal{L}$ is a matrix group of order 4, whereas $SL(2, \mathbb{C})$ is a matrix group of order 2. The collection of all $n \times n$ invertible matrices (with either real or complex entries) clearly also constitutes a matrix group and is called *the general linear group* of order n and written either $GL(n, \mathbb{R})$ or $GL(n, \mathbb{C})$ depending on whether the entries are real or complex. Observe that a matrix group of order n necessarily contains the $n \times n$ identity matrix $I_n = I$ since, for any G in the group, $GG^{-1} = I$. If $\mathcal{G}$ is a matrix group and $\mathcal{G}'$ is a subset of $\mathcal{G}$, then $\mathcal{G}'$ is called a *subgroup* of $\mathcal{G}$ if it is closed under the formation of products and inverses, i.e., if it is itself a matrix group. For example, the set $\mathcal{R}$ of rotations in $\mathcal{L}$ is a subgroup of $\mathcal{L}$ (Exercise 1.3.7), SU_2 is a subgroup of $SL(2,\mathbb{C})$ (Exercise 1.7.6) and, of course, any matrix group is a subgroup of some general linear group. A *homomorphism* from one matrix group $\mathcal{G}$ to another $\mathcal{H}$ is a map $D: \mathcal{G} \to \mathcal{H}$ that preserves matrix multiplication, i.e., satisfies $D(G_1 G_2) = D(G_1)\, D(G_2)$ whenever G_1 and G_2 are in $\mathcal{G}$. As is customary we shall often write the image of G under D as D_G rather than $D(G)$ and denote the action of D on G by $G \to D_G$. If $\mathcal{G}$ has order n and $\mathcal{H}$ has order m, then D necessarily carries I_n onto I_m since $D(I_n) = D(I_n I_n) = D(I_n)D(I_n)$ so that $D(I_n)(D(I_n))^{-1} = D(I_n)D(I_n)(D(I_n))^{-1}$ and therefore $I_m = D(I_n)\, I_m = D(I_n)$.

Exercise 3.1.2. Show that a homomorphism $D: \mathcal{G} \to \mathcal{H}$ preserves inverses, i.e., that $D(G^{-1}) = (D(G))^{-1}$ for all G in $\mathcal{G}$.

Exercise 3.1.3. Show that if $D: \mathcal{G} \to \mathcal{H}$ is a homomorphism, then its image $D(\mathcal{G}) = \{D(G): G \in \mathcal{G}\}$ is a subgroup of $\mathcal{H}$.

A homomorphism of one matrix group $\mathcal{G}$ into another $\mathcal{H}$ is also called a (*finite dimensional*) *representation* of $\mathcal{G}$. For reasons that will become clear shortly, we will be particularly concerned with the representations of $\mathcal{L}$ and $SL(2,\mathbb{C})$. If $\mathcal{H}$ is of order m and V_m is an m-dimensional vector space (over $\mathbb{C}$ if the entries in $\mathcal{H}$ are complex, but otherwise arbitrary), then the elements of $\mathcal{H}$ can, by selecting a basis for V_m, be regarded as linear transformations or, equivalently, as change of basis matrices on V_m. In this case the elements of V_m are called *carriers* of the representation. $\mathcal{M}$ itself may be regarded as a space of carriers for the representation $D: \mathcal{L} \to GL(4,\mathbb{R})$ of $\mathcal{L}$ corresponding to (3.1.1), i.e., the identity representation $\Lambda \to D_\Lambda = \Lambda$. Similarly, the vector space of linear transformations from $\mathcal{M}$ to $\mathcal{M}$ and that of bilinear forms on $\mathcal{M}$ act as carriers for the representations $[\Lambda^a{}_b] \to [\Lambda^a{}_\alpha \Lambda_b{}^\beta]$ and $[\Lambda^a{}_b] \to [\Lambda_a{}^\alpha \Lambda_b{}^\beta]$ corresponding to (3.1.2) and (3.1.3), respectively. It is rather inconvenient, however, to have different representations of $\mathcal{L}$ acting on carriers of such diverse type (vectors, linear transformations, bilinear forms) and we shall see presently that this can be avoided.

The picture we see emerging here from these few examples is really quite general. Suppose that we have under consideration some geometrical or physical quantity that is described in each admissible basis/frame by m uniquely determined numbers and suppose furthermore that these sets of numbers corresponding to different bases are related by *linear* transformation laws that depend on the Lorentz transformation relating the bases (there are objects of interest that do not satisfy this linearity requirement, but we shall have no occasion to consider them). In each basis we may write the m numbers that describe our object as a column matrix $T = \mathrm{col}[T_1 \cdots T_m]$. Then, associated with every $\Lambda \in \mathcal{L}$ there will be an $m \times m$ matrix D_Λ whose entries depend on those of Λ and with the property that $\hat{T} = D_\Lambda T$ if $\{e_a\}$ and $\{\hat{e}_a\}$ are related by Λ. Since the numbers describing the object in each basis are uniquely determined, the association $\Lambda \to D_\Lambda$ must carry the identity onto the identity and satisfy $\Lambda_1 \Lambda_2 \to D_{\Lambda_1 \Lambda_2} = D_{\Lambda_1} D_{\Lambda_2}$, i.e., must be a representation of the Lorentz group. Thus, the representations of the Lorentz group are precisely the (linear) transformation laws relating the components of physical and geometrical objects of interest in Minkowski spacetime. The objects themselves are the carriers of these representations. Of course, an $m \times m$ matrix can be thought of as acting on any m-dimensional vector space so the precise mathematical nature of these carriers is, to a large extent, arbitrary. We shall find next, however, that one particularly natural choice recommends itself.

We denote by $\mathcal{M}^*$ the dual of the vector space $\mathcal{M}$, i.e., the set of all real-valued linear functionals on $\mathcal{M}$. Thus, $\mathcal{M}^* = \{f : \mathcal{M} \to \mathbb{R} : f(\alpha u + \beta v) = \alpha f(u) + \beta f(v)\ \forall\ u, v \in \mathcal{M}$ and $\alpha, \beta \in \mathbb{R}\}$. The elements of $\mathcal{M}^*$ are called *covectors*. The vector space structure of $\mathcal{M}^*$ is defined in the obvious way, i.e., if f and g are in $\mathcal{M}^*$ and α and β are in $\mathbb{R}$, then $\alpha f + \beta g$ is defined by $(\alpha f + \beta g)(u) = \alpha f(u) + \beta g(u)$. If $\{e_a\}$ is an admissible basis for $\mathcal{M}$, its dual basis $\{e^a\}$ for $\mathcal{M}^*$ is defined by the requirement that $e^a(e_b) = \delta^a_b$ for $a, b = 1, 2, 3, 4$. Let $\{\hat{e}_a\}$ be another admissible basis for $\mathcal{M}$ and $\{\hat{e}^a\}$ its dual basis. If Λ is the element of $\mathcal{L}$ relating $\{e_a\}$ and $\{\hat{e}_a\}$, then

$$\hat{e}_a = \Lambda_a{}^\alpha e_\alpha, \quad a = 1, 2, 3, 4, \tag{3.1.4}$$

and

$$\hat{e}^a = \Lambda^a{}_\alpha e^\alpha, \quad a = 1, 2, 3, 4. \tag{3.1.5}$$

We prove (3.1.5) by showing that the left- and right-hand sides agree on the basis $\{\hat{e}_b\}$ [(3.1.4) is just (1.2.15)]. Of course, $\hat{e}^a(\hat{e}_b) = \delta^a_b$. But also $\Lambda^a{}_\alpha e^\alpha(\hat{e}_b) = \Lambda^a{}_\alpha e^\alpha(\Lambda_b{}^\beta e_\beta) = \Lambda^a{}_\alpha \Lambda_b{}^\beta e^\alpha(e_\beta) = \Lambda^a{}_\alpha \Lambda_b{}^\beta \delta^\alpha_\beta = \Lambda^a{}_\alpha \Lambda_b{}^\alpha = \delta^a_b$ since $[\Lambda^a{}_\alpha]$ and $[\Lambda_b{}^\beta]$ are inverses.

Recall that each $v \in \mathcal{M}$ gives rise, via the Lorentz inner product, to a $v^* \in \mathcal{M}^*$ defined by $v^*(u) = v \cdot u$ for all $u \in \mathcal{M}$. Moreover, if $v = v^a e_a$, then $v^* = v_a e^a$, where $v_a = \eta_{a\alpha} v^\alpha$ since $v_a = v^*(e_a) = v \cdot e_a = (v^\alpha e_\alpha) \cdot e_a = v^\alpha(e_\alpha \cdot e_a) = \eta_{a\alpha} v^\alpha$. Moreover, relative to another basis, $v^* = \hat{v}_a \hat{e}^a = \hat{v}_a(\Lambda^a{}_\alpha e^\alpha) = (\Lambda^a{}_\alpha \hat{v}_a)e^\alpha$ so $v_\alpha = \Lambda^a{}_\alpha \hat{v}_a$ and, applying the inverse, $\hat{v}_a = \Lambda_a{}^\alpha v_\alpha$.

With this we can show that all of the representations of $\mathcal{L}$ considered thus far can, in a very natural way, be regarded as acting on vector spaces of multilinear functionals (defined shortly). Consider first the collection $\mathcal{T}_2^0$ of bilinear forms $L : \mathcal{M} \times \mathcal{M} \to \mathbb{R}$ on $\mathcal{M}$. If $L,\ T \in \mathcal{T}_2^0$ and $\alpha \in \mathbb{R}$, then the definitions $(L+T)(u,v) = L(u,v) + T(u,v)$ and $(\alpha L)(u,v) = \alpha L(u,v)$ are easily seen to give $\mathcal{T}_2^0$ the structure of a real vector space. For any two elements f and g in $\mathcal{M}^*$ we define their *tensor product* $f \otimes g \colon \mathcal{M} \times \mathcal{M} \to \mathbb{R}$ by $f \otimes g\,(u,v) = f(u)\,g(v)$. Then $f \otimes g \in \mathcal{T}_2^0$.

Exercise 3.1.4. Show that, if $\{e^a\}$ is the dual of an admissible basis, then $\{e^a \otimes e^b \colon a,b = 1,2,3,4\}$ is a basis for $\mathcal{T}_2^0$ and that, for any $L \in \mathcal{T}_2^0$,

$$L = L(e_a, e_b)\, e^a \otimes e^b = L_{ab}\, e^a \otimes e^b\,. \tag{3.1.6}$$

Now, in another basis, $L(\hat{e}_a, \hat{e}_b) = L(\Lambda_a{}^\alpha\, e_\alpha, \Lambda_b{}^\beta\, e_\beta) = \Lambda_a{}^\alpha\, \Lambda_b{}^\beta\, L(e_\alpha, e_\beta)$ so

$$\hat{L}_{ab} = \Lambda_a{}^\alpha\, \Lambda_b{}^\beta\, L_{\alpha\beta}\,. \tag{3.1.7}$$

Thus, components relative to bases of the form $\{e^a \otimes e^b \colon a,b = 1,2,3,4\}$ for $\mathcal{T}_2^0$ transform under the representation $[\Lambda^a{}_b] \to [\Lambda_a{}^\alpha\, \Lambda_b{}^\beta]$ of (3.1.3) and we may therefore regard the bilinear forms in $\mathcal{T}_2^0$ as the carriers of this representation. Elements of $\mathcal{T}_2^0$ are called *world tensors of contravariant rank* 0 *and covariant rank* 2 (we will discuss the terminology shortly).

Next we consider the representation $[\Lambda^a{}_b] \to [\Lambda^a{}_\alpha\, \Lambda_b{}^\beta]$ of $\mathcal{L}$ appropriate to (3.1.2). Let $\mathcal{T}_1^1$ denote the set of all real-valued functions $L : \mathcal{M}^* \times \mathcal{M} \to \mathbb{R}$ that are linear in each variable, i.e., satisfy $L(\alpha f + \beta g, u) = \alpha L(f,u) + \beta L(g,u)$ and $L(f, \alpha u + \beta v) = \alpha L(f,u) + \beta L(f,v)$ whenever $\alpha, \beta \in \mathbb{R}$, $f, g \in \mathcal{M}^*$ and $u, v \in \mathcal{M}$. The vector space structure of $\mathcal{T}_1^1$ is defined in the obvious way: If $L, T \in \mathcal{T}_1^1$ and $\alpha, \beta \in \mathbb{R}$, then $\alpha L + \beta T \in \mathcal{T}_1^1$ is defined by $(\alpha L + \beta T)(f,u) = \alpha L(f,u) + \beta T(f,u)$. For $u \in \mathcal{M}$ and $f \in \mathcal{M}^*$ we define $u \otimes f : \mathcal{M}^* \times \mathcal{M} \to \mathbb{R}$ by $u \otimes f\ (g,v) = g(u)\, f(v)$. Again, it is easy to see that $u \otimes f \in \mathcal{T}_1^1$, that $\{e_a \otimes e^b \colon a,b = 1,2,3,4\}$ is a basis for $\mathcal{T}_1^1$ and that, for any L in $\mathcal{T}_1^1$,

$$L = L(e^a, e_b)\, e_a \otimes e^b = L^a{}_b\, e_a \otimes e^b\,. \tag{3.1.8}$$

In another basis, $L(\hat{e}^a, \hat{e}_b) = L(\Lambda^a{}_\alpha\, e^\alpha, \Lambda_b{}^\beta\, e_\beta) = \Lambda^a{}_\alpha\, \Lambda_b{}^\beta\, L(e^\alpha, e_\beta) = \Lambda^a{}_\alpha\, \Lambda_b{}^\beta\, L^\alpha{}_\beta$ so

$$\hat{L}^a{}_b = \Lambda^a{}_\alpha\, \Lambda_b{}^\beta\, L^\alpha{}_\beta\,. \tag{3.1.9}$$

Thus, components relative to bases of the form $\{e_a \otimes e^b \colon a,b = 1,2,3,4\}$ transform under the representation $[\Lambda^a{}_b] \to [\Lambda^a{}_\alpha\, \Lambda_b{}^\beta]$ of (3.1.2) so that the elements of $\mathcal{T}_1^1$ are a natural choice for the carriers of this representation. The elements of $\mathcal{T}_1^1$ are called world tensors of *contravariant rank* 1 *and covariant rank* 1.

The appropriate generalization of these ideas should by now be clear. Let $r \geq 0$ and $s \geq 0$ be integers. Denote by $\mathcal{T}_s^r$ the set of all real-valued functions defined on

$$\underbrace{\mathcal{M}^* \times \cdots \times \mathcal{M}^*}_{r \text{ factors}} \times \underbrace{\mathcal{M} \times \cdots \times \mathcal{M}}_{s \text{ factors}}$$

that are linear in each variable separately (these are called *multilinear functionals*). $\mathcal{T}_s^r$ is made into a real vector space by the obvious pointwise definitions of addition and scalar multiplication. If $u_1, \ldots, u_r \in \mathcal{M}$ and $f_1, \ldots, f_s \in \mathcal{M}^*$ one defines $u_1 \otimes \cdots \otimes u_r \otimes f_1 \otimes \cdots \otimes f_s$ in $\mathcal{T}_s^r$ by

$$\begin{aligned} u_1 \otimes \cdots \otimes u_r \otimes f_1 \otimes \cdots \otimes f_s \, (g_1, \ldots, g_r, v_1, \ldots, v_s) \\ = g_1(u_1) \cdots g_r(u_r) \cdot f_1(v_1) \cdots f_s(v_s) \end{aligned}$$

and finds that the set of $e_{a_1} \otimes \cdots \otimes e_{a_r} \otimes e^{b_1} \otimes \cdots \otimes e^{b_s}$, $a_1, \ldots, a_r = 1,2,3,4$ and $b_1, \ldots, b_s = 1,2,3,4$, form a basis for $\mathcal{T}_s^r$. Moreover, if $L \in \mathcal{T}_s^r$, then

$$\begin{aligned} L &= L\left(e^{a_1}, \ldots, e^{a_r}, e_{b_1}, \ldots, e_{b_s}\right) e_{a_1} \otimes \cdots \otimes e_{a_r} \otimes e^{b_1} \otimes \cdots \otimes e^{b_s} \\ &= L^{a_1 \cdots a_r}{}_{b_1 \cdots b_s} \, e_{a_1} \otimes \cdots \otimes e_{a_r} \otimes e^{b_1} \otimes \cdots \otimes e^{b_s}. \end{aligned} \tag{3.1.10}$$

Relative to another basis,

$$\begin{aligned} &L(\hat{e}^{a_1}, \ldots, \hat{e}^{a_r}, \hat{e}_{b_1}, \ldots, \hat{e}_{b_s}) \\ &\quad = L(\Lambda^{a_1}{}_{\alpha_1} e^{\alpha_1}, \ldots, \Lambda^{a_r}{}_{\alpha_r} e^{\alpha_r}, \Lambda_{b_1}{}^{\beta_1} e_{\beta_1}, \ldots, \Lambda_{b_s}{}^{\beta_s} e_{\beta_s}) \\ &\quad = \Lambda^{a_1}{}_{\alpha_1} \cdots \Lambda^{a_r}{}_{\alpha_r} \Lambda_{b_1}{}^{\beta_1} \cdots \Lambda_{b_s}{}^{\beta_s} L(e^{\alpha_1}, \ldots, e^{\alpha_r}, e_{\beta_1}, \ldots, e_{\beta_s}) \end{aligned}$$

so

$$\hat{L}^{a_1 \cdots a_r}{}_{b_1 \cdots b_s} = \Lambda^{a_1}{}_{\alpha_1} \cdots \Lambda^{a_r}{}_{\alpha_r} \Lambda_{b_1}{}^{\beta_1} \cdots \Lambda_{b_s}{}^{\beta_s} L^{\alpha_1 \cdots \alpha_r}{}_{\beta_1 \cdots \beta_s}. \tag{3.1.11}$$

The elements of $\mathcal{T}_s^r$ are called *world tensors of contravariant rank* r *and covariant rank* s. "Contravariant rank r" refers to the r indices $a_1, \ldots, a_r$ that are written as superscripts in the expression for the components and which appear in the transformation law attached to an entry in Λ (rather than Λ^{-1}). Covariant indices are written as subscripts in the components and transform under Λ^{-1}. An element of $\mathcal{T}_s^r$ has 4^{r+s} components and if these are written as a column matrix, then the transformation law (3.1.11) can be written as a matrix product thus giving rise to an assignment

$$[\Lambda^a{}_b] \longrightarrow \left[\Lambda^{a_1}{}_{\alpha_1} \cdots \Lambda^{a_r}{}_{\alpha_r} \Lambda_{b_1}{}^{\beta_1} \cdots \Lambda_{b_s}{}^{\beta_s}\right]$$

to each element of $\mathcal{L}$ of a $4^{r+s} \times 4^{r+s}$ matrix which can be shown to be a representation of $\mathcal{L}$ and is called the *world tensor representation of contravariant rank* r *and covariant rank* s. Notice that even the identity representation of $\mathcal{L}$ corresponding to (3.1.1) is included in this scheme (with $r = 1$ and $s = 0$). The carriers, however, are now viewed as linear functionals on $\mathcal{M}^*$, i.e., we are employing the standard isomorphism of $\mathcal{M}$ onto $\mathcal{M}^{**}$ [$x \in \mathcal{M} \to x^{**} \in \mathcal{M}^{**}$ defined by $x^{**}(f) = f(x)$ for all $f \in \mathcal{M}^*$]. The elements of $\mathcal{T}_0^1$ are sometimes called *contravariant vectors,* whereas those of $\mathcal{T}_1^0$ are *covariant vectors* or *covectors.*

World tensors were introduced by Minkowski in 1908 as a language in which to express Lorentz invariant relationships. Any assertion that two world tensors L and T are equal would be checked in a given admissible basis/frame by comparing their components $L^{a_1\cdots a_r}{}_{b_1\cdots b_s}$ and $T^{a_1\cdots a_r}{}_{b_1\cdots b_s}$ in that basis and, if these are indeed found to be equal in one basis, then the components in any other basis must necessarily also be equal since they both transform to the new basis under (3.1.11). World tensor equations are true in all admissible frames if and only if they are true in any one admissible frame, i.e., they are Lorentz invariant. World tensors were introduced, in analogy with the 3-vectors of classical mechanics, to serve as the basic "building blocks" from which to construct the laws of relativistic (i.e., Lorentz invariant) physics. So admirably suited were they to this task that it was not until attempts got under way to reconcile the principles of relativistic and quantum mechanics that it was found that there were not enough "building blocks". The reason for this can be traced to the fact that the underlying physically significant quantities in quantum mechanics (e.g., wave functions) are described by *complex* numbers ψ, whereas the result of a specific measurement carried out on a quantum mechanical system is a *real* number that depends only on quantities of the form $\psi\bar{\psi}$ and these last quantities are insensitive to changes in sign, i.e., $(-\psi)(\overline{-\psi}) = \psi\bar{\psi}$. Consequently, ψ and $-\psi$ give rise to precisely the same predictions as to the result of any experiment and so must represent the same state of the system. As a result, transforming the state's description in one admissible frame to that in another (related to it by Λ) can be accomplished by either one of two matrices $\pm D_\Lambda$. As we shall see in Section 3.5 this ambiguity in the sign is often an essential feature of the situation and cannot be consistently removed by making one choice or the other. This fact leads directly to the notion of what Penrose [**PR**] has called a "spinorial object" and which we shall discuss in some detail in Appendix B. For the present we will only take these remarks as motivation for introducing what are called "two-valued representations" of the Lorentz group (intuitively, assignments $\Lambda \to \pm D_\Lambda$ of two component transformation matrices, differing only by sign, to each $\Lambda \in \mathcal{L}$).

In Section 1.7 we constructed a mapping of $SL(2, \mathbb{C})$ onto $\mathcal{L}$ called the spinor map which we now designate

$$\text{Spin}: SL(2,\mathbb{C}) \longrightarrow \mathcal{L}.$$

Spin was a homomorphism of the matrix group $SL(2,\mathbb{C})$ onto the matrix group $\mathcal{L}$ that mapped the unitary subgroup SU_2 of $SL(2,\mathbb{C})$ onto the rotation subgroup $\mathcal{R}$ of $\mathcal{L}$ and was precisely two-to-one, carrying $\pm G$ in $SL(2,\mathbb{C})$ onto the same element of $\mathcal{L}$ [which we denote either $\text{Spin}(G) = \text{Spin}(-G)$ or $\Lambda_G = \Lambda_{-G}$]. Next we observe that any representation $\tilde{D}: \mathcal{L} \to \mathcal{H}$ "lifts" to a representation of $SL(2,\mathbb{C})$. More precisely, we define $D: SL(2,\mathbb{C}) \to \mathcal{H}$ by $D = \tilde{D} \circ \text{Spin}$. Of course, D has the property that, for every $G \in SL(2,\mathbb{C})$, $D_{-G} = \tilde{D}(\text{Spin}(-G)) = \tilde{D}(\text{Spin}(G)) = D_G$. Conversely, suppose $D: SL(2,\mathbb{C}) \to \mathcal{H}$ is a representation of $SL(2,\mathbb{C})$ with the property that $D_{-G} = D_G$ for every $G \in SL(2,\mathbb{C})$. We define $\tilde{D}: \mathcal{L} \to \mathcal{H}$ as follows: Let $\Lambda \in \mathcal{L}$. Then there exists a $G \in SL(2,\mathbb{C})$ such that $\Lambda_G = \Lambda$. Define $\tilde{D}(\Lambda) = \tilde{D}(\Lambda_G) = D_G$. Then $\tilde{D}$ is a representation of $\mathcal{L}$ since $\tilde{D}(\Lambda_1 \Lambda_2) = \tilde{D}(\Lambda_{G_1} \Lambda_{G_2}) = \tilde{D}(\Lambda_{G_1 G_2}) = D_{G_1 G_2} = D_{G_1} D_{G_2} = \tilde{D}(\Lambda_1)\, \tilde{D}(\Lambda_2)$. Thus, *there is a one-to-one correspondence between the representations of* $\mathcal{L}$ *and the representations of* $SL(2,\mathbb{C})$ *that satisfy* $D_{-G} = D_G$ *for all* $G \in SL(2,\mathbb{C})$.

Before proceeding with the discussion of those representations of $SL(2,\mathbb{C})$ for which $D_{-G} \neq D_G$ we introduce a few more definitions. Thus, we let $\mathcal{G}$ and $\mathcal{H}$ be arbitrary matrix groups and $D: \mathcal{G} \to \mathcal{H}$ a representation of $\mathcal{G}$. If the order of $\mathcal{H}$ is m, we let V_m stand for any space of carriers for D. A subspace S of V_m is said to be *invariant under* D if each D_G, thought of as a linear transformation of V_m, carries S into itself, i.e., satisfies $D_G S \subseteq S$. For example, V_m itself and the trivial subspace $\{0\}$ of V_m are obviously invariant under any D. If $\{0\}$ and V_m are the only subspaces of V_m that are invariant under D, then D is said to be *irreducible;* otherwise, D is *reducible.* It can be shown (see [**GMS**]) that all of the representations of $SL(2,\mathbb{C})$ can be constructed from those that are irreducible. Finally, two representations $D^{(1)}: \mathcal{G} \to \mathcal{H}^1$ and $D^{(2)}: \mathcal{G} \to \mathcal{H}^2$, where $\mathcal{H}^1$ and $\mathcal{H}^2$ have the same order, are said to be *equivalent* if there exists an invertible matrix P such that

$$D_G^{(2)} = P^{-1}\, D_G^{(1)} P$$

for all $G \in \mathcal{G}$. This is clearly equivalent to the requirement that, if V_m is a space of carriers for both $D^{(1)}$ and $D^{(2)}$, then there exist bases $\{v_a^{(1)}\}$ and $\{v_a^{(2)}\}$ for V_m such that, for every $G \in \mathcal{G}$, the linear transformation whose matrix relative to $\{v_a^{(1)}\}$ is $D_G^{(1)}$ has matrix $D_G^{(2)}$ relative to $\{v_a^{(2)}\}$.

Theorem 3.1.1. (*Schur's Lemma*) *Let* $\mathcal{G}$ *and* $\mathcal{H}$ *be matrix groups of order* n *and* m *respectively and* $D: \mathcal{G} \to \mathcal{H}$ *an irreducible representation* *of $\mathcal{G}$. If A is an $m \times m$ matrix which commutes with every D_G, i.e., $AD_G = D_G A$ for every $G \in \mathcal{G}$, then A is a multiple of the identity matrix, i.e., $A = \lambda I$ for some (in general, complex) number λ.*

Proof: We select a space V_m of carriers and regard A and all the D_G as linear transformations on V_m. Let $S = \ker A$. Then S is a subspace of V_m. For each $s \in S$, $As = 0$ implies $A(D_G\, s) = D_G(As) = D_G(0) = 0$ so $D_G\, s \in S$, i.e., S is invariant under D. Since D is irreducible, either $S = V_m$ or $S = \{0\}$. If $S = V_m$, then $A = 0 = 0 \cdot I$ and we are done. If $S = \{0\}$, then A is invertible and so has a nonzero (complex) eigenvalue λ. Notice that $(A - \lambda I)D_G = AD_G - (\lambda I)D_G = D_G\, A - D_G(\lambda I) = D_G(A - \lambda I)$ so $A - \lambda I$ commutes with every D_G. The argument given above shows that $A - \lambda I$ is either 0 or invertible. But λ is an eigenvalue of A so $A - \lambda I$ is not invertible and therefore $A - \lambda I = 0$ as required. ∎

Corollary 3.1.2. *Let $\mathcal{G}$ be a matrix group that contains $-G$ for every $G \in \mathcal{G}$ and $D: \mathcal{G} \to \mathcal{H}$ an irreducible representation of $\mathcal{G}$. Then*

$$D_{-G} = \pm D_G. \tag{3.1.12}$$

Proof: $-G = (-I)G$ so $D_{-G} = D_{(-I)G} = D_{-I}\, D_G$ and it will suffice to show that $D_{-I} = \pm I$. For any $G' \in \mathcal{G}$, $D_{-I}\, D_{G'} = D_{(-I)G'} = D_{G'(-I)} = D_{G'}\, D_{-I}$ so D_{-I} commutes with each $D_{G'}$, $G' \in \mathcal{G}$. By Schur's Lemma, $D_{-I} = \lambda I$ for some λ. But $D_{-I}\, D_{-I} = D_{(-I)(-I)} = D_I = I$ so $(\lambda I)(\lambda I) = \lambda^2 I = I$. Thus, $\lambda^2 = 1$ so $\lambda = \pm 1$ and $D_{-I} = \pm I$. ∎

Since $SL(2, \mathbb{C})$ clearly contains $-G$ for every $G \in SL(2, \mathbb{C})$, we find that every irreducible representation D of $SL(2, \mathbb{C})$ satisfies either $D_{-G} = D_G$ or $D_{-G} = -D_G$. As we have seen, those of the first type give representations of the Lorentz group. Although those that satisfy $D_{-G} = -D_G$ cannot legitimately be regarded as representations of $\mathcal{L}$ (not being single-valued), it has become customary to refer to such a representation of $SL(2, \mathbb{C})$ as a *two-valued representation of* $\mathcal{L}$ and we shall adhere to the custom.

The problem of determining the finite-dimensional, irreducible representations of $SL(2, \mathbb{C})$ is thus seen to be a matter of considerable interest in mathematical physics. As it happens, these representations are well-known and rather easy to describe. Moreover, such a description is well worth the effort required to produce it since it leads inevitably to the notion of a "spinor", which will be our major concern in this chapter.

In order to enumerate these representations of $SL(2, \mathbb{C})$ it will be convenient to reverse our usual procedure and specify first a space of carriers and a basis and then describe the linear transformations whose matrices relative to this basis will constitute our representations. If $m \geq 0$ and $n \geq 0$ are integers we denote by P_{mn} the vector space of all polynomials in z and $\bar{z}$ with complex coefficients and of degree at most m in z and at most n in $\bar{z}$, i.e.,

$$P_{mn} = \{p(z,\bar{z}) = p_{00} + p_{10}z + p_{01}\bar{z} + p_{11}z\bar{z} + \cdots + p_{mn}z^m\bar{z}^n = p_{rs}z^r\bar{z}^s : \ p_{rs} \in \mathbb{C}\},$$

with the usual coefficientwise addition and scalar multiplication, i.e., $p(z,\bar{z}) + q(z,\bar{z}) = [p_{00} + p_{10}z + \cdots + p_{mn}\, z^m\bar{z}^n] + [q_{00} + q_{10}z + \cdots + q_{mn}z^m\bar{z}^n] = [p_{00} + q_{00}] + [p_{10} + q_{10}]z + \cdots + [p_{mn} + q_{mn}]z^m\bar{z}^n$ and $\alpha p(z,\bar{z}) = (\alpha p_{00}) + (\alpha p_{10})z + \cdots + (\alpha p_{mn})z^m\bar{z}^n$. The basis implicit here is $\{1, z, \bar{z}, z\,\bar{z}, \ldots, z^m\bar{z}^n\}$ so $\dim P_{mn} = (m+1)(n+1)$. Now, for each $G = \left[\begin{smallmatrix} a & b \\ c & d \end{smallmatrix}\right] \in SL(2,\mathbb{C})$ we define $D_G^{(\frac{m}{2},\frac{n}{2})} : P_{mn} \to P_{mn}$ by

$$D_G^{(\frac{m}{2},\frac{n}{2})}\,(p(z,\bar{z})) = D_G^{(\frac{m}{2},\frac{n}{2})}\,(p_{rs}\,z^r\,\bar{z}^s) = (bz+d)^m\,(\bar{b}\bar{z}+\bar{d})^n\,p(w,\bar{w}),$$

where

$$w = \frac{az+c}{bz+d}.$$

Then $D_G^{(\frac{m}{2},\frac{n}{2})}$ is clearly linear in $p(z,\bar{z})$ and maps P_{mn} to P_{mn}. Although algebraically a bit messy it is straightforward to show that $D_G^{(\frac{m}{2},\frac{n}{2})}$ has the properties required to determine a representation of $SL(2,\mathbb{C})$. We leave the manual labor to the reader.

Exercise 3.1.5. Show that $D_I^{(\frac{m}{2},\frac{n}{2})}$ is the identity transformation on P_{mn} and that if G_1 and G_2 are in $SL(2,\mathbb{C})$, then

$$D_{G_1G_2}^{(\frac{m}{2},\frac{n}{2})} = D_{G_1}^{(\frac{m}{2},\frac{n}{2})} \circ D_{G_2}^{(\frac{m}{2},\frac{n}{2})}.$$

Thus, the matrices of the linear transformations $D_G^{(\frac{m}{2},\frac{n}{2})}$ relative to the basis $\{1, z, \bar{z}, \ldots, z^m\bar{z}^n\}$ for P_{mn} constitute a representation of $SL(2,\mathbb{C})$ which we also denote

$$G \longrightarrow D_G^{(\frac{m}{2},\frac{n}{2})}$$

and call the *spinor representation* of *type* (m,n). Although it is by no means obvious the spinor representations are all irreducible and, in fact, exhaust all of the finite-dimensional, irreducible representations of $SL(2,\mathbb{C})$ (we refer the interested reader to [**GMS**] for a proof of Theorem 3.1.3).

Theorem 3.1.3. *For all* $m,n = 0,1,2,\ldots,$ *the spinor representation* $D^{(\frac{m}{2},\frac{n}{2})}$ *of* $SL(2,\mathbb{C})$ *is irreducible and every finite-dimensional irreducible representation of* $SL(2,\mathbb{C})$ *is equivalent to some* $D^{(\frac{m}{2},\frac{n}{2})}$.

We consider a few specific examples. First suppose $m = 1$ and $n = 0$: $\mathcal{P}_{10} = \{p(z,\bar{z}) = p_{00} + p_{10}z : p_{rs} \in \mathbb{C}\}$. For $G = \left[\begin{smallmatrix} a & b \\ c & d \end{smallmatrix}\right] \in SL(2,\mathbb{C})$,

$$\begin{aligned} D_G^{(\frac{1}{2},0)}\,(p(z,\bar{z})) &= (bz+d)^1(\bar{b}\bar{z}+\bar{d})^0 p(w,\bar{w}) \\ &= (bz+d)\left(p_{00} + p_{10}\left(\frac{az+c}{bz+d}\right)\right) \\ &= (bz+d)p_{00} + (az+c)p_{10} \\ &= (cp_{10} + dp_{00}) + (ap_{10} + bp_{00})z \\ &= \hat{p}_{00} + \hat{p}_{10}z\,, \end{aligned}$$

where

$$\begin{bmatrix} \hat{p}_{10} \\ \hat{p}_{00} \end{bmatrix} = \begin{bmatrix} a & b \\ c & d \end{bmatrix} \begin{bmatrix} p_{10} \\ p_{00} \end{bmatrix}.$$

Thus, the representation $G \to D_G^{(\frac{1}{2},0)}$ is given by

$$\begin{bmatrix} a & b \\ c & d \end{bmatrix} \longrightarrow D^{(\frac{1}{2},0)}_{\left[\begin{smallmatrix} a & b \\ c & d \end{smallmatrix}\right]} = \begin{bmatrix} a & b \\ c & d \end{bmatrix},$$

i.e., $D^{(\frac{1}{2},0)}$ is the *identity representation* of $SL(2,\mathbb{C})$.

Exercise 3.1.6. Show in the same way that $D^{(0,\frac{1}{2})}$ is the *conjugation representation*

$$\begin{bmatrix} a & b \\ c & d \end{bmatrix} \longrightarrow D^{(0,\frac{1}{2})}_{\left[\begin{smallmatrix} a & b \\ c & d \end{smallmatrix}\right]} = \begin{bmatrix} \bar{a} & \bar{b} \\ \bar{c} & \bar{d} \end{bmatrix}.$$

Exercise 3.1.7. Show that $D^{(\frac{1}{2},0)}$ and $D^{(0,\frac{1}{2})}$ are *not* equivalent representations of $SL(2,\mathbb{C})$, i.e., that there does not exist an invertible matrix P such that $P^{-1}GP = \bar{G}$ for all $G \in SL(2,\mathbb{C})$. *Hint:* Let $G = \left[\begin{smallmatrix} i & 0 \\ 0 & -i \end{smallmatrix}\right]$ and $P = \left[\begin{smallmatrix} 0 & i \\ -i & 0 \end{smallmatrix}\right]$. Show that $P^{-1}GP = \bar{G}$ and that P and its nonzero scalar multiples are the only matrices for which this is true. Now find a $G' \in SL(2,\mathbb{C})$ for which $P^{-1}G'P \neq \overline{G'}$.

Before working out another example we include a few more observations about $D^{(\frac{1}{2},0)}$ and $D^{(0,\frac{1}{2})}$. First note that if $G \to D_G$ is any representation of $SL(2,\mathbb{C})$, then the assignment $G \to (D_G^{-1})^T = (D_G^T)^{-1}$ of the transposed inverse of D_G to each G is also a representation of $SL(2,\mathbb{C})$ since $I \to (D_I^{-1})^T = (I^{-1})^T = I^T = I$ and $G_1G_2 \to ((D_{G_1G_2})^{-1})^T = ((D_{G_1}D_{G_2})^{-1})^T = (D_{G_2}^{-1}D_{G_1}^{-1})^T = (D_{G_1}^{-1})^T(D_{G_2}^{-1})^T$ (note that inversion or transposition alone would not accomplish this since each reverses products). Applying this, in particular, to the identity representation $D^{(\frac{1}{2},0)}$ gives

$$G = \begin{bmatrix} a & b \\ c & d \end{bmatrix} \longrightarrow (G^{-1})^T = (G^T)^{-1} = \begin{bmatrix} d & -c \\ -b & a \end{bmatrix}.$$

Letting $\epsilon = \left[\begin{smallmatrix} 0 & -1 \\ 1 & 0 \end{smallmatrix}\right]$ it is easily checked that $\epsilon^{-1} = -\epsilon$ and

$$\left(G^{-1}\right)^T = \epsilon^{-1} G \epsilon .$$

Thus, $G \to (G^{-1})^T = (G^T)^{-1}$ *is* equivalent to $D^{(\frac{1}{2},0)}$ and we shall denote it $\tilde{D}^{(\frac{1}{2},0)}$. Similarly, one can define a representation $\tilde{D}^{(0,\frac{1}{2})}$ equivalent to conjugation by

$$G \longrightarrow \left(\bar{G}^{-1}\right)^T = \left(\bar{G}^T\right)^{-1} = \epsilon^{-1} \bar{G} \epsilon .$$

These equivalent versions of $D^{(\frac{1}{2},0)}$ and $D^{(0,\frac{1}{2})}$ as well as analogous versions of $D^{(\frac{m}{2},\frac{n}{2})}$ are often convenient and we shall return to them in the next section.

Now let $m = n = 1$. Then $P_{11} = \{p(z,\bar{z}) = p_{00} + p_{10} z + p_{01} \bar{z} + p_{11} z\bar{z} : p_{rs} \in \mathbb{C}\}$ and for each $G = \left[\begin{smallmatrix} a & b \\ c & d \end{smallmatrix}\right] \in SL(2,\mathbb{C})$ one has

$$D_G^{(\frac{1}{2},\frac{1}{2})}(p(z,\bar{z})) = (bz+d)^1 (\bar{b}\bar{z}+\bar{d})^1 (p_{00} + p_{10} w + p_{01} \bar{w} + p_{11} w \bar{w}),$$

where $w = \frac{az+c}{bz+d}$. Multiplying out and rearranging yields

$$D_G^{(\frac{1}{2},\frac{1}{2})}(p(z,\bar{z})) = \hat{p}_{00} + \hat{p}_{10} z + \hat{p}_{01} \bar{z} + \hat{p}_{11} z\bar{z} ,$$

where

$$\begin{bmatrix} \hat{p}_{11} \\ \hat{p}_{10} \\ \hat{p}_{01} \\ \hat{p}_{00} \end{bmatrix} = \begin{bmatrix} a\bar{a} & a\bar{b} & \bar{a}b & b\bar{b} \\ a\bar{c} & a\bar{d} & b\bar{c} & b\bar{d} \\ \bar{a}c & \bar{b}c & \bar{a}d & \bar{b}d \\ c\bar{c} & c\bar{d} & \bar{c}d & d\bar{d} \end{bmatrix} \begin{bmatrix} p_{11} \\ p_{10} \\ p_{01} \\ p_{00} \end{bmatrix}, \qquad (3.1.13)$$

so that

$$D^{(\frac{1}{2},\frac{1}{2})}_{\left[\begin{smallmatrix} a & b \\ c & d \end{smallmatrix}\right]} = \begin{bmatrix} a\bar{a} & a\bar{b} & \bar{a}b & b\bar{b} \\ a\bar{c} & a\bar{d} & b\bar{c} & b\bar{d} \\ \bar{a}c & \bar{b}c & \bar{a}d & \bar{b}d \\ c\bar{c} & c\bar{d} & \bar{c}d & d\bar{d} \end{bmatrix} .$$

Proceeding in this manner with the notation currently at our disposal would soon become algebraically unmanageable. For this reason we now introduce new and powerful notational devices that will constitute the language in which the remainder of the chapter will be written. First we rephrase the example of $D^{(\frac{1}{2},\frac{1}{2})}$ in these new terms. We begin by rewriting each $p(z,\bar{z})$ as a sum of terms of the form

$$\phi_{A\dot{X}} z^A \bar{z}^{\dot{X}},$$

where $A = 1,0$ and $\dot{X} = \dot{1},\dot{0}$ (the dot is used only to indicate a power of $\bar{z}$ rather than z and $\dot{1},\dot{0}$ are treated exactly as if they were $1,0$, i.e., $\bar{z}^{\dot{0}} = 1$, $\bar{z}^{\dot{1}} = \bar{z}$, $\dot{0} + \dot{1} = \dot{1}$, etc.). Thus,

$$p_{00} + p_{10} z + p_{01} \bar{z} + p_{11} z\bar{z} = \phi_{0\dot{0}} z^0 \bar{z}^{\dot{0}} + \phi_{1\dot{0}} z^1 \bar{z}^{\dot{0}} + \phi_{0\dot{1}} z^0 \bar{z}^{\dot{1}} + \phi_{1\dot{1}} z^1 \bar{z}^{\dot{1}},$$

where $\phi_{0\dot{0}} = p_{00}$, $\phi_{1\dot{0}} = p_{10}$, $\phi_{1\dot{1}} = p_{11}$. With the summation convention (over $A = 1, 0$, $\dot{X} = \dot{1}, \dot{0}$),

$$p(z, \bar{z}) = \phi_{A\dot{X}} z^A \bar{z}^{\dot{X}}.$$

To set up another application of the summation convention we henceforth denote the entries in $G \in SL(2, \mathbb{C})$ as

$$G = \left[G_A{}^B\right] = \begin{bmatrix} G_1{}^1 & G_1{}^0 \\ G_0{}^1 & G_0{}^0 \end{bmatrix}$$

and write the conjugate $\bar{G}$ of G as

$$\bar{G} = \left[\bar{G}_{\dot{X}}{}^{\dot{Y}}\right] = \begin{bmatrix} \bar{G}_{\dot{1}}{}^{\dot{1}} & \bar{G}_{\dot{1}}{}^{\dot{0}} \\ \bar{G}_{\dot{0}}{}^{\dot{1}} & \bar{G}_{\dot{0}}{}^{\dot{0}} \end{bmatrix}.$$

Convention: Henceforth, conjugating a term with undotted indices dots them all and introduces a bar, whereas conjugating a term with dotted indices undots them and removes the bar. Whenever possible we will select undotted index names from the beginning of the alphabet $(A, B, C, \ldots)$ and dotted indices from the end $(\ldots, \dot{X}, \dot{Y}, \dot{Z})$.

Now, if we let $D_G^{(\frac{1}{2},\frac{1}{2})}(\phi_{A\dot{X}} z^A \bar{z}^{\dot{X}}) = \hat{\phi}_{A\dot{X}} z^A \bar{z}^{\dot{X}}$ we find from (3.1.13) that

$$\begin{bmatrix} \hat{\phi}_{1\dot{1}} \\ \hat{\phi}_{1\dot{0}} \\ \hat{\phi}_{0\dot{1}} \\ \hat{\phi}_{0\dot{0}} \end{bmatrix} = \begin{bmatrix} G_1{}^1\bar{G}_{\dot{1}}{}^{\dot{1}} & G_1{}^1\bar{G}_{\dot{1}}{}^{\dot{0}} & G_1{}^0\bar{G}_{\dot{1}}{}^{\dot{1}} & G_1{}^0\bar{G}_{\dot{1}}{}^{\dot{0}} \\ G_1{}^1\bar{G}_{\dot{0}}{}^{\dot{1}} & G_1{}^1\bar{G}_{\dot{0}}{}^{\dot{0}} & G_1{}^0\bar{G}_{\dot{0}}{}^{\dot{1}} & G_1{}^0\bar{G}_{\dot{0}}{}^{\dot{0}} \\ G_0{}^1\bar{G}_{\dot{1}}{}^{\dot{1}} & G_0{}^1\bar{G}_{\dot{1}}{}^{\dot{0}} & G_0{}^0\bar{G}_{\dot{1}}{}^{\dot{1}} & G_0{}^0\bar{G}_{\dot{1}}{}^{\dot{0}} \\ G_0{}^1\bar{G}_{\dot{0}}{}^{\dot{1}} & G_0{}^1\bar{G}_{\dot{0}}{}^{\dot{0}} & G_0{}^0\bar{G}_{\dot{0}}{}^{\dot{1}} & G_0{}^0\bar{G}_{\dot{0}}{}^{\dot{0}} \end{bmatrix} \begin{bmatrix} \phi_{1\dot{1}} \\ \phi_{1\dot{0}} \\ \phi_{0\dot{1}} \\ \phi_{0\dot{0}} \end{bmatrix} \tag{3.1.14}$$

which all collapses quite nicely with the summation convention to

$$\hat{\phi}_{A\dot{X}} = G_A{}^B \bar{G}_{\dot{X}}{}^{\dot{Y}} \phi_{B\dot{Y}}, \quad A = 1, 0, \ \dot{X} = \dot{1}, \dot{0}. \tag{3.1.15}$$

For $D^{(\frac{1}{2},0)}$ we would write $p_{00} + p_{10} z = \phi_0 z^0 + \phi_1 z^1 = \phi_A z^A$ and $D_G^{(\frac{1}{2},0)}(\phi_A z^A) = \hat{\phi}_A z^A$, where

$$\hat{\phi}_A = G_A{}^B \phi_B, \quad A = 1, 0. \tag{3.1.16}$$

Similarly, for $D^{(0,\frac{1}{2})}$, $p_{00} + p_{01}\bar{z} = \phi_{\dot{0}} \bar{z}^{\dot{0}} + \phi_{\dot{1}} \bar{z}^{\dot{1}} = \phi_{\dot{X}} \bar{z}^{\dot{X}}$ and $D_G^{(0,\frac{1}{2})}(\phi_{\dot{X}} \bar{z}^{\dot{X}}) = \hat{\phi}_{\dot{X}} \bar{z}^{\dot{X}}$, where

$$\hat{\phi}_{\dot{X}} = \bar{G}_{\dot{X}}{}^{\dot{Y}} \phi_{\dot{Y}}, \quad \dot{X} = \dot{1}, \dot{0}. \tag{3.1.17}$$

Notice that the 4×4 matrix in (3.1.14) is precisely $D_G^{(\frac{1}{2},\frac{1}{2})}$ and that analogous statements would be true of (3.1.16) and (3.1.17) if these were written as matrix products. The situation changes somewhat for larger m and n so we wish to treat one more example before describing the general case. Thus, we let $m = 2$ and $n = 1$. An element $p(z,\bar{z}) = p_{00} + p_{10}z + \cdots + p_{21}z^2\bar{z}$ is to be written as a sum of terms of the form $\phi_{A_1A_2\dot{X}}z^{A_1}z^{A_2}\bar{z}^{\dot{X}}$. For example, the constant term p_{00} is written $\phi_{00\dot{0}}z^0z^0\bar{z}^{\dot{0}}$ so $\phi_{00\dot{0}} = p_{00}$ and $p_{10}z$ becomes $\phi_{10\dot{0}}z^1z^0\bar{z}^{\dot{0}} + \phi_{01\dot{0}}z^0z^1\bar{z}^{\dot{0}}$ and we take $\phi_{10\dot{0}} = \phi_{01\dot{0}} = \frac{1}{2}p_{10}$, and so on. The result is

$$\begin{aligned} p(z,\bar{z}) \;=\; & \phi_{00\dot{0}}z^0z^0\bar{z}^{\dot{0}} + \phi_{10\dot{0}}z^1z^0\bar{z}^{\dot{0}} + \phi_{01\dot{0}}z^0z^1\bar{z}^{\dot{0}} \\ & + \phi_{00\dot{1}}z^0z^0\bar{z}^{\dot{1}} + \phi_{10\dot{1}}z^1z^0\bar{z}^{\dot{1}} + \phi_{01\dot{1}}z^0z^1\bar{z}^{\dot{1}} \\ & + \phi_{11\dot{0}}z^1z^1\bar{z}^{\dot{0}} + \phi_{11\dot{1}}z^1z^1\bar{z}^{\dot{1}}, \end{aligned}$$

where we take

$$\begin{aligned} &\phi_{00\dot{0}} = p_{00}, && \phi_{10\dot{0}} = \phi_{01\dot{0}} = \tfrac{1}{2}p_{10}, \quad \phi_{00\dot{1}} = p_{01}, \\ &\phi_{10\dot{1}} = \phi_{01\dot{1}} = \tfrac{1}{2}p_{11}, && \phi_{11\dot{0}} = p_{20}, \quad \phi_{11\dot{1}} = p_{21}, \end{aligned}$$

so that, in particular, $\phi_{A_1A_2\dot{X}}$ is *symmetric in* A_1 *and* A_2, *i.e.*, $\phi_{A_2A_1\dot{X}} = \phi_{A_1A_2\dot{X}}$ *for* $A_1, A_2 = 1, 0$. Thus, with the summation convention,

$$p(z,\bar{z}) = \phi_{A_1A_2\dot{X}}z^{A_1}z^{A_2}\bar{z}^{\dot{X}} = \phi_{A_1A_2\dot{X}}z^{A_1+A_2}\bar{z}^{\dot{X}}.$$

Now,

$$D_G^{(\frac{2}{2},\frac{1}{2})}(p(z,\bar{z})) = (G_1{}^0z + G_0{}^0)^2(\bar{G}_{\dot{1}}{}^{\dot{0}}\bar{z} + \bar{G}_{\dot{0}}{}^{\dot{0}})^1(\phi_{A_1A_2\dot{X}}w^{A_1+A_2}\bar{w}^{\dot{X}}),$$

where

$$w = \frac{G_1{}^1z + G_0{}^1}{G_1{}^0z + G_0{}^0},$$

so

$$\begin{aligned} D_G^{(\frac{2}{2},\frac{1}{2})}(p(z,\bar{z})) = \; & \phi_{A_1A_2\dot{X}}(G_1{}^1z + G_0{}^1)^{A_1+A_2}(G_1{}^0z + G_0{}^0)^{2-A_1-A_2} \\ & \cdot(\bar{G}_{\dot{1}}{}^{\dot{1}}\bar{z} + \bar{G}_{\dot{0}}{}^{\dot{1}})^{\dot{X}}(\bar{G}_{\dot{1}}{}^{\dot{0}}\bar{z} + \bar{G}_{\dot{0}}{}^{\dot{0}})^{1-\dot{X}} \\ = \; & \phi_{00\dot{0}}(G_1{}^0z + G_0{}^0)(G_1{}^0z + G_0{}^0)(\bar{G}_{\dot{1}}{}^{\dot{0}}\bar{z} + \bar{G}_{\dot{0}}{}^{\dot{0}}) \\ & + \phi_{10\dot{0}}(G_1{}^1z + G_0{}^1)(G_1{}^0z + G_0{}^0)(\bar{G}_{\dot{1}}{}^{\dot{0}}\bar{z} + \bar{G}_{\dot{0}}{}^{\dot{0}}) \\ & + \phi_{01\dot{0}}(G_1{}^1z + G_0{}^1)(G_1{}^0z + G_0{}^0)(\bar{G}_{\dot{1}}{}^{\dot{0}}\bar{z} + \bar{G}_{\dot{0}}{}^{\dot{0}}) \\ & + \phi_{11\dot{0}}(G_1{}^1z + G_0{}^1)(G_1{}^1z + G_0{}^1)(\bar{G}_{\dot{1}}{}^{\dot{0}}\bar{z} + \bar{G}_{\dot{0}}{}^{\dot{0}}) \\ & + \phi_{00\dot{1}}(G_1{}^0z + G_0{}^0)(G_1{}^0z + G_0{}^0)(\bar{G}_{\dot{1}}{}^{\dot{1}}\bar{z} + \bar{G}_{\dot{0}}{}^{\dot{1}}) \\ & + \phi_{10\dot{1}}(G_1{}^1z + G_0{}^1)(G_1{}^0z + G_0{}^0)(\bar{G}_{\dot{1}}{}^{\dot{1}}\bar{z} + \bar{G}_{\dot{0}}{}^{\dot{1}}) \\ & + \phi_{01\dot{1}}(G_1{}^1z + G_0{}^1)(G_1{}^0z + G_0{}^0)(\bar{G}_{\dot{1}}{}^{\dot{1}}\bar{z} + \bar{G}_{\dot{0}}{}^{\dot{1}}) \\ & + \phi_{11\dot{1}}(G_1{}^1z + G_0{}^1)(G_1{}^1z + G_0{}^1)(\bar{G}_{\dot{1}}{}^{\dot{1}}\bar{z} + \bar{G}_{\dot{0}}{}^{\dot{1}}). \end{aligned}$$

Multiplying out and collecting terms gives $D_G^{(\frac{2}{2},\frac{1}{2})}(\phi_{A_1A_2\dot{X}}z^{A_1+A_2}\bar{z}^{\dot{X}}) = \hat{\phi}_{A_1A_2\dot{X}}z^{A_1+A_2}\bar{z}^{\dot{X}}$, where

$$\hat{\phi}_{A_1A_2\dot{X}} = G_{A_1}{}^{B_1}G_{A_2}{}^{B_2}\bar{G}_{\dot{X}}{}^{\dot{Y}}\,\phi_{B_1B_2\dot{Y}}, \quad A_1, A_2 = 1, 0, \;\; \dot{X} = \dot{1}, \dot{0}. \quad (3.1.18)$$

Exercise 3.1.8. Write out all terms in the expansion of $D_G^{(\frac{2}{2},\frac{1}{2})}(p(z,\bar{z}))$ that contain $z^2\bar{z}$ and show that they can be written in the form

$$G_1{}^{B_1}G_1{}^{B_2}\bar{G}_{\dot{1}}{}^{\dot{Y}}\,\phi_{B_1B_2\dot{Y}}z^{B_1}z^{B_2}\bar{z}^{\dot{Y}}$$

and so verify (3.1.18) for $A_1 = A_2 = 1$, $\dot{X} = \dot{1}$. Similarly, find the constant term and the terms with $z, z^2, \bar{z}$ and $z\bar{z}$ to verify (3.1.18) for all A_1, A_2 and $\dot{X}$.

Now observe that by writing the $\hat{\phi}_{A_1A_2\dot{X}}$ as a column matrix $[\hat{\phi}_{A_1A_2\dot{X}}] = \text{col}[\hat{\phi}_{11\dot{1}}\;\hat{\phi}_{10\dot{1}}\;\hat{\phi}_{01\dot{1}}\;\hat{\phi}_{00\dot{1}}\;\hat{\phi}_{11\dot{0}}\;\hat{\phi}_{10\dot{0}}\;\hat{\phi}_{01\dot{0}}\;\hat{\phi}_{00\dot{0}}]$, and similarly for the $\phi_{B_1B_2\dot{Y}}$, (3.1.18) can be written as a matrix product

$$\begin{bmatrix} \hat{\phi}_{11\dot{1}} \\ \hat{\phi}_{10\dot{1}} \\ \vdots \\ \hat{\phi}_{00\dot{0}} \end{bmatrix} = \begin{bmatrix} G_1{}^1G_1{}^1\bar{G}_{\dot{1}}{}^{\dot{1}} & G_1{}^1G_1{}^0\bar{G}_{\dot{1}}{}^{\dot{1}} & G_1{}^0G_1{}^1\bar{G}_{\dot{1}}{}^{\dot{1}} & \cdots \\ G_1{}^1G_0{}^1\bar{G}_{\dot{1}}{}^{\dot{1}} & G_1{}^1G_0{}^0\bar{G}_{\dot{1}}{}^{\dot{1}} & G_1{}^0G_0{}^1\bar{G}_{\dot{1}}{}^{\dot{1}} & \cdots \\ \vdots & \vdots & \vdots & \\ G_0{}^1G_0{}^1\bar{G}_{\dot{0}}{}^{\dot{1}} & G_0{}^1G_0{}^0\bar{G}_{\dot{0}}{}^{\dot{1}} & G_0{}^0G_0{}^1\bar{G}_{\dot{0}}{}^{\dot{1}} & \cdots \end{bmatrix} \begin{bmatrix} \phi_{11\dot{1}} \\ \phi_{10\dot{1}} \\ \vdots \\ \phi_{00\dot{0}} \end{bmatrix}.$$

Unlike (3.1.14), however, the 8×8 coefficient matrix here is *not* $D_G^{(\frac{2}{2},\frac{1}{2})}$. Indeed, the representation $G \to [G_{A_1}{}^{B_1}G_{A_2}{}^{B_2}\bar{G}_{\dot{X}}{}^{\dot{Y}}]$ which assigns this matrix to G is not even equivalent to the spinor representation of type $(2,1)$ since the latter has order $(2+1)(1+1) = 6$, not 8. The reason is that, in writing the elements of P_{21} in the form $\phi_{A_1A_2\dot{X}}z^{A_1}z^{A_2}\bar{z}^{\dot{X}}$, we are not finding components relative to the basis $\{1, z, \bar{z}, \ldots, z^2\bar{z}\}$ since, for example, $z^1z^0\bar{z}^{\dot{1}}$ and $z^0z^1\bar{z}^{\dot{1}}$ are both $z\bar{z}$. Nevertheless, it is the transformation law (3.1.18) that is of most interest to us.

The general case proceeds in much the same way. P_{mn} consists of all $p(z,\bar{z}) = p_{00} + p_{10}z + \cdots + p_{mn}z^m\bar{z}^n = p_{rs}z^r\bar{z}^s$, $r = 0, \ldots, m$, $s = 0, \ldots, n$. Each of these is written as a sum of terms of the form

$$\phi_{A_1\cdots A_m\dot{X}_1\cdots\dot{X}_n}z^{A_1}\cdots z^{A_m}\bar{z}^{\dot{X}_1}\cdots\bar{z}^{\dot{X}_n},$$

where $A_1, \ldots, A_m = 1, 0$ and $\dot{X}_1, \ldots, \dot{X}_n = \dot{1}, \dot{0}$, and $\phi_{A_1\cdots A_m\dot{X}_1\cdots\dot{X}_n}$ is completely symmetric in $A_1, \ldots, A_m$ (i.e., $\phi_{A_1\cdots A_i\cdots A_j\cdots A_m\dot{X}_1\cdots\dot{X}_n} = \phi_{A_1\cdots A_j\cdots A_i\cdots A_m\dot{X}_1\cdots\dot{X}_n}$ for all i and j) and completely symmetric in the

$\dot{X}_1, \ldots, \dot{X}_n$. For example,

$$\begin{aligned} p_{22} z^2 \bar{z}^2 &= \phi_{110\cdots 0\dot{1}\dot{1}\dot{0}\cdots\dot{0}} \underbrace{z^1 z^1 z^0 \cdots z^0}_{m} \underbrace{\bar{z}^{\dot{1}} \bar{z}^{\dot{1}} \bar{z}^{\dot{0}} \cdots \bar{z}^{\dot{0}}}_{n} \\ &\quad + \phi_{1010\cdots 0\dot{1}\dot{0}\dot{1}\dot{0}\cdots\dot{0}} z^1 z^0 z^1 z^0 \cdots z^0 \bar{z}^{\dot{1}} \bar{z}^{\dot{0}} \bar{z}^{\dot{1}} \bar{z}^{\dot{0}} \cdots \bar{z}^{\dot{0}} \\ &\quad + \cdots + \phi_{00\cdots 011\dot{0}\dot{0}\cdots\dot{0}\dot{1}\dot{1}} z^0 z^0 \cdots z^0 z^1 z^1 \bar{z}^{\dot{0}} \bar{z}^{\dot{0}} \cdots \bar{z}^{\dot{0}} \bar{z}^{\dot{1}} \bar{z}^{\dot{1}}. \end{aligned}$$

There are $\binom{m}{2}\binom{n}{2}$ terms in the sum so we may take

$$\phi_{A_1 \cdots A_m \dot{X}_1 \cdots \dot{X}_n} = \frac{1}{\binom{m}{2}\binom{n}{2}} p_{22},$$

where $A_1 + \cdots + A_m = 2$ and $\dot{X}_1 + \cdots + \dot{X}_n = \dot{2}$. Similarly, for each $0 \le r \le m$ and $0 \le s \le n$, if $A_1, \ldots, A_m$ take the values 1 and 0 with $A_1 + \cdots + A_m = r$ and $\dot{X}_1, \ldots, \dot{X}_n$ take the values $\dot{1}$ and $\dot{0}$ with $\dot{X}_1 + \cdots + \dot{X}_n = \dot{s}$, we define

$$\phi_{A_1 \cdots A_m \dot{X}_1 \cdots \dot{X}_n} = \frac{1}{\binom{m}{r}\binom{n}{s}} p_{rs}.$$

Then

$$p_{rs} z^r \bar{z}^s = \sum_{\substack{A_1 + \cdots + A_m = r \\ \dot{X}_1 + \cdots + \dot{X}_n = \dot{s} \\ A_i = 1, 0 \\ \dot{X}_i = \dot{1}, \dot{0}}} \phi_{A_1 \cdots A_m \dot{X}_1 \cdots \dot{X}_n} z^{A_1} \cdots z^{A_m} \bar{z}^{\dot{X}_1} \cdots \bar{z}^{\dot{X}_n},$$

where there is *no sum* on the left. Summing over all $r = 0, \ldots, m$ and $s = 0, \ldots, n$ and using the summation convention on both sides gives

$$\begin{aligned} p(z, \bar{z}) = p_{rs} z^r \bar{z}^s &= \phi_{A_1 \cdots A_m \dot{X}_1 \cdots \dot{X}_n} z^{A_1} \cdots z^{A_m} \bar{z}^{\dot{X}_1} \cdots \bar{z}^{\dot{X}_n} \\ &= \phi_{A_1 \cdots A_m \dot{X}_1 \cdots \dot{X}_n} z^{A_1 + \cdots + A_m} \bar{z}^{\dot{X}_1 + \cdots + \dot{X}_n}. \end{aligned}$$

Again we observe that the ϕ's are symmetric in $A_1, \ldots, A_m$ and symmetric in $\dot{X}_1, \ldots, \dot{X}_n$. Applying the transformation $D_G^{(\frac{m}{2}, \frac{n}{2})}$ to $p(z, \bar{z})$ yields

$$\begin{aligned} &D_G^{(\frac{m}{2}, \frac{n}{2})} \left(\phi_{A_1 \cdots A_m \dot{X}_1 \cdots \dot{X}_n} z^{A_1 + \cdots + A_m} \bar{z}^{\dot{X}_1 + \cdots + \dot{X}_n} \right) \\ &\quad = \hat{\phi}_{A_1 \cdots A_m \dot{X}_1 \cdots \dot{X}_n} z^{A_1 + \cdots + A_m} \bar{z}^{\dot{X}_1 + \cdots + \dot{X}_n}, \end{aligned} \tag{3.1.19}$$

where

$$\hat{\phi}_{A_1 \cdots A_m \dot{X}_1 \cdots \dot{X}_n} = G_{A_1}{}^{B_1} \cdots G_{A_m}{}^{B_m} \bar{G}_{\dot{X}_1}{}^{\dot{Y}_1} \cdots \bar{G}_{\dot{X}_n}{}^{\dot{Y}_n} \phi_{B_1 \cdots B_m \dot{Y}_1 \cdots \dot{Y}_n}, \tag{3.1.20}$$

and the sum is over all $r = 0, \ldots, m$, $B_1 + \cdots + B_m = r$ with $B_1, \ldots, B_m = 1, 0$ and all $s = 0, \ldots, n$, $\dot{Y}_1 + \cdots + \dot{Y}_n = \dot{s}$ with $\dot{Y}_1, \ldots, \dot{Y}_n = \dot{1}, \dot{0}$.

The transformation law (3.1.20) is typical of a certain type of "spinor" [those of "valence" $\left(\begin{smallmatrix} 0 & 0 \\ m & n \end{smallmatrix}\right)$] that we will define in the next section. For the precise definition we wish to follow a procedure analogous to that employed in our definition of a world tensor. The idea there was that the underlying group being represented $(\mathcal{L})$ was the group of matrices of orthogonal transformations of $\mathcal{M}$ relative to orthonormal bases and that a world tensor could be identified with a multilinear functional on $\mathcal{M}$ and its dual. By analogy we would like to regard the elements of $SL(2, \mathbb{C})$ as matrices of the structure preserving maps of some "inner-product-like" space $\mathfrak{B}$ and identify "spinors" as multilinear functionals. This is, indeed, possible, although we will have to stretch our notion of "inner product" a bit.

Since the elements of $SL(2, \mathbb{C})$ are 2×2 complex matrices, the space $\mathfrak{B}$ we seek must be a 2-dimensional vector space over $\mathbb{C}$. Observe that if $\left[\begin{smallmatrix} \phi_1 \\ \phi_0 \end{smallmatrix}\right]$ and $\left[\begin{smallmatrix} \psi_1 \\ \psi_0 \end{smallmatrix}\right]$ are two ordered pairs of complex numbers and $G = [G_A{}^B]$ is in $SL(2, \mathbb{C})$ and if we define $\left[\begin{smallmatrix} \hat{\phi}_1 \\ \hat{\phi}_0 \end{smallmatrix}\right]$ and $\left[\begin{smallmatrix} \hat{\psi}_1 \\ \hat{\psi}_0 \end{smallmatrix}\right]$ by

$$\begin{bmatrix} \hat{\phi}_1 \\ \hat{\phi}_0 \end{bmatrix} = \begin{bmatrix} G_1{}^1 & G_1{}^0 \\ G_0{}^1 & G_0{}^0 \end{bmatrix} \begin{bmatrix} \phi_1 \\ \phi_0 \end{bmatrix} = \begin{bmatrix} G_1{}^1 \phi_1 + G_1{}^0 \phi_0 \\ G_0{}^1 \phi_1 + G_0{}^0 \phi_0 \end{bmatrix}$$

and similarly for $\left[\begin{smallmatrix} \hat{\psi}_1 \\ \hat{\psi}_0 \end{smallmatrix}\right]$, then

$$\begin{aligned} \begin{vmatrix} \hat{\phi}_1 & \hat{\psi}_1 \\ \hat{\phi}_0 & \hat{\psi}_0 \end{vmatrix} &= \left| \begin{bmatrix} G_1{}^1 & G_1{}^0 \\ G_0{}^1 & G_0{}^0 \end{bmatrix} \begin{bmatrix} \phi_1 & \psi_1 \\ \phi_0 & \psi_0 \end{bmatrix} \right| \\ &= \begin{vmatrix} G_1{}^1 & G_1{}^0 \\ G_0{}^1 & G_0{}^0 \end{vmatrix} \begin{vmatrix} \phi_1 & \psi_1 \\ \phi_0 & \psi_0 \end{vmatrix} \\ &= \begin{vmatrix} \phi_1 & \psi_1 \\ \phi_0 & \psi_0 \end{vmatrix} \end{aligned}$$

and so

$$\hat{\phi}_1 \hat{\psi}_0 - \hat{\phi}_0 \hat{\psi}_1 = \phi_1 \psi_0 - \phi_0 \psi_1. \tag{3.1.21}$$

Conversely, if (3.1.21) is satisfied, then G must be in $SL(2, \mathbb{C})$. Thus, if we define on the vector space

$$\mathbb{C}^2 = \left\{ \phi = \begin{bmatrix} \phi_1 \\ \phi_0 \end{bmatrix} : \phi_A \in \mathbb{C} \text{ for } A = 1, 0 \right\}$$

a mapping

$$<,> : \mathbb{C}^2 \times \mathbb{C}^2 \longrightarrow \mathbb{C}$$

by

$$< \phi, \psi > = \phi_1 \psi_0 - \phi_0 \psi_1,$$

then the elements of $SL(2, \mathbb{C})$ are precisely the matrices that preserve $< , >$.

Exercise 3.1.9. Verify the following properties of $< , >$.

1. $< , >$ is bilinear, i.e., $< \phi, a\psi + b\xi > = a < \phi, \psi > + b < \phi, \xi >$ and $< a\phi + b\psi, \xi > = a < \phi, \xi > + b < \psi, \xi >$ for all $a, b \in \mathbb{C}$ and $\phi, \psi, \xi \in \mathbb{C}^2$.
2. $< , >$ is skew-symmetric, i.e., $< \psi, \phi > = - < \phi, \psi >$.
3. $< \phi, \psi > \xi + < \xi, \phi > \psi + < \psi, \xi > \phi = 0$ for all $\phi, \psi, \xi \in \mathbb{C}^2$.

With these observations as motivation we proceed in the next section with an abstract definition of the underlying 2-dimensional complex vector space ß whose multilinear functionals are "spinors".

3.2 Spin Space

Spin space is a vector space ß over the complex numbers on which is defined a map $< , > : ß \times ß \to \mathbb{C}$ which satisfies:

1. there exist ϕ and ψ in ß such that $< \phi, \psi > \neq 0$,
2. $< \psi, \phi > = - < \phi, \psi >$ for all $\phi, \psi \in ß$,
3. $< a\phi + b\psi, \xi > = a < \phi, \xi > + b < \psi, \xi >$ for all $\phi, \psi, \xi \in ß$ and all $a, b \in \mathbb{C}$,
4. $< \phi, \psi > \xi + < \xi, \phi > \psi + < \psi, \xi > \phi = 0$ for all $\phi, \psi, \xi \in ß$.

An element of ß is called a *spin vector*. The existence of a vector space of the type described was established in Exercise 3.1.9.

Lemma 3.2.1. *Each of the following holds in spin space.*

(a) $< \phi, \phi > = 0$ *for every* $\phi \in ß$.

(b) $< , >$ *is bilinear, i.e., in addition to #3 in the definition we have* $< \phi, a\psi + b\xi > = a < \phi, \psi > + b < \phi, \xi >$ *for all* $\phi, \psi, \xi \in ß$ *and all* $a, b \in \mathbb{C}$.

(c) Any ϕ *and* ψ *in* ß *which satisfy* $< \phi, \psi > \neq 0$ *form a basis for* ß. *In particular,* $\dim ß = 2$.

(d) There exists a basis $\{s^1, s^0\}$ *for* ß *which satisfies* $< s^1, s^0 > = 1 = - < s^0, s^1 >$ *(any such basis is called a spin frame for* ß*).*

(e) *If* $\{s^1, s^0\}$ *is a spin frame and* $\phi = \phi_1 s^1 + \phi_0 s^0 = \phi_A s^A$, *then* $\phi_1 = <\phi, s^0>$ *and* $\phi_0 = -<\phi, s^1>$.

(f) *If* $\{s^1, s^0\}$ *is a spin frame and* $\phi = \phi_A s^A$ *and* $\psi = \psi_A s^A$, *then*

$$<\phi, \psi> = \begin{vmatrix} \phi_1 & \psi_1 \\ \phi_0 & \psi_0 \end{vmatrix} = \phi_1 \psi_0 - \phi_0 \psi_1.$$

(g) ϕ *and* ψ *in* $\mathfrak{B}$ *are linearly independent if and only if* $<\phi, \psi> \neq 0$.

(h) *If* $\{s^1, s^0\}$ *and* $\{\hat{s}^1, \hat{s}^0\}$ *are two spin frames with* $s^1 = G_1{}^1 \hat{s}^1 + G_0{}^1 \hat{s}^0 = G_A{}^1 \hat{s}^A$ *and* $s^0 = G_1{}^0 \hat{s}^1 + G_0{}^0 \hat{s}^0 = G_A{}^0 \hat{s}^A$, *i.e.,*

$$s^B = G_A{}^B \hat{s}^A, \quad B = 1, 0, \tag{3.2.1}$$

then $G = [G_A{}^B] = \begin{bmatrix} G_1{}^1 & G_1{}^0 \\ G_0{}^1 & G_0{}^0 \end{bmatrix}$ *is in* $SL(2, \mathbb{C})$.

(i) *If* $\{s^1, s^0\}$ *and* $\{\hat{s}^1, \hat{s}^0\}$ *are two spin frames and* $\phi = \phi_A s^A = \hat{\phi}_A \hat{s}^A$, *then*

$$\begin{bmatrix} \hat{\phi}_1 \\ \hat{\phi}_0 \end{bmatrix} = \begin{bmatrix} G_1{}^1 & G_1{}^0 \\ G_0{}^1 & G_0{}^0 \end{bmatrix} \begin{bmatrix} \phi_1 \\ \phi_0 \end{bmatrix},$$

i.e.,

$$\hat{\phi}_A = G_A{}^B \phi_B, \quad A = 1, 0, \tag{3.2.2}$$

where the $G_A{}^B$ *are given by (3.2.1).*

(j) *A linear transformation* $T: \mathfrak{B} \to \mathfrak{B}$ *preserves* $<\ ,\ >$ *(i.e., satisfies* $<T\phi, T\psi> = <\phi, \psi>$ *for all* $\phi, \psi \in \mathfrak{B}$*) if and only if the matrix of* T *relative to any spin frame is in* $SL(2, \mathbb{C})$.

Proof:

Exercise 3.2.1. Prove (a) and (b).

(c) From (a) and (b) it follows that $<\lambda\phi, \phi> = <\phi, \lambda\phi> = 0$ for all $\lambda \in \mathbb{C}$ and all $\phi \in \mathfrak{B}$. Consequently, if $<\phi, \psi> \neq 0$, neither ϕ nor ψ can be a multiple of the other, i.e., they are linearly independent. Moreover, for any $\xi \in \mathfrak{B}$, #4 gives $<\phi, \psi> \xi = -<\xi, \phi> \psi - <\psi, \xi> \phi$ so, since $<\phi, \psi> \neq 0$, ξ is a linear combination of ϕ and ψ so $\{\phi, \psi\}$ is a basis for $\mathfrak{B}$.

(d) Suppose $<\phi,\psi>\neq 0$. By switching names if necessary and using #2 we may assume $<\phi,\psi>>0$. By (c), ϕ and ψ form a basis for $\mathfrak{B}$ and therefore so do $s^1=<\phi,\psi>^{-\frac{1}{2}}\phi$ and $s^0=<\phi,\psi>^{-\frac{1}{2}}\psi$. But then bilinearity of $<\, ,\,>$ gives $<s^1,s^0>=1$ and so, by #2, $<s^0,s^1>=-1$.

(e) $\phi=\phi_1 s^1+\phi_0 s^0 \Rightarrow\ <\phi,s^0>=\phi_1<s^1,s^0>+\phi_0<s^0,s^0>=\phi_1$, and, similarly, $<\phi,s^1>=-\phi_0$.

(f) $<\phi,\psi>=<\phi_1 s^1+\phi_0 s^0,\psi_1 s^1+\psi_0 s^0>=\phi_1\psi_1<s^1,s^1>+\phi_1\psi_0<s^1,s^0>+\phi_0\psi_1<s^0,s^1>+\phi_0\psi_0<s^0,s^0>=\phi_1\psi_0-\phi_0\psi_1$.

(g) $<\phi,\psi>\neq 0$ implies ϕ and ψ linearly independent by (c). For the converse suppose $<\phi,\psi>=0$. If $\phi=0$ they are obviously dependent so assume $\phi\neq 0$. Select a spin frame $\{s^1,s^0\}$ and set $\phi=\phi_A s^A$ and $\psi=\psi_A s^A$. Suppose $\phi_1\neq 0$ (the proof is analogous if $\phi_0\neq 0$). By (f), $<\phi,\psi>=0$ implies $\phi_1\psi_0-\phi_0\psi_1=0$ so $\psi_0=(\phi_0/\phi_1)\psi_1$ and therefore

$$\begin{bmatrix}\psi_1\\ \psi_0\end{bmatrix}=\frac{\psi_1}{\phi_1}\begin{bmatrix}\phi_1\\ \phi_0\end{bmatrix},$$

so $\psi=(\psi_1/\phi_1)\phi$ and ϕ and ψ are linearly dependent.

(h) $<s^1,s^0>=1$ implies $1=<G_1{}^1\hat{s}^1+G_0{}^1\hat{s}^0,G_1{}^0\hat{s}^1+G_0{}^0\hat{s}^0>=G_1{}^1G_1{}^0<\hat{s}^1,\hat{s}^1>+G_1{}^1G_0{}^0<\hat{s}^1,\hat{s}^0>+G_0{}^1G_1{}^0<\hat{s}^0,\hat{s}^1>+G_0{}^1G_0{}^0<\hat{s}^0,\hat{s}^0>=G_1{}^1G_0{}^0-G_0{}^1G_1{}^0=\det G$ as required.

(i) $\hat{\phi}_1\hat{s}^1+\hat{\phi}_0\hat{s}^0=\phi_1 s^1+\phi_0 s^0=\phi_1(G_1{}^1\hat{s}^1+G_0{}^1\hat{s}^0)+\phi_0(G_1{}^0\hat{s}^1+G_0{}^0\hat{s}^0)=(G_1{}^1\phi_1+G_1{}^0\phi_0)\hat{s}^1+(G_0{}^1\phi_1+G_0{}^0\phi_0)\hat{s}^0$ so the result follows by equating components.

(j) Let $T:\mathfrak{B}\rightarrow\mathfrak{B}$ be a linear transformation and $\{s^1,s^0\}$ a spin frame. Let $[T_A{}^B]$ be the matrix of T relative to $\{s^1,s^0\}$. Then, for all ϕ and ψ in $\mathfrak{B}$, $T\phi=T_A{}^B\phi_B$, $T\psi=T_A{}^B\psi_B$ and

$$<\phi,\psi>=\begin{vmatrix}\phi_1 & \psi_1\\ \phi_0 & \psi_0\end{vmatrix}.$$

Now compute

$$<T\phi,T\psi>\ =\begin{vmatrix}T_1{}^1\phi_1+T_1{}^0\phi_0 & T_1{}^1\psi_1+T_1{}^0\psi_0\\ T_0{}^1\phi_1+T_0{}^0\phi_0 & T_0{}^1\psi_1+T_0{}^0\psi_0\end{vmatrix}$$

$$=\left|\begin{bmatrix}T_1{}^1 & T_1{}^0\\ T_0{}^1 & T_0{}^0\end{bmatrix}\begin{bmatrix}\phi_1 & \psi_1\\ \phi_0 & \psi_0\end{bmatrix}\right|$$

$$= \begin{vmatrix} T_1{}^1 & T_1{}^0 \\ T_0{}^1 & T_0{}^0 \end{vmatrix} \begin{vmatrix} \phi_1 & \psi_1 \\ \phi_0 & \psi_0 \end{vmatrix}.$$

Thus, $< T\phi, T\psi > = < \phi, \psi >$ if and only if $\det\,[T_A{}^B] = 1$, i.e., if and only if $[T_A{}^B] \in SL(2, \mathbb{C})$. ■

Comparing (3.2.2) and (3.1.16) we see that spin vectors are a natural choice as carriers for the identity representation $D^{(\frac{1}{2},0)}$ of $SL(2, \mathbb{C})$. To find an equally natural choice for the carrier space of the equivalent representation $\tilde{D}^{(\frac{1}{2},0)}$ we denote by $\mathfrak{B}^*$ the dual of the vector space $\mathfrak{B}$ and by $\{s_1, s_0\}$ the basis for $\mathfrak{B}^*$ dual to the spin frame $\{s^1, s^0\}$. Thus,

$$s_A(s^B) = \delta_A^B, \quad A, B = 1, 0. \tag{3.2.3}$$

The elements of $\mathfrak{B}^*$ are called *spin covectors*. For each $\phi \in \mathfrak{B}$ we define $\phi^* \in \mathfrak{B}^*$ by

$$\phi^*(\psi) = < \phi, \psi >$$

for every $\psi \in \mathfrak{B}$ [ϕ^* is linear by (b) of Lemma 3.2.1].

Lemma 3.2.2. *Every element of $\mathfrak{B}^*$ is ϕ^* for some $\phi \in \mathfrak{B}$.*

Proof: Let $f \in \mathfrak{B}^*$. Select a spin frame $\{s^1, s^0\}$ and define $\phi \in \mathfrak{B}$ by $\phi = f(s^0)s^1 - f(s^1)s^0$. Then, for every $\psi \in \mathfrak{B}$, $\phi^*(\psi) = < \phi, \psi > = < f(s^0)s^1 - f(s^1)s^0, \psi > = f(s^0) < s^1, \psi > - f(s^1) < s^0, \psi > = -f(s^0) < \psi, s^1 > + f(s^1) < \psi, s^0 > = f(s^1)\psi_1 + f(s^0)\psi_0$. But $f(\psi) = f(\psi_1 s^1 + \psi_0 s^0) = \psi_1 f(s^1) + \psi_0 f(s^0) = \phi^*(\psi)$ so $f = \phi^*$. ■

Now, for every $\phi \in \mathfrak{B}$, we may write $\phi = \phi_A s^A$ and $\phi^* = \phi^A s_A$ for some constants ϕ_A and ϕ^A, $A = 1, 0$. By (3.2.3), $\phi^*(s^1) = (\phi^A s_A)(s^1) = \phi^1 s_1(s^1) = \phi^1$ and, similarly, $\phi^*(s^0) = \phi^0$. On the other hand, $\phi^*(s^1) = < \phi, s^1 > = < \phi_1 s^1 + \phi_0 s^0, s^1 > = -\phi_0$ and, similarly, $\phi^*(s^0) = \phi_1$ so we find that

$$\begin{cases} \phi^1 = -\phi_0 \\ \phi^0 = \phi_1. \end{cases} \tag{3.2.4}$$

Now, if $\{\hat{s}^1, \hat{s}^0\}$ is another spin frame with $s^B = G_A{}^B \hat{s}^A$ as in (3.2.1), then, by (i) of Lemma 3.2.1, we have

$$\begin{bmatrix} \hat{\phi}_1 \\ \hat{\phi}_0 \end{bmatrix} = \begin{bmatrix} G_1{}^1 & G_1{}^0 \\ G_0{}^1 & G_0{}^0 \end{bmatrix} \begin{bmatrix} \phi_1 \\ \phi_0 \end{bmatrix}$$

for every $\phi = \phi_A s^A = \hat{\phi}_A \hat{s}^A$ in $\mathfrak{B}$. Letting

$$\begin{bmatrix} \mathcal{G}^1{}_1 & \mathcal{G}^1{}_0 \\ \mathcal{G}^0{}_1 & \mathcal{G}^0{}_0 \end{bmatrix} = \begin{bmatrix} G_0{}^0 & -G_0{}^1 \\ -G_1{}^0 & G_1{}^1 \end{bmatrix} = \left([G_A{}^B]^{-1} \right)^T,$$

we find that

$$\begin{bmatrix} \mathcal{G}^1{}_1 & \mathcal{G}^1{}_0 \\ \mathcal{G}^0{}_1 & \mathcal{G}^0{}_0 \end{bmatrix} \begin{bmatrix} \phi^1 \\ \phi^0 \end{bmatrix} = \begin{bmatrix} G_0{}^0 & -G_0{}^1 \\ -G_1{}^0 & G_1{}^1 \end{bmatrix} \begin{bmatrix} -\phi_0 \\ \phi_1 \end{bmatrix}$$

$$= \begin{bmatrix} -G_0{}^B\phi_B \\ G_1{}^B\phi_B \end{bmatrix} = \begin{bmatrix} -\hat{\phi}_0 \\ \hat{\phi}_1 \end{bmatrix}$$

$$= \begin{bmatrix} \hat{\phi}^1 \\ \hat{\phi}^0 \end{bmatrix},$$

so

$$\hat{\phi}^A = \mathcal{G}^A{}_B\phi^B, \qquad A = 1, 0. \tag{3.2.5}$$

Consequently, spin covectors have components relative to dual spin frames that transform under $\tilde{D}^{(\frac{1}{2},0)}$ so $\mathfrak{B}^*$ is a natural choice for a space of carriers of this representation of $SL(2, \mathbb{C})$.

Exercise 3.2.2. Verify that $\mathcal{G}^A{}_C G_B{}^C = G_C{}^A \mathcal{G}^C{}_B = \delta^A_B$ and show that

$$\hat{s}^A = \mathcal{G}^A{}_B s^B \tag{3.2.6}$$

and

$$\hat{s}_A = G_A{}^B s_B \tag{3.2.7}$$

and therefore

$$s_B = \mathcal{G}^A{}_B \hat{s}_A. \tag{3.2.8}$$

Exercise 3.2.3. For each $\phi \in \mathfrak{B}$ define $\phi^{**} : \mathfrak{B}^* \to \mathbb{C}$ by $\phi^{**}(f) = f(\phi)$ for each $f \in \mathfrak{B}^*$. Show that ϕ^{**} is a linear functional on $\mathfrak{B}^*$, i.e., $\phi^{**} \in (\mathfrak{B}^*)^*$, and that the map $\phi \to \phi^{**}$ is an isomorphism of $\mathfrak{B}$ onto $(\mathfrak{B}^*)^*$.

From Exercise 3.2.3 we conclude that, just as the elements of $\mathfrak{B}^*$ are linear functionals on $\mathfrak{B}$, so we can regard the elements of $\mathfrak{B}$ as linear functionals on $\mathfrak{B}^*$. Of course, the transformation law for components relative to a double dual basis for $(\mathfrak{B}^*)^*$ is the same as that for the spin frame it came from since one takes the transposed inverse twice. The point is that we now have carrier spaces for $D^{(\frac{1}{2},0)}$ and $\tilde{D}^{(\frac{1}{2},0)}$ that are both spaces of linear functionals (on $\mathfrak{B}^*$ and $\mathfrak{B}$, respectively).

Next consider a bilinear functional on, say, $\mathfrak{B}^* \times \mathfrak{B}^* : \xi : \mathfrak{B}^* \times \mathfrak{B}^* \to \mathbb{C}$. If $\{s^1, s^0\}$ is a spin frame and $\{s_1, s_0\}$ its dual, then for all $\phi^* = \phi^A s_A$ and $\psi^* = \psi^A s_A$ in $\mathfrak{B}^*$ we have $\xi(\phi^*, \psi^*) = \xi(\phi^A s_A, \psi^B s_B) = \xi(s_A, s_B)\phi^A \psi^B$. Letting $\xi_{AB} = \xi(s_A, s_B)$ we find that $\xi(\phi^*, \psi^*) = \xi_{AB}\phi^A\psi^B$. Now, if $\{\hat{s}^1, \hat{s}^0\}$ is another spin frame with dual $\{\hat{s}_1, \hat{s}_0\}$, then $\hat{\xi}_{AB} = \xi(\hat{s}_A, \hat{s}_B) = \xi(G_A{}^C s_C, G_B{}^D s_D) = G_A{}^C G_B{}^D \xi(s_C, s_D) = G_A{}^C G_B{}^D \xi_{CD}$ which we write as

$$\hat{\xi}_{A_1 A_2} = G_{A_1}{}^{B_1} G_{A_2}{}^{B_2} \xi_{B_1 B_2}, \quad A_1, A_2 = 1, 0, \tag{3.2.9}$$

and recognize as being the transformation law (3.1.20) with $m = 2$ and $n = 0$. Multilinear functionals on larger products $\mathfrak{B}^* \times \mathfrak{B}^* \times \cdots \times \mathfrak{B}^*$ will, in the same way, have components which transform according to (3.1.20) for larger m and $n = 0$ (we will consider nonzero n shortly). For a bilinear $\xi : \mathfrak{B} \times \mathfrak{B} \to \mathbb{C}$ we find that $\xi(\phi, \psi) = \xi(\phi_A s^A, \psi_B s^B) = \xi(s^A, s^B)\phi_A \psi_B = \xi^{AB}\phi_A \psi_B$ and, in another spin frame,

$$\hat{\xi}^{C_1 C_2} = \mathcal{G}^{C_1}{}_{D_1} \mathcal{G}^{C_2}{}_{D_2} \xi^{D_1 D_2}, \quad C_1, C_2 = 1, 0, \tag{3.2.10}$$

and similarly for larger products.

Exercise 3.2.4. Verify (3.2.10). Also show that if $\xi : \mathfrak{B}^* \times \mathfrak{B} \to \mathbb{C}$ is bilinear, $\{s^A\}$ and $\{\hat{s}^A\}$ are spin frames with duals $\{s_A\}$ and $\{\hat{s}_A\}$, $\xi_A{}^C = \xi(s_A, s^C)$ and $\hat{\xi}_A{}^C = \xi(\hat{s}_A, \hat{s}^C)$, then, for any $\phi = \phi_A s^A = \hat{\phi}_A \hat{s}^A \in \mathfrak{B}$ and $\psi^* = \psi^A s_A = \hat{\psi}^A \hat{s}_A \in \mathfrak{B}^*$ we have $\xi(\psi^*, \phi) = \xi_A{}^C \psi^A \phi_C = \hat{\xi}_A{}^C \hat{\psi}^A \hat{\phi}_C$ and

$$\hat{\xi}_{A_1}{}^{C_1} = G_{A_1}{}^{B_1} \mathcal{G}^{C_1}{}_{D_1} \xi_{B_1}{}^{D_1}, \quad A_1, C_1 = 1, 0. \tag{3.2.11}$$

All of this will be generalized shortly in our definition of a "spinor", but first we must construct carriers for $D^{(0,\frac{1}{2})}$ and $\tilde{D}^{(0,\frac{1}{2})}$. For this we shall require a copy $\bar{\mathfrak{B}}$ of $\mathfrak{B}$ that is distinct from $\mathfrak{B}$. For example, we might take $\bar{\mathfrak{B}} = \mathfrak{B} \times \{1\}$ so that each element of $\bar{\mathfrak{B}}$ is of the form $(\phi, 1)$ for some $\phi \in \mathfrak{B}$. We denote by $\bar{\phi}$ the element $(\phi, 1) \in \bar{\mathfrak{B}}$. Thus,

$$\bar{\mathfrak{B}} = \{\bar{\phi} : \phi \in \mathfrak{B}\}.$$

We define the linear space structure on $\bar{\mathfrak{B}}$ as follows: For $\bar{\phi}, \bar{\psi}$ and $\bar{\xi}$ in $\bar{\mathfrak{B}}$ and $c \in \mathbb{C}$ we have

$$\bar{\phi} + \bar{\psi} = \overline{\phi + \psi}$$

and

$$c\bar{\phi} = \overline{\bar{c}\phi}$$

(this last being equivalent to $\bar{\lambda}\bar{\phi} = \overline{\lambda\phi}$, where $\bar{c}$ and $\bar{\lambda}$ are the usual conjugates of the complex numbers c and λ). Thus, the map $\phi \to \bar{\phi}$ of $\mathfrak{B}$

to $\bar{\mathfrak{B}}$, which is obviously bijective, is a *conjugate* (or *anti-*) *isomorphism*, i.e., satisfies

$$\phi + \psi \longrightarrow \bar{\phi} + \bar{\psi}$$

and

$$c\phi \longrightarrow \bar{c}\bar{\phi}.$$

The elements of $\bar{\mathfrak{B}}$ are called *conjugate spin vectors.*

Let $\{s^1, s^0\}$ be a spin frame in $\mathfrak{B}$ and denote by $\bar{s}^{\dot{1}}$ and $\bar{s}^{\dot{0}}$ the images of s^1 and s^0 respectively under $\phi \to \bar{\phi}$. Then $\{\bar{s}^{\dot{1}}, \bar{s}^{\dot{0}}\}$ is a basis for $\bar{\mathfrak{B}}$. Moreover, if $\phi = \phi_1 s^1 + \phi_0 s^0$ is in $\mathfrak{B}$, then $\bar{\phi} = \bar{\phi}_{\dot{1}}\bar{s}^{\dot{1}} + \bar{\phi}_{\dot{0}}\bar{s}^{\dot{0}}$ (recall our notational conventions from Section 3.1 concerning dotted indices, bars, etc.). Now, if $\{\hat{s}^1, \hat{s}^0\}$ is another spin frame, related to $\{s^A\}$ by (3.2.6), and $\{\bar{\hat{s}}^{\dot{1}}, \bar{\hat{s}}^{\dot{0}}\}$ is its image under $\phi \to \bar{\phi}$, then $\hat{s}^1 = \mathcal{G}^1{}_B s^B = \mathcal{G}^1{}_1 s^1 + \mathcal{G}^1{}_0 s^0$ implies $\bar{\hat{s}}^{\dot{1}} = \bar{\mathcal{G}}^{\dot{1}}{}_{\dot{1}}\bar{s}^{\dot{1}} + \bar{\mathcal{G}}^{\dot{1}}{}_{\dot{0}}\bar{s}^{\dot{0}}$ and similarly for $\bar{\hat{s}}^{\dot{0}}$ so

$$\bar{\hat{s}}^{\dot{X}} = \bar{\mathcal{G}}^{\dot{X}}{}_{\dot{Y}}\bar{s}^{\dot{Y}}, \quad \dot{X} = \dot{1}, \dot{0}, \tag{3.2.12}$$

and so

$$\bar{s}^{\dot{Y}} = \bar{G}^{\dot{Y}}{}_{\dot{X}}\bar{\hat{s}}^{\dot{X}}, \quad \dot{Y} = \dot{1}, \dot{0}. \tag{3.2.13}$$

It follows that if $\bar{\phi} = \bar{\phi}_{\dot{Y}}\bar{s}^{\dot{Y}} = \bar{\hat{\phi}}_{\dot{X}}\bar{\hat{s}}^{\dot{X}}$, then

$$\bar{\hat{\phi}}_{\dot{X}} = \bar{G}^{\dot{Y}}{}_{\dot{X}}\bar{\phi}_{\dot{Y}}, \quad \dot{X} = \dot{1}, \dot{0}, \tag{3.2.14}$$

and

$$\bar{\phi}_{\dot{Y}} = \bar{\mathcal{G}}^{\dot{X}}{}_{\dot{Y}}\bar{\hat{\phi}}_{\dot{X}}, \quad \dot{Y} = \dot{1}, \dot{0}. \tag{3.2.15}$$

The elements of the dual $\bar{\mathfrak{B}}^*$ of $\bar{\mathfrak{B}}$ are called *conjugate spin covectors* and the bases dual to $\{\bar{s}^{\dot{X}}\}$ and $\{\bar{\hat{s}}^{\dot{X}}\}$ are denoted $\{\bar{s}_{\dot{X}}\}$ and $\{\bar{\hat{s}}_{\dot{X}}\}$ respectively. Just as before we have

$$\bar{\hat{s}}_{\dot{X}} = \bar{G}^{\dot{Y}}{}_{\dot{X}}\bar{s}_{\dot{Y}}, \quad \dot{X} = \dot{1}, \dot{0}, \tag{3.2.16}$$

and

$$\bar{s}_{\dot{Y}} = \bar{\mathcal{G}}^{\dot{X}}{}_{\dot{Y}}\bar{\hat{s}}_{\dot{X}}, \quad \dot{Y} = \dot{1}, \dot{0}. \tag{3.2.17}$$

For each $\phi^* = \phi^A s_A = \hat{\phi}^A \hat{s}_A \in \mathfrak{B}^*$ we define $\bar{\phi}^* \in \bar{\mathfrak{B}}^*$ by $\bar{\phi}^* = \bar{\phi}^{\dot{X}} \bar{s}_{\dot{X}} = \bar{\hat{\phi}}^{\dot{X}} \bar{\hat{s}}_{\dot{X}}$. Then

$$\bar{\hat{\phi}}^{\dot{X}} = \bar{\mathcal{G}}^{\dot{X}}{}_{\dot{Y}} \bar{\phi}^{\dot{Y}}, \quad \dot{X} = \dot{1}, \dot{0}, \tag{3.2.18}$$

and

$$\bar{\phi}^{\dot{Y}} = \bar{G}_{\dot{X}}{}^{\dot{Y}} \bar{\hat{\phi}}^{\dot{X}}, \quad \dot{Y} = \dot{1}, \dot{0}. \tag{3.2.19}$$

Before giving the general definitions we once again illustrate with a specific example. Thus, consider a multilinear functional $\xi \colon \mathfrak{B} \times \bar{\mathfrak{B}} \times \mathfrak{B}^* \times \bar{\mathfrak{B}}^* \to \mathbb{C}$. If $\phi = \phi_A s^A$, $\bar{\psi} = \bar{\psi}_{\dot{X}} \bar{s}^{\dot{X}}$, $\zeta = \zeta^B s_B$ and $\bar{\nu} = \bar{\nu}^{\dot{Y}} \bar{s}_{\dot{Y}}$ are in $\mathfrak{B}$, $\bar{\mathfrak{B}}$, $\mathfrak{B}^*$ and $\bar{\mathfrak{B}}^*$ respectively, then

$$\begin{aligned}
\xi(\phi, \bar{\psi}, \zeta, \bar{\nu}) &= \xi\left(\phi_A s^A, \bar{\psi}_{\dot{X}} \bar{s}^{\dot{X}}, \zeta^B s_B, \bar{\nu}^{\dot{Y}} \bar{s}_{\dot{Y}}\right) \\
&= \xi\left(s^A, \bar{s}^{\dot{X}}, s_B, \bar{s}_{\dot{Y}}\right) \phi_A \bar{\psi}_{\dot{X}} \zeta^B \bar{\nu}^{\dot{Y}} \\
&= \xi^{A\dot{X}}{}_{B\dot{Y}} \phi_A \bar{\psi}_{\dot{X}} \zeta^B \bar{\nu}^{\dot{Y}},
\end{aligned}$$

where $\xi^{A\dot{X}}{}_{B\dot{Y}} = \xi(s^A, \bar{s}^{\dot{X}}, s_B, \bar{s}_{\dot{Y}})$ are the components of ξ relative to the spin frame $\{s^1, s^0\}$ (and the related bases for $\bar{\mathfrak{B}}$, $\mathfrak{B}^*$ and $\bar{\mathfrak{B}}^*$). In another spin frame $\{\hat{s}^1, \hat{s}^0\}$ we have $\xi(\phi, \bar{\psi}, \zeta, \bar{\nu}) = \hat{\xi}^{A\dot{X}}{}_{B\dot{Y}} \hat{\phi}_A \bar{\hat{\psi}}_{\dot{X}} \hat{\zeta}^B \bar{\hat{\nu}}^{\dot{Y}}$, where

$$\begin{aligned}
\hat{\xi}^{A\dot{X}}{}_{B\dot{Y}} &= \xi\left(\hat{s}^A, \bar{\hat{s}}^{\dot{X}}, \hat{s}_B, \bar{\hat{s}}_{\dot{Y}}\right) \\
&= \xi\left(\mathcal{G}^A{}_{A_1} s^{A_1}, \bar{\mathcal{G}}^{\dot{X}}{}_{\dot{X}_1} \bar{s}^{\dot{X}_1}, G_B{}^{B_1} s_{B_1}, \bar{G}_{\dot{Y}}{}^{\dot{Y}_1} \bar{s}_{\dot{Y}_1}\right) \\
&= \mathcal{G}^A{}_{A_1} \bar{\mathcal{G}}^{\dot{X}}{}_{\dot{X}_1} G_B{}^{B_1} \bar{G}_{\dot{Y}}{}^{\dot{Y}_1} \xi^{A_1 \dot{X}_1}{}_{B_1 \dot{Y}_1}
\end{aligned}$$

which, as we shall see, is the transformation law for the components relative to a spin frame of a "spinor of valence $\left(\begin{smallmatrix} 1 & 1 \\ 1 & 1 \end{smallmatrix}\right)$". With this we are finally prepared to present the general definitions.

A *spinor* of *valence* $\left(\begin{smallmatrix} r & s \\ m & n \end{smallmatrix}\right)$, also called a spinor with m *undotted lower indices, n dotted lower indices, r undotted upper indices,* and s *dotted upper indices* is a multilinear functional

$$\xi : \underbrace{\mathfrak{B} \times \cdots \times \mathfrak{B}}_{r \text{ factors}} \times \underbrace{\bar{\mathfrak{B}} \times \cdots \times \bar{\mathfrak{B}}}_{s \text{ factors}} \times \underbrace{\mathfrak{B}^* \times \cdots \times \mathfrak{B}^*}_{m \text{ factors}} \times \underbrace{\bar{\mathfrak{B}}^* \times \cdots \times \bar{\mathfrak{B}}^*}_{n \text{ factors}} \longrightarrow \mathbb{C}.$$

If $\{s^1, s^0\}$ is a spin frame (with associated bases $\{\bar{s}^{\dot{1}}, \bar{s}^{\dot{0}}\}$, $\{s_1, s_0\}$ and $\{\bar{s}_{\dot{1}}, \bar{s}_{\dot{0}}\}$ for $\bar{\mathfrak{B}}$, $\mathfrak{B}^*$ and $\bar{\mathfrak{B}}^*$), then the *components of ξ relative to* $\{s^A\}$

are defined by

$$\xi^{A_1\cdots A_r\dot{X}_1\cdots\dot{X}_s}{}_{B_1\cdots B_m\dot{Y}_1\cdots\dot{Y}_n} = \xi\left(s^{A_1},\ldots,s^{A_r},\bar{s}^{\dot{X}_1},\ldots,\bar{s}^{\dot{X}_s},s_{B_1},\ldots,\right.$$
$$\left. s_{B_m},\bar{s}_{\dot{Y}_1},\ldots,\bar{s}_{\dot{Y}_n}\right), \tag{3.2.20}$$
$$A_1,\ldots,A_r,B_1,\ldots,B_m = 1,0,$$
$$\dot{X}_1,\ldots,\dot{X}_s,\dot{Y}_1,\ldots,\dot{Y}_n = \dot{1},\dot{0}.$$

Exercise 3.2.5. Show that, if $\{\hat{s}^1,\hat{s}^0\}$ is another spin frame, then

$$\hat{\xi}^{A_1\cdots A_r\dot{X}_1\cdots\dot{X}_s}{}_{B_1\cdots B_m\dot{Y}_1\cdots\dot{Y}_n} = \mathcal{G}^{A_1}{}_{C_1}\cdots\mathcal{G}^{A_r}{}_{C_r}\bar{\mathcal{G}}^{\dot{X}_1}{}_{\dot{U}_1}\cdots\bar{\mathcal{G}}^{\dot{X}_s}{}_{\dot{U}_s}G_{B_1}{}^{D_1}\cdots$$
$$G_{B_m}{}^{D_m}\bar{G}_{\dot{Y}_1}{}^{\dot{V}_1}\cdots\bar{G}_{\dot{Y}_n}{}^{\dot{V}_n}\xi^{C_1\cdots C_r\dot{U}_1\cdots\dot{U}_s}{}_{D_1\cdots D_m\dot{V}_1\cdots\dot{V}_n}. \tag{3.2.21}$$

It is traditional, particularly in the physics literature, to define a "spinor with r contravariant and m covariant undotted indices and s contravariant and n covariant dotted indices" to be an assignment of $2^{r+m+s+n}$ complex numbers $\{\xi^{A_1\cdots A_r\dot{X}_1\cdots\dot{X}_s}{}_{B_1\cdots B_m\dot{Y}_1\cdots\dot{Y}_n}\}$ to each spin frame (or, rather, an assignment of two such sets of numbers $\{\pm\,\xi^{A_1\cdots A_r\dot{X}_1\cdots\dot{X}_s}{}_{B_1\cdots B_m\dot{Y}_1\cdots\dot{Y}_n}\}$ to each admissible basis for $\mathcal{M}$) which transform according to (3.2.21) under a change of basis. Although our approach is more in keeping with the "coordinate-free" fashion that is currently in vogue, most calculations are, in fact, performed in terms of components and the transformation law (3.2.21). Observe also that, when $r = s = 0$, (3.2.21) coincides with the transformation law (3.2.20) for the carriers of the representation $D^{(\frac{m}{2},\frac{n}{2})}$ of $SL(2,\mathbb{C})$. There is a difference, however, in that the $\phi_{A_1\cdots A_m\dot{X}_1\cdots\dot{X}_n}$ constructed in Section 3.1 are symmetric in $A_1,\ldots,A_m$ and symmetric in $\dot{X}_1,\ldots,\dot{X}_n$ and no such symmetry assumption is made in the definition of a spinor of valence $\left(\begin{smallmatrix}0 & 0\\ m & n\end{smallmatrix}\right)$. The representations of $SL(2,\mathbb{C})$ corresponding to the transformation laws (3.2.21) will, in general, be *reducible,* unlike the irreducible spinor representations of Section 3.1. One final remark on the ordering of indices is apropos. The position of an index in (3.2.20) indicates the "slot" in ξ into which the corresponding basis element is to be inserted for evaluation. For two indices of the same type (both upper and undotted, both lower and dotted, etc.) the order in which the indices appear is crucial since, for example, there is no reason to suppose that $\xi(s^1,s^0,\ldots)$ and $\xi(s^0,s^1,\ldots)$ are the same. However, since slots corresponding to different types of indices accept different sorts of objects (e.g., spin vectors and conjugate spin covectors) there is no reason to insist upon any relative ordering of different types of indices and we shall not do so. Thus, for example, $\xi^{A_1A_2\dot{X}_1}{}_{B_1} = \xi^{A_1\dot{X}_1A_2}{}_{B_1} = \xi^{A_1}{}_{B_1}{}^{A_2\dot{X}_1}$ etc., but these need not be the same as $\xi^{A_2A_1\dot{X}_1}{}_{B_1}$, etc.

3.3 Spinor Algebra

In this section we collect together the basic algebraic and computational tools that will be used in the remainder of the chapter. We begin by introducing a matrix that will figure prominantly in many of the calculations that are before us. Thus, we let ϵ denote the 2×2 matrix defined by

$$\epsilon = \begin{bmatrix} 0 & -1 \\ 1 & 0 \end{bmatrix} = \begin{bmatrix} \epsilon_{11} & \epsilon_{10} \\ \epsilon_{01} & \epsilon_{00} \end{bmatrix} = [\epsilon_{AB}] .$$

Depending on the context and the requirements of the summation convention we will also denote the entries of ϵ in any of the following ways:

$$\epsilon = [\epsilon_{AB}] = [\epsilon^{AB}] = [\bar{\epsilon}_{\dot{X}\dot{Y}}] = [\bar{\epsilon}^{\dot{X}\dot{Y}}] .$$

Observe that $\epsilon^{-1} = -\epsilon$. Moreover, if ϕ and ψ are two spin vectors and $\{s^1, s^0\}$ is a spin frame with $\phi = \phi_A s^A$ and $\psi = \psi_B s^B$, then, with the summation convention, $\epsilon^{AB}\psi_A\phi_B = \epsilon^{10}\psi_1\phi_0 + \epsilon^{01}\psi_0\phi_1 = \phi_1\psi_0 - \phi_0\psi_1 = < \phi, \psi >$. Also let $\phi^* = \phi^A s_A$ and $\psi^* = \psi^A s_A$ be the corresponding spin covectors.

Exercise 3.3.1. Verify each of the following:

$$< \phi, \psi > \; = \; \epsilon^{AB}\psi_A\phi_B = -\epsilon^{AB}\phi_A\psi_B, \tag{3.3.1}$$

$$\phi^A \; = \; \epsilon^{AB}\phi_B = -\phi_B\epsilon^{BA}, \tag{3.3.2}$$

$$\phi_A \; = \; \phi^B\epsilon_{BA} = -\epsilon_{AB}\phi^B, \tag{3.3.3}$$

$$\phi^A\psi_A \; = \; < \phi, \psi > = -\phi_A\psi^A, \tag{3.3.4}$$

$$\epsilon^{AC}\epsilon_{BC} \; = \; \delta^A_B = \epsilon^{CA}\epsilon_{CB}, \tag{3.3.5}$$

$$\left(\epsilon^{CB}\phi_B\right)\epsilon_{CA} \; = \; \phi_A \text{ and } \epsilon^{AC}\left(\phi^B\epsilon_{BC}\right) = \phi^A, \tag{3.3.6}$$

$$\epsilon^{AB}\epsilon_{AB} \; = \; 2 = \epsilon_{AB}\epsilon^{AB}. \tag{3.3.7}$$

Of course, each of the identities (3.3.1)-(3.3.7) has an obvious "barred and dotted" version, e.g., (3.3.6) would read $(\bar{\epsilon}^{\dot{Z}\dot{Y}}\bar{\phi}_{\dot{Y}})\bar{\epsilon}_{\dot{Z}\dot{X}} = \bar{\phi}_{\dot{X}}$. In addition to these we record several more identities that will be used repeatedly in the sequel.

$$\epsilon_{AB}\epsilon_{CD} + \epsilon_{AC}\epsilon_{DB} + \epsilon_{AD}\epsilon_{BC} \; = \; 0, \quad A, B, C, D = 1, 0. \tag{3.3.8}$$

To prove (3.3.8) we suppose first that $A=1$. Thus, we consider $\epsilon_{1B}\epsilon_{CD}+\epsilon_{1C}\epsilon_{DB}+\epsilon_{1D}\epsilon_{BC}$. If $B=1$ this becomes $\epsilon_{1C}\epsilon_{D1}+\epsilon_{1D}\epsilon_{1C}$. $C=1$ or $D=1$ gives 0 for both terms. For $C=0$ and $D=0$ we obtain $\epsilon_{10}\epsilon_{01}+\epsilon_{10}\epsilon_{10}=(-1)(1)+(-1)(-1)=0$. On the other hand, if $B=0$ we have $\epsilon_{10}\epsilon_{CD}+\epsilon_{1C}\epsilon_{D0}+\epsilon_{1D}\epsilon_{0C}$. $C=D$ gives 0 for each term. For $C=0$ and $D=1$ we obtain $\epsilon_{10}\epsilon_{01}+\epsilon_{10}\epsilon_{10}+\epsilon_{11}\epsilon_{00}=(-1)(1)+(-1)(-1)+0=0$. If $C=1$ and $D=0$, $\epsilon_{10}\epsilon_{10}+\epsilon_{11}\epsilon_{00}+\epsilon_{10}\epsilon_{01}=(-1)(-1)+0+(-1)(1)=0$. Thus, (3.3.8) is proved if $A=1$ and the argument is the same if $A=0$. Next we show that if $G=[G_A{}^B]\in SL(2,\mathbb{C})$, then

$$G_A{}^{A_1}G_B{}^{B_1}\epsilon_{A_1B_1}=\epsilon_{AB},\quad A,B=1,0. \tag{3.3.9}$$

This follows from

$$\begin{aligned}
G_A{}^{A_1}G_B{}^{B_1}\epsilon_{A_1B_1} &= G_A{}^1G_B{}^0\epsilon_{10}+G_A{}^0G_B{}^1\epsilon_{01}\\
&= G_A{}^0G_B{}^1-G_A{}^1G_B{}^0\\
&= \begin{cases} 0, & \text{if } A=B\\ \det G, & \text{if } A=0,\ B=1\\ -\det G, & \text{if } A=1,\ B=0\end{cases}\\
&= \epsilon_{AB}\,(\det G)\\
&= \epsilon_{AB}
\end{aligned}$$

since $\det G=1$. Similarly, if $\mathcal{G}=[\mathcal{G}^A{}_B]=([G_A{}^B]^{-1})^T$,

$$\mathcal{G}^A{}_{A_1}\mathcal{G}^B{}_{B_1}\epsilon^{A_1B_1}=\epsilon^{AB},\quad A,B=1,0, \tag{3.3.10}$$

and both (3.3.9) and (3.3.10) have barred and dotted versions. Observe that the bilinear form $<\ ,\ >:\ \mathfrak{B}\times\mathfrak{B}\to\mathbb{C}$ is, according to our definitions in Section 3.2, a spinor of valence $\left(\begin{smallmatrix}2&0\\0&0\end{smallmatrix}\right)$ and has components in *any* spin frame given by $<s^A,s^B>=-\epsilon^{AB}$ and that (3.2.10), with $\hat{\epsilon}^{AB}=\epsilon^{AB}$, simply confirms the appropriate transformation law. In the same way, (3.3.9) asserts that the ϵ_{AB} can be regarded as the (constant) components of a spinor of valence $\left(\begin{smallmatrix}0&0\\2&0\end{smallmatrix}\right)$, whereas the barred and dotted versions of these make similar assertions about the $\bar{\epsilon}^{\dot{X}\dot{Y}}$ and $\bar{\epsilon}_{\dot{X}\dot{Y}}$.

Exercise 3.3.2. Write out explicitly the bilinear forms (spinors) whose components relative to every spin frame are ϵ_{AB}, $\bar{\epsilon}^{\dot{X}\dot{Y}}$ and $\bar{\epsilon}_{\dot{X}\dot{Y}}$.

The first equality in (3.3.2) asserts that, given a spin vector ϕ and the corresponding spin covector ϕ^*, then, relative to any spin frame, the components of $\phi^*=\phi^A s_A$ are mechanically retrievable from those of

$\phi = \phi_B s^B$ by forming the sum $\epsilon^{AB}\phi_B$. This process is called *raising the index* of ϕ_B. Similarly, obtaining the ϕ_A from the ϕ^B according to (3.3.3) by computing $\phi^B \epsilon_{BA}$ is termed *lowering the index* of ϕ^B. Due to the skew-symmetry of the ϵ_{AB} care must be exercised in arranging the order of the factors and the placement of the indices when carrying out these processes. As an aid to the memory, one "raises on the left and lowers on the right" with the summed indices "adjacent and descending to the right". The equalities in (3.3.6) assert that these two operations are consistent, i.e., that lowering a raised index or vice versa returns the original component. The operations of raising and lowering indices extend easily to higher valence spinors. Consider, for example, a spinor ξ of valence $\left(\begin{smallmatrix} 2 & 0 \\ 1 & 0 \end{smallmatrix}\right)$. In each spin frame ξ has components $\xi^{AB}{}_C$ and we now define numbers $\xi_A{}^B{}_C$ in this frame by

$$\xi_A{}^B{}_C = \xi^{A_1 B}{}_C \, \epsilon_{A_1 A}.$$

In another spin frame we have $\hat{\xi}_A{}^B{}_C = \hat{\xi}^{A_1 B}{}_C \epsilon_{A_1 A}$ and we now show that

$$\hat{\xi}_A{}^B{}_C = G_A{}^{A_1} \mathcal{G}^B{}_{B_1} G_C{}^{C_1} \xi_{A_1}{}^{B_1}{}_{C_1}, \tag{3.3.11}$$

so that the $\xi_A{}^B{}_C$ transform as the components of a spinor of valence $\left(\begin{smallmatrix} 1 & 0 \\ 2 & 0 \end{smallmatrix}\right)$. This last spinor we shall say is obtained from ξ by "lowering the first (undotted) upper index". To prove (3.3.11) we use (3.3.9) and the transformation law for the $\xi^{AB}{}_C$ as follows:

$$\begin{aligned}
\hat{\xi}_A{}^B{}_C &= \hat{\xi}^{A_1 B}{}_C \epsilon_{A_1 A} = \left(\mathcal{G}^{A_1}{}_{A_2} \mathcal{G}^B{}_{B_1} G_C{}^{C_1} \xi^{A_2 B_1}{}_{C_1}\right)\left(G_{A_1}{}^{A_3} G_A{}^{A_4} \epsilon_{A_3 A_4}\right) \\
&= \left(\mathcal{G}^{A_1}{}_{A_2} G_{A_1}{}^{A_3}\right)\left(\mathcal{G}^B{}_{B_1} G_C{}^{C_1} G_A{}^{A_4}\right)\left(\xi^{A_2 B_1}{}_{C_1} \epsilon_{A_3 A_4}\right) \\
&= \delta^{A_3}_{A_2} \left(\mathcal{G}^B{}_{B_1} G_C{}^{C_1} G_A{}^{A_4}\right)\left(\xi^{A_2 B_1}{}_{C_1} \epsilon_{A_3 A_4}\right) \\
&= \left(\mathcal{G}^B{}_{B_1} G_C{}^{C_1} G_A{}^{A_4}\right)\left(\xi^{A_3 B_1}{}_{C_1} \epsilon_{A_3 A_4}\right) \\
&= \mathcal{G}^B{}_{B_1} G_C{}^{C_1} G_A{}^{A_4} \xi_{A_4}{}^{B_1}{}_{C_1} \\
&= G_A{}^{A_4} \mathcal{G}^B{}_{B_1} G_C{}^{C_1} \xi_{A_4}{}^{B_1}{}_{C_1} \\
&= G_A{}^{A_1} \mathcal{G}^B{}_{B_1} G_C{}^{C_1} \xi_{A_1}{}^{B_1}{}_{C_1}
\end{aligned}$$

as required.

Exercise 3.3.3. With ξ as above, let $\xi^{ABC} = \epsilon^{CC_1} \xi^{AB}{}_{C_1}$ in each spin frame. Show that $\hat{\xi}^{ABC} = \mathcal{G}^A{}_{A_1} \mathcal{G}^B{}_{B_1} \mathcal{G}^C{}_{C_1} \xi^{A_1 B_1 C_1}$ and conclude that the ξ^{ABC} determine a spinor of valence $\left(\begin{smallmatrix} 3 & 0 \\ 0 & 0 \end{smallmatrix}\right)$.

The calculations in these last examples make it clear that a spinor of any valence can have any one of its lower (upper) indices raised (lowered) to yield a spinor with one more upper (lower) index. Applying this to the constant spinors ϵ_{AB} and ϵ^{AB} and using (3.3.5) yields the following useful identities.

$$\epsilon_A{}^B = \epsilon^{BC}\epsilon_{AC} = \delta_A^B \qquad (3.3.12)$$

and

$$\epsilon^A{}_B = \epsilon^{AC}\epsilon_{CB} = -\delta_B^A. \qquad (3.3.13)$$

We derive a somewhat less obvious identity by beginning with (3.3.8) and first raising C.

$$\begin{aligned}
\epsilon_{AB}\epsilon_{CD} + \epsilon_{AC}\epsilon_{DB} + \epsilon_{AD}\epsilon_{BC} &= 0, \\
\epsilon_{AB}\left(\epsilon^{CE}\epsilon_{ED}\right) + \left(\epsilon^{CE}\epsilon_{AE}\right)\epsilon_{DB} + \epsilon_{AD}\left(\epsilon^{CE}\epsilon_{BE}\right) &= 0, \\
\epsilon_{AB}\epsilon^C{}_D + \epsilon_A{}^C\epsilon_{DB} + \epsilon_{AD}\epsilon_B{}^C &= 0, \\
-\epsilon_{AB}\delta_D^C + \delta_A^C\epsilon_{DB} + \epsilon_{AD}\delta_B^C &= 0.
\end{aligned}$$

Now, raise D.

$$\begin{aligned}
-\epsilon_{AB}\left(\epsilon^{DE}\delta_E^C\right) + \delta_A^C\left(\epsilon^{DE}\epsilon_{EB}\right) + \left(\epsilon^{DE}\epsilon_{AE}\right)\delta_B^C &= 0, \\
-\epsilon_{AB}\epsilon^{DC} + \delta_A^C\epsilon^D{}_B + \epsilon_A{}^D\delta_B^C &= 0, \\
-\epsilon_{AB}\epsilon^{DC} - \delta_A^C\delta_B^D + \delta_A^D\delta_B^C &= 0.
\end{aligned}$$

Using $\epsilon^{DC} = -\epsilon^{CD}$ we finally obtain

$$\epsilon_{AB}\epsilon^{CD} = \delta_A^C\delta_B^D - \delta_A^D\delta_B^C, \quad A,B,C,D = 1,0. \qquad (3.3.14)$$

It will also be useful to introduce, for each $[G_A{}^B] = \begin{bmatrix} G_1{}^1 & G_1{}^0 \\ G_0{}^1 & G_0{}^0 \end{bmatrix}$ in $SL(2,\mathbb{C})$, an associated matrix $[G^A{}_B] = \begin{bmatrix} G^1{}_1 & G^0{}_1 \\ G^1{}_0 & G^0{}_0 \end{bmatrix}$, where

$$G^A{}_B = \epsilon^{AA_1}G_{A_1}{}^{B_1}\epsilon_{B_1B}, \quad A,B = 1,0.$$

Exercise 3.3.4. Show that

$$\begin{bmatrix} G^1{}_1 & G^0{}_1 \\ G^1{}_0 & G^0{}_0 \end{bmatrix} = \begin{bmatrix} -G_0{}^0 & G_1{}^0 \\ G_0{}^1 & -G_1{}^1 \end{bmatrix} = -\begin{bmatrix} \mathcal{G}^1{}_1 & \mathcal{G}^0{}_1 \\ \mathcal{G}^1{}_0 & \mathcal{G}^0{}_0 \end{bmatrix} \qquad (3.3.15)$$

and that

$$G^A{}_{A_1}G^B{}_{B_1}\epsilon^{A_1B_1} = \epsilon^{AB}, \quad A,B = 1,0. \qquad (3.3.16)$$

As usual, all of these have obvious barred and dotted versions.

We shall denote by $\mathfrak{B}^{rs}_{mn}$ the set of all spinors of valence $\binom{r\ s}{m\ n}$. Being a collection of multilinear functionals, $\mathfrak{B}^{rs}_{mn}$ admits a natural "pointwise" vector space structure. Specifically, if $\xi,\ \zeta \in \mathfrak{B}^{rs}_{mn}$ and $\overset{1}{\phi},\ldots,\overset{r}{\phi} \in \mathfrak{B}$, $\overset{1}{\bar{\psi}}, \ldots, \overset{s}{\bar{\psi}} \in \bar{\mathfrak{B}}$, $\overset{1}{\mu},\ldots,\overset{m}{\mu} \in \mathfrak{B}^*$ and $\overset{1}{\bar{\nu}},\ldots,\overset{n}{\bar{\nu}} \in \bar{\mathfrak{B}}^*$, then

$$(\xi+\zeta)\left(\overset{1}{\phi},\ldots,\overset{n}{\bar{\nu}}\right) = \xi\left(\overset{1}{\phi},\ldots,\overset{n}{\bar{\nu}}\right) + \zeta\left(\overset{1}{\phi},\ldots,\overset{n}{\bar{\nu}}\right)$$

and, for $\alpha \in \mathbb{C}$,

$$(\alpha\xi)\left(\overset{1}{\phi},\ldots,\overset{n}{\bar{\nu}}\right) = \alpha\left(\xi\left(\overset{1}{\phi},\ldots,\overset{n}{\bar{\nu}}\right)\right).$$

If $\{s^1, s^0\}$ is a spin frame, $\{s_1, s_0\}$ its dual basis for $\mathfrak{B}^*$, $\{\bar{s}^{\dot{1}}, \bar{s}^{\dot{0}}\}$ the corresponding conjugate basis for $\bar{\mathfrak{B}}$ and $\{\bar{s}_{\dot{1}}, \bar{s}_{\dot{0}}\}$ its dual, we define $s_{A_1} \otimes \cdots \otimes s_{A_r} \otimes \bar{s}_{\dot{X}_1} \otimes \cdots \otimes \bar{s}_{\dot{X}_s} \otimes s^{B_1} \otimes \cdots \otimes s^{B_m} \otimes \bar{s}^{\dot{Y}_1} \otimes \cdots \otimes \bar{s}^{\dot{Y}_n}$, abbreviated $s_{A_1} \otimes \cdots \otimes \bar{s}^{\dot{Y}_n}$, by

$$\begin{aligned} s_{A_1} \otimes \cdots \otimes \bar{s}^{\dot{Y}_n}\left(\overset{1}{\phi},\ldots,\overset{n}{\bar{\nu}}\right) &= s_{A_1}\left(\overset{1}{\phi}\right)\cdots \bar{s}^{\dot{Y}_n}\left(\overset{n}{\bar{\nu}}\right) \\ &= \overset{1}{\phi}_{A_1} \cdots \overset{n}{\bar{\nu}}_{\dot{Y}_n}. \end{aligned}$$

Thus, for example, in $\mathfrak{B}^{11}_{10}$ we have $s_1 \otimes \bar{s}_{\dot{0}} \otimes s^0$ defined by $s_1 \otimes \bar{s}_{\dot{0}} \otimes s^0\,(\phi, \bar{\psi}, \mu) = s_1(\phi)\bar{s}_{\dot{0}}(\bar{\psi})s^0(\mu) = \phi_1\bar{\psi}_{\dot{0}}\mu^0$, where $\phi = \phi_A s^A$, $\bar{\psi} = \bar{\psi}_{\dot{X}}\bar{s}^{\dot{X}}$ and $\mu = \mu^A s_A$. For $\xi \in \mathfrak{B}^{rs}_{mn}$ we define

$$\xi^{A_1\cdots\dot{X}_s}{}_{B_1\cdots\dot{Y}_n} = \xi\left(s^{A_1},\ldots,\bar{s}^{\dot{X}_s}, s_{B_1},\ldots,\bar{s}_{\dot{Y}_n}\right) \tag{3.3.17}$$

for $A_i, B_i = 1, 0$ and $\dot{X}_i, \dot{Y}_i = \dot{1}, \dot{0}$.

Exercise 3.3.5. Show that the elements $s_{A_1} \otimes \cdots \otimes \bar{s}^{\dot{Y}_n}$ of $\mathfrak{B}^{rs}_{mn}$ are linearly independent and that any $\xi \in \mathfrak{B}^{rs}_{mn}$ can be written

$$\xi = \xi^{A_1\cdots\dot{X}_s}{}_{B_1\cdots\dot{Y}_n}\, s_{A_1} \otimes \cdots \otimes \bar{s}_{\dot{X}_s} \otimes s^{B_1} \otimes \cdots \otimes \bar{s}^{\dot{Y}_n}.$$

From Exercise 3.3.5 we conclude that the $s_{A_1} \otimes \cdots \otimes \bar{s}^{\dot{Y}_n}$ form a basis for $\mathfrak{B}^{rs}_{mn}$ which therefore has dimension $2^{r+s+m+n}$. For each $\xi \in \mathfrak{B}^{rs}_{mn}$ the numbers $\xi^{A_1\cdots\dot{X}_s}{}_{B_1\cdots\dot{Y}_n}$ defined by (3.3.17) are the components of ξ relative to the basis $\{s_{A_1} \otimes \cdots \otimes \bar{s}^{\dot{Y}_n}\}$ and, in terms of them, the linear operations on $\mathfrak{B}^{rs}_{mn}$ can be expressed as

$$(\xi+\zeta)^{A_1\cdots\dot{X}_s}{}_{B_1\cdots\dot{Y}_n} = \xi^{A_1\cdots\dot{X}_s}{}_{B_1\cdots\dot{Y}_n} + \zeta^{A_1\cdots\dot{X}_s}{}_{B_1\cdots\dot{Y}_n}$$

and

$$(\alpha\xi)^{A_1\cdots\dot{X}_s}{}_{B_1\cdots\dot{Y}_n} = \alpha\xi^{A_1\cdots\dot{X}_s}{}_{B_1\cdots\dot{Y}_n}.$$

The next algebraic operation on spinors that we must consider is a generalization of the procedure we just employed to construct a basis for $\mathfrak{B}^{rs}_{mn}$ from a spin frame. Suppose ξ is a spinor of valence $\left(\begin{smallmatrix} r_1 & s_1 \\ m_1 & n_1 \end{smallmatrix}\right)$ and ζ is a spinor of valence $\left(\begin{smallmatrix} r_2 & s_2 \\ m_2 & n_2 \end{smallmatrix}\right)$. The *outer product* of ξ and ζ is the spinor $\xi \otimes \zeta$ of valence $\left(\begin{smallmatrix} r_1+r_2 & s_1+s_2 \\ m_1+m_2 & n_1+n_2 \end{smallmatrix}\right)$ defined as follows. If $\overset{1}{\phi}, \ldots, \overset{r_1+r_2}{\phi} \in \mathfrak{B}$, $\overset{1}{\bar{\psi}}, \ldots, \overset{s_1+s_2}{\bar{\psi}} \in \bar{\mathfrak{B}}$, $\overset{1}{\mu}, \ldots, \overset{m_1+m_2}{\mu} \in \mathfrak{B}^*$ and $\overset{1}{\bar{\nu}}, \ldots, \overset{n_1+n_2}{\bar{\nu}} \in \bar{\mathfrak{B}}^*$, then

$$\begin{aligned}
&(\xi \otimes \zeta)\left(\overset{1}{\phi}, \ldots, \overset{r_1+r_2}{\phi}, \overset{1}{\bar{\psi}}, \ldots, \overset{s_1+s_2}{\bar{\psi}}, \overset{1}{\mu}, \ldots, \overset{m_1+m_2}{\mu}, \overset{1}{\bar{\nu}}, \ldots, \overset{n_1+n_2}{\bar{\nu}}\right) \\
&= \xi\left(\overset{1}{\phi}, \ldots, \overset{r_1}{\phi}, \overset{1}{\bar{\psi}}, \ldots, \overset{s_1}{\bar{\psi}}, \overset{1}{\mu}, \ldots, \overset{m_1}{\mu}, \overset{1}{\bar{\nu}}, \ldots, \overset{n_1}{\bar{\nu}}\right) \zeta\left(\overset{r_1+1}{\phi}, \ldots, \overset{r_1+r_2}{\phi}, \overset{s_1+1}{\bar{\psi}},\right. \\
&\qquad \left.\ldots, \overset{s_1+s_2}{\bar{\psi}}, \overset{m_1+1}{\mu}, \ldots, \overset{m_1+m_2}{\mu}, \overset{n_1+1}{\bar{\nu}}, \ldots, \overset{n_1+n_2}{\bar{\nu}}\right).
\end{aligned}$$

It follows immediately from the definition that, in terms of components,

$$\begin{aligned}
&(\xi \otimes \zeta)^{A_1\cdots A_{r_1+r_2}\dot{X}_1\cdots\dot{X}_{s_1+s_2}}{}_{B_1\cdots B_{m_1+m_2}\dot{Y}_1\cdots\dot{Y}_{n_1+n_2}} = \\
&(\xi^{A_1\cdots A_{r_1}\dot{X}_1\cdots\dot{X}_{s_1}}{}_{B_1\cdots B_{m_1}\dot{Y}_1\cdots\dot{Y}_{n_1}})(\zeta^{A_{r_1+1}\cdots A_{r_1+r_2}\dot{X}_{s_1+1}\cdots\dot{X}_{s_1+s_2}}{}_{B_{m_1+1}\cdots B_{m_1+m_2}\dot{Y}_{n_1+1}\cdots\dot{Y}_{n_1+n_2}}).
\end{aligned}$$

Moreover, outer multiplication is clearly associative $[(\xi\otimes\zeta)\otimes v = \xi\otimes(\zeta\otimes v)]$ and distributive $[\xi\otimes(\zeta+v) = \xi\otimes\zeta+\xi\otimes v$ and $(\xi+\zeta)\otimes v = \xi\otimes v+\zeta\otimes v]$, but is not commutative. For example, if $\{s^1, s^0\}$ is a spin frame, then $s^1 \otimes s^0$ does not equal $s^0 \otimes s^1$ since $s^1 \otimes s^0(\phi^*, \psi^*) = \phi^1\psi^0$, but $s^0 \otimes s^1(\phi^*, \psi^*) = \phi^0\psi^1$ and these are generally not the same.

Next we consider a spinor ξ of valence $\left(\begin{smallmatrix} r & s \\ m & n \end{smallmatrix}\right)$ and two integers k and l with $1 \le k \le r$ and $1 \le l \le m$. Then the *contraction of* ξ *in the indices* A_k *and* B_l is the spinor $\mathcal{C}_{kl}(\xi)$ of valence $\left(\begin{smallmatrix} r-1 & s \\ m-1 & n \end{smallmatrix}\right)$ whose components relative to any spin frame are obtained by equating A_k and B_l in those of ξ and summing as indicated, i.e., if

$$\xi = \xi^{A_1\cdots A_k\cdots A_r\dot{X}_1\cdots\dot{X}_s}{}_{B_1\cdots B_l\cdots B_m\dot{Y}_1\cdots\dot{Y}_n}\, s_{A_1} \otimes \cdots \otimes \bar{s}^{\dot{Y}_n},$$

then

$$\mathcal{C}_{kl}(\xi) = \xi^{A_1\cdots A\cdots A_r\dot{X}_1\cdots\dot{X}_s}{}_{B_1\cdots A\cdots B_m\dot{Y}_1\cdots\dot{Y}_n}\, s_{A_1} \otimes \cdots \otimes \bar{s}^{\dot{Y}_n}, \tag{3.3.18}$$

where, in this last expression, it is understood that s_{A_k} and s^{B_l} are missing in $s_{A_1} \otimes \cdots \otimes \bar{s}^{\dot{Y}_n}$. Thus, for example, if ξ is of valence $\binom{1\ 1}{1\ 0}$ with

$$\xi = \xi^{A_1 \dot{X}_1}{}_{B_1} s_{A_1} \otimes \bar{s}_{\dot{X}_1} \otimes s^{B_1}$$

and $k = l = 1$, then $\mathcal{C}_{11}(\xi)$ is the spinor of valence $\binom{0\ 1}{0\ 0}$ given by

$$\mathcal{C}_{11}(\xi) = \xi^{A\dot{X}_1}{}_A \bar{s}_{\dot{X}_1} = \left(\xi^{1\dot{X}_1}{}_1 + \xi^{0\dot{X}_1}{}_0\right) \bar{s}_{\dot{X}_1}.$$

Unlike our previous definitions that were coordinate-free, contractions are defined in terms of components and so it is not immediately apparent that we have defined a spinor at all. We must verify that the components of $\mathcal{C}_{kl}(\xi)$ as defined by (3.3.18) transform correctly, i.e., as the components of a spinor of valence $\binom{r-1\ \ s}{m-1\ \ n}$. But this is clearly the case since, in a new spin frame,

$$\begin{aligned}
&\hat{\xi}^{A_1\cdots A\cdots A_r\dot{X}_1\cdots\dot{X}_s}{}_{B_1\cdots A\cdots B_m\dot{Y}_1\cdots\dot{Y}_n} \\
&= \mathcal{G}^{A_1}{}_{C_1}\cdots\mathcal{G}^{A}{}_{C_k}\cdots\mathcal{G}^{A_r}{}_{C_r}\bar{\mathcal{G}}^{\dot{X}_1}{}_{\dot{U}_1}\cdots\bar{\mathcal{G}}^{\dot{X}_s}{}_{\dot{U}_s} G_{B_1}{}^{D_1}\cdots G_A{}^{D_l}\cdots \\
&\quad G_{B_m}{}^{D_m}\bar{G}_{\dot{Y}_1}{}^{\dot{V}_1}\cdots\bar{G}_{\dot{Y}_n}{}^{\dot{V}_n}\xi^{C_1\cdots C_k\cdots C_r\dot{U}_1\cdots\dot{U}_s}{}_{D_1\cdots D_l\cdots D_m\dot{Y}_1\cdots\dot{Y}_n} \\
&= \left(\mathcal{G}^{A}{}_{C_k}G_A{}^{D_l}\right)\mathcal{G}^{A_1}{}_{C_1}\cdots\bar{G}_{\dot{Y}_n}{}^{\dot{V}_n}\xi^{C_1\cdots C_k\cdots C_r\dot{U}_1\cdots\dot{U}_s}{}_{D_1\cdots D_l\cdots D_m\dot{Y}_1\cdots\dot{Y}_n} \\
&= \delta^{D_l}_{C_k}\mathcal{G}^{A_1}{}_{C_1}\cdots\bar{G}_{\dot{Y}_n}{}^{\dot{V}_n}\xi^{C_1\cdots C_k\cdots C_r\dot{U}_1\cdots\dot{U}_s}{}_{D_1\cdots D_l\cdots D_m\dot{Y}_1\cdots\dot{Y}_n} \\
&= \mathcal{G}^{A_1}{}_{C_1}\cdots\bar{G}_{\dot{Y}_n}{}^{\dot{V}_n}\xi^{C_1\cdots A\cdots C_r\dot{U}_1\cdots\dot{U}_s}{}_{D_1\cdots A\cdots D_m\dot{Y}_1\cdots\dot{Y}_n}.
\end{aligned}$$

One can, in the same way, contract a spinor ξ of valence $\binom{r\ s}{m\ n}$ in two dotted indices $\dot{k}$ and $\dot{l}$, one upper and one lower, to obtain a spinor $\mathcal{C}_{\dot{k}\dot{l}}(\xi)$ of valence $\binom{r\ \ s-1}{m\ \ n-1}$. Observe that the processes of raising and lowering indices discussed earlier are actually outer products (with an ϵ spinor) followed by a contraction.

Exercise 3.3.6. Let ϕ be a spinor of valence $\binom{0\ 0}{2\ 0}$ and denote its components in a spin frame by ϕ_{AB}. Show that

1. $\phi_1{}^1 = -\phi_{10},\ \phi_0{}^0 = \phi_{01},\ \phi_1{}^0 = \phi_{11},\ \phi_0{}^1 = -\phi_{00},$

2. $\phi^{11} = \phi_{00},\ \phi^{00} = \phi_{11},\ \phi^{10} = -\phi_{01},\ \phi^{01} = -\phi_{10},$

3. $\phi_{AB}\phi^{AB} = 2\ \det\begin{bmatrix}\phi_{11} & \phi_{10}\\ \phi_{01} & \phi_{00}\end{bmatrix} = 2\ \det\begin{bmatrix}\phi_1{}^1 & \phi_1{}^0\\ \phi_0{}^1 & \phi_0{}^0\end{bmatrix},$

4. $\phi_{AC}\phi_B{}^C = \begin{cases} 0\quad, & A = B\\ \det[\phi_{AB}], & A = 0,\ B = 1\\ -\det[\phi_{AB}], & A = 1,\ B = 0.\end{cases}$

Let ξ denote a spinor with the same number of dotted and undotted indices, say, of valence $\left[\begin{smallmatrix} r & r \\ 0 & 0 \end{smallmatrix}\right]$. We define a new spinor denoted $\bar{\xi}$ and called the *conjugate* of ξ by specifying that its components $\bar{\xi}^{\dot{A}_1\cdots\dot{A}_r X_1\cdots X_r}$ in any spin frame are given by

$$\bar{\xi}^{\dot{A}_1\cdots\dot{A}_r X_1\cdots X_r} = \overline{\xi^{A_1\cdots A_r \dot{X}_1\cdots\dot{X}_r}}$$

(here we must depart from our habit of selecting dotted/undotted indices from the end/beginning of the alphabet). Thus, for example, if ξ has components $\xi^{A\dot{X}}$, then the components of $\bar{\xi}$ are given by $\bar{\xi}^{\dot{0}1} = \overline{\xi^{0\dot{1}}}$, $\bar{\xi}^{\dot{1}1} = \overline{\xi^{1\dot{1}}}$, etc.

Exercise 3.3.7. Show that we have actually defined a spinor of the required type by verifying the appropriate transformation law, i.e.,

$$\hat{\bar{\xi}}^{\dot{A}_1\cdots\dot{A}_r X_1\cdots X_r} = \bar{\mathcal{G}}^{\dot{A}_1}{}_{\dot{C}_1}\cdots\bar{\mathcal{G}}^{\dot{A}_r}{}_{\dot{C}_r}\,\mathcal{G}^{X_1}{}_{U_1}\cdots\mathcal{G}^{X_r}{}_{U_r}\,\bar{\xi}^{\dot{C}_1\cdots\dot{C}_r U_1\cdots U_r}.$$

Entirely analogous definitions and results apply regardless of the positions (upper or lower) of the indices, provided only that the number of dotted indices is the same as the number of undotted indices. We shall say that such a spinor ξ is *Hermitian* if $\bar{\xi} = \xi$. Thus, for example, if ξ is of valence $\left(\begin{smallmatrix} r & r \\ 0 & 0 \end{smallmatrix}\right)$, then it is Hermitian if $\bar{\xi}^{A_1\cdots A_r\dot{X}_1\cdots\dot{X}_r} = \xi^{A_1\cdots A_r\dot{X}_1\cdots\dot{X}_r}$ for all $A_1,\ldots,A_r$, $\dot{X}_1,\ldots,\dot{X}_r$, i.e., if

$$\overline{\xi^{\dot{A}_1\cdots\dot{A}_r X_1\cdots X_r}} = \xi^{A_1\cdots A_r\dot{X}_1\cdots\dot{X}_r},$$

e.g., if $r = 1$, $\overline{\xi^{\dot{0}1}} = \xi^{0\dot{1}}$, $\overline{\xi^{0\dot{0}}} = \xi^{\dot{0}0}$, etc.

As a multilinear functional a spinor ξ operates on four distinct types of objects (elements of $\mathfrak{B}$, $\bar{\mathfrak{B}}$, $\mathfrak{B}^*$ and $\bar{\mathfrak{B}}^*$) and, if the valence is $\left(\begin{smallmatrix} r & s \\ m & n \end{smallmatrix}\right)$, has $r+s+m+n$ "slots" (variables) into which these objects are inserted for evaluation, each slot corresponding to an index position in our notation for ξ's components. If ξ has the property that, for two such slots of the same type, $\xi(\ldots,p,\ldots,q,\ldots) = \xi(\ldots,q,\ldots,p,\ldots)$ for all p and q of the appropriate type, then ξ is said to be *symmetric* in these two variables (if $\xi(\ldots,p,\ldots,q,\ldots) = -\xi(\ldots,q,\ldots,p,\ldots)$, it is *skew-symmetric*). It follows at once from the definition that ξ is symmetric (skew-symmetric) in the variables p and q if and only if the components of ξ in every spin frame are unchanged (change sign) when the corresponding indices are interchanged. We will be particularly interested in the case of spinors with just two indices. Thus, for example, a spinor ϕ of valence $\left(\begin{smallmatrix} 0 & 0 \\ 2 & 0 \end{smallmatrix}\right)$ is symmetric (in its only two variables) if and only if, in every spin frame, $\phi_{BA} = -\phi_{AB}$ for all $A, B = 1, 0$; ϕ is skew-symmetric if $\phi_{BA} = -\phi_{AB}$ for all A and B. On the other hand, an arbitrary spinor ξ of valence $\left(\begin{smallmatrix} 0 & 0 \\ 2 & 0 \end{smallmatrix}\right)$ has a *symmetrization* whose components in each spin frame are given by

$$\xi_{(AB)} = \tfrac{1}{2}\left(\xi_{AB} + \xi_{BA}\right)$$

and a *skew-symmetrization* given by

$$\xi_{[AB]} = \frac{1}{2}\left(\xi_{AB} - \xi_{BA}\right).$$

The symmetrization (skew-symmetrization) of ξ clearly defines a spinor, also of valence $\left(\begin{smallmatrix}0 & 0\\ 2 & 0\end{smallmatrix}\right)$, that is symmetric (skew-symmetric).

Exercise 3.3.8. Let α and β be two spin vectors. The outer product $\alpha \otimes \beta$ is a spinor of valence $\left(\begin{smallmatrix}0 & 0\\ 2 & 0\end{smallmatrix}\right)$ whose components in any spin frame are given by $\alpha_A \beta_B$, $A, B = 1, 0$. Let ϕ be the symmetrization of $\alpha \otimes \beta$ so that

$$\phi_{AB} = \alpha_{(A}\beta_{B)} = \frac{1}{2}\left(\alpha_A\beta_B + \alpha_B\beta_A\right).$$

Show that $\phi^{AB} = \frac{1}{2}(\alpha^A\beta^B + \alpha^B\beta^A)$ and that

$$\phi_{AB}\phi^{AB} = -\frac{1}{2} < \alpha, \beta >^2 .$$

3.4 Spinors and World Vectors

In this section we will establish a correspondence between spinors of valence $\left(\begin{smallmatrix}1 & 1\\ 0 & 0\end{smallmatrix}\right)$ and vectors in Minkowski spacetime (also called *world vectors*). This correspondence, which we have actually seen before (in Section 1.7), is most easily phrased in terms of the Pauli spin matrices.

Exercise 3.4.1. Let $\sigma_1 = \left[\begin{smallmatrix}0 & 1\\ 1 & 0\end{smallmatrix}\right]$, $\sigma_2 = \left[\begin{smallmatrix}0 & i\\ -i & 0\end{smallmatrix}\right]$, $\sigma_3 = \left[\begin{smallmatrix}1 & 0\\ 0 & -1\end{smallmatrix}\right]$, and $\sigma_4 = \left[\begin{smallmatrix}1 & 0\\ 0 & 1\end{smallmatrix}\right]$. Verify the following *commutation relations*:

$$\begin{aligned}
{\sigma_1}^2 = {\sigma_2}^2 = {\sigma_3}^2 = {\sigma_4}^2 &= \sigma_4,\\
\sigma_1\sigma_2 = -\sigma_2\sigma_1 &= -i\sigma_3,\\
\sigma_1\sigma_3 = -\sigma_3\sigma_1 &= i\sigma_2,\\
\sigma_2\sigma_3 = -\sigma_3\sigma_2 &= -i\sigma_1.
\end{aligned}$$

For what follows it will be convenient to introduce a factor of $\frac{1}{\sqrt{2}}$ and some rather peculiar looking indices, the significance of which will become clear shortly. Thus, for each $A = 1, 0$ and $\dot{X} = \dot{1}, \dot{0}$, we define matrices

$$\sigma_a{}^{A\dot{X}} = \begin{bmatrix} \sigma_a{}^{1\dot{1}} & \sigma_a{}^{1\dot{0}} \\ \sigma_a{}^{0\dot{1}} & \sigma_a{}^{0\dot{0}} \end{bmatrix}, \qquad a = 1, 2, 3, 4,$$

by

$$\sigma_a{}^{A\dot{X}} = \frac{1}{\sqrt{2}}\sigma_a, \qquad a = 1, 2, 3, 4.$$

Thus,

$$\sigma_1{}^{A\dot{X}} = \frac{1}{\sqrt{2}}\begin{bmatrix} 0 & 1 \\ 1 & 0 \end{bmatrix}, \qquad \sigma_2{}^{A\dot{X}} = \frac{1}{\sqrt{2}}\begin{bmatrix} 0 & i \\ -i & 0 \end{bmatrix},$$

$$\sigma_3{}^{A\dot{X}} = \frac{1}{\sqrt{2}}\begin{bmatrix} 1 & 0 \\ 0 & -1 \end{bmatrix}, \qquad \sigma_4{}^{A\dot{X}} = \frac{1}{\sqrt{2}}\begin{bmatrix} 1 & 0 \\ 0 & 1 \end{bmatrix}.$$

We again adopt the convention that the relative position of dotted and undotted indices is immaterial so $\sigma_a{}^{A\dot{X}} = \sigma_a{}^{\dot{X}A}$. Undotted indices indicate rows; dotted indices number the columns. Observe that each of these is a Hermitian matrix, i.e., equals its conjugate transpose (Section 1.7).

Now we describe a procedure for taking a vector $v \in \mathcal{M}$, an admissible basis $\{e_a\}$ for $\mathcal{M}$ and a spin frame $\{s^A\}$ and constructing from them a spinor V of valence $\binom{1\,1}{0\,0}$. We do this by specifying the components of V in every spin frame and verifying that they have the correct transformation law. We begin by writing $v = v^a e_a$. Define the components $V^{A\dot{X}}$ of V relative to $\{s^A\}$ by

$$V^{A\dot{X}} = \sigma_a{}^{A\dot{X}} v^a, \qquad A = 1, 0, \ \dot{X} = \dot{1}, \dot{0}. \tag{3.4.1}$$

Thus,

$$\begin{aligned} V^{1\dot{1}} &= \tfrac{1}{\sqrt{2}}\left(v^3 + v^4\right), \\ V^{1\dot{0}} &= \tfrac{1}{\sqrt{2}}\left(v^1 + iv^2\right), \\ V^{0\dot{1}} &= \tfrac{1}{\sqrt{2}}\left(v^1 - iv^2\right), \\ V^{0\dot{0}} &= \tfrac{1}{\sqrt{2}}\left(-v^3 + v^4\right) \end{aligned} \tag{3.4.2}$$

(cf. Exercise 1.7.1). Now, suppose $\{\hat{s}^1, \hat{s}^0\}$ is another spin frame, related to $\{s^A\}$ by (3.2.1) $(s^B = G_A{}^B \hat{s}^A)$ and (3.2.6) $(\hat{s}^A = \mathcal{G}^A{}_B s^B)$. We define the components $\hat{V}^{A\dot{X}}$ of V relative to $\{\hat{s}^A\}$ as follows: Let $\Lambda = \Lambda_{\mathcal{G}} =$ Spin$(\mathcal{G})$ be the element of $\mathcal{L}$ that $\mathcal{G}$ maps onto under the spinor map and $\hat{v}^a = \Lambda^a{}_b v^b$, $a = 1, 2, 3, 4$. Now let

$$\hat{V}^{A\dot{X}} = \sigma_a{}^{A\dot{X}} \hat{v}^a, \qquad A = 1, 0, \ \dot{X} = \dot{1}, \dot{0}. \tag{3.4.3}$$

That we have actually defined a spinor of valence $\binom{1\,1}{0\,0}$ is not obvious, of course, since it is not clear that the $V^{A\dot{X}}$ transform correctly. To show this we must prove that

$$\hat{V}^{A\dot{X}} = \mathcal{G}^A{}_B \bar{\mathcal{G}}^{\dot{X}}{}_{\dot{Y}} V^{B\dot{Y}}, \qquad A = 1, 0, \ \dot{X} = \dot{1}, \dot{0}. \tag{3.4.4}$$

For this we temporarily denote the right-hand side of (3.4.4) by $\tilde{V}^{A\dot{X}}$, i.e., $\tilde{V}^{A\dot{X}} = \mathcal{G}^A{}_B \bar{\mathcal{G}}^{\dot{X}}{}_{\dot{Y}} V^{B\dot{Y}}$. Writing this as a matrix product gives

$$\begin{bmatrix} \tilde{V}^{1\dot{1}} \\ \tilde{V}^{1\dot{0}} \\ \tilde{V}^{0\dot{1}} \\ \tilde{V}^{0\dot{0}} \end{bmatrix} = \begin{bmatrix} \mathcal{G}^1{}_1\bar{\mathcal{G}}^{\dot{1}}{}_{\dot{1}} & \mathcal{G}^1{}_1\bar{\mathcal{G}}^{\dot{1}}{}_{\dot{0}} & \mathcal{G}^1{}_0\bar{\mathcal{G}}^{\dot{1}}{}_{\dot{1}} & \mathcal{G}^1{}_0\bar{\mathcal{G}}^{\dot{1}}{}_{\dot{0}} \\ \mathcal{G}^1{}_1\bar{\mathcal{G}}^{\dot{0}}{}_{\dot{1}} & \mathcal{G}^1{}_1\bar{\mathcal{G}}^{\dot{0}}{}_{\dot{0}} & \mathcal{G}^1{}_0\bar{\mathcal{G}}^{\dot{0}}{}_{\dot{1}} & \mathcal{G}^1{}_0\bar{\mathcal{G}}^{\dot{0}}{}_{\dot{0}} \\ \mathcal{G}^0{}_1\bar{\mathcal{G}}^{\dot{1}}{}_{\dot{1}} & \mathcal{G}^0{}_1\bar{\mathcal{G}}^{\dot{1}}{}_{\dot{0}} & \mathcal{G}^0{}_0\bar{\mathcal{G}}^{\dot{1}}{}_{\dot{1}} & \mathcal{G}^0{}_0\bar{\mathcal{G}}^{\dot{1}}{}_{\dot{0}} \\ \mathcal{G}^0{}_1\bar{\mathcal{G}}^{\dot{0}}{}_{\dot{1}} & \mathcal{G}^0{}_1\bar{\mathcal{G}}^{\dot{0}}{}_{\dot{0}} & \mathcal{G}^0{}_0\bar{\mathcal{G}}^{\dot{0}}{}_{\dot{1}} & \mathcal{G}^0{}_0\bar{\mathcal{G}}^{\dot{0}}{}_{\dot{0}} \end{bmatrix} \begin{bmatrix} V^{1\dot{1}} \\ V^{1\dot{0}} \\ V^{0\dot{1}} \\ V^{0\dot{0}} \end{bmatrix}. \qquad (3.4.5)$$

But if we let

$$\mathcal{G} = \begin{bmatrix} \mathcal{G}^1{}_1 & \mathcal{G}^1{}_0 \\ \mathcal{G}^0{}_1 & \mathcal{G}^0{}_0 \end{bmatrix} = \begin{bmatrix} \alpha & \beta \\ \gamma & \delta \end{bmatrix},$$

then (3.4.5) becomes

$$\begin{bmatrix} \tilde{V}^{1\dot{1}} \\ \tilde{V}^{1\dot{0}} \\ \tilde{V}^{0\dot{1}} \\ \tilde{V}^{0\dot{0}} \end{bmatrix} = \begin{bmatrix} \alpha\bar{\alpha} & \alpha\bar{\beta} & \bar{\alpha}\beta & \beta\bar{\beta} \\ \alpha\bar{\gamma} & \alpha\bar{\delta} & \beta\bar{\gamma} & \beta\bar{\delta} \\ \bar{\alpha}\gamma & \bar{\beta}\gamma & \bar{\alpha}\delta & \bar{\beta}\delta \\ \gamma\bar{\gamma} & \gamma\bar{\delta} & \bar{\gamma}\delta & \delta\bar{\delta} \end{bmatrix} \begin{bmatrix} V^{1\dot{1}} \\ V^{1\dot{0}} \\ V^{0\dot{1}} \\ V^{0\dot{0}} \end{bmatrix}. \qquad (3.4.6)$$

Now, using (3.4.2) and the corresponding equalities for $\hat{V}^{A\dot{X}}$ it follows from Exercise 1.7.2 (with the appropriate notational changes) that the right-hand side of (3.4.6) is equal to

$$\frac{1}{\sqrt{2}} \begin{bmatrix} \hat{v}^3 + \hat{v}^4 \\ \hat{v}^1 + i\hat{v}^2 \\ \hat{v}^1 - i\hat{v}^2 \\ -\hat{v}^3 + \hat{v}^4 \end{bmatrix} = \begin{bmatrix} \hat{V}^{1\dot{1}} \\ \hat{V}^{1\dot{0}} \\ \hat{V}^{0\dot{1}} \\ \hat{V}^{0\dot{0}} \end{bmatrix},$$

where the $\hat{v}^a$ are the images of the v^a under $\Lambda = \Lambda_{\mathcal{G}}$. Substituting this into (3.4.6) then gives $[\tilde{V}^{A\dot{X}}] = [\hat{V}^{A\dot{X}}]$ and this proves (3.4.4). Observe that, since

$$\begin{bmatrix} G^1{}_1 & G^1{}_0 \\ G^0{}_1 & G^0{}_0 \end{bmatrix} = - \begin{bmatrix} \mathcal{G}^1{}_1 & \mathcal{G}^1{}_0 \\ \mathcal{G}^0{}_1 & \mathcal{G}^0{}_0 \end{bmatrix},$$

(3.4.4) can also be written as

$$\hat{V}^{A\dot{X}} = G^A{}_B \bar{G}^{\dot{X}}{}_{\dot{Y}} V^{B\dot{Y}}, \quad A = 1, 0, \;\; \dot{X} = \dot{1}, \dot{0}. \qquad (3.4.7)$$

We conclude then that the procedure we have described does indeed define a spinor of valence $\left(\begin{smallmatrix} 1 & 1 \\ 0 & 0 \end{smallmatrix}\right)$ which we shall call the *spinor equivalent of* $v \in \mathcal{M}$ (somewhat imprecisely since V depends not only on v, but also on the initial choices of $\{e_a\}$ and $\{s^A\}$). Observe that the conjugate $\bar{V}$ of V has components $\bar{V}^{A\dot{X}} = \overline{V^{\dot{A}X}} = \overline{\sigma_a{}^{\dot{A}X} v^a} = \overline{\sigma_a{}^{\dot{A}X}}\,\overline{v^a} = \sigma_a{}^{A\dot{X}} v^a = V^{A\dot{X}}$ since the matrices $\sigma_a{}^{A\dot{X}}$ are Hermitian and the v^a are real. Thus, the spinor equivalent of any world vector is a Hermitian spinor.

With (3.4.4) we can now justify the odd arrangement of indices in the symbols $\sigma_a{}^{A\dot{X}}$ by showing that the $\sigma_a{}^{A\dot{X}}$ are constant under the combined effect of a $\mathcal{G} \in SL(2,\mathbb{C})$ and the corresponding $\Lambda = \Lambda_{\mathcal{G}}$ in $\mathcal{L}$, i.e., that

$$\Lambda_a{}^b \mathcal{G}^A{}_B \bar{\mathcal{G}}^{\dot{X}}{}_{\dot{Y}} \sigma_b{}^{B\dot{Y}} = \sigma_a{}^{A\dot{X}}, \quad a = 1,2,3,4, \;\; A = 1,0, \;\; \dot{X} = \dot{1},\dot{0} \quad (3.4.8)$$

(one might say that the $\sigma_a{}^{A\dot{X}}$ are the components of a constant "spinor-covector"). To see this we select an arbitrary admissible basis and spin frame. Fix A and $\dot{X}$. Now let $v = v^a e_a$ be an arbitrary vector in $\mathcal{M}$. Then $V^{A\dot{X}} = \sigma_a{}^{A\dot{X}} v^a$. In another spin frame, related to the original by $\mathcal{G}$, we have

$$\hat{V}^{A\dot{X}} = \sigma_a{}^{A\dot{X}} \hat{v}^a.$$

But also,

$$\begin{aligned}
\hat{V}^{A\dot{X}} &= \mathcal{G}^A{}_B \bar{\mathcal{G}}^{\dot{X}}{}_{\dot{Y}} V^{B\dot{Y}} = \mathcal{G}^A{}_B \bar{\mathcal{G}}^{\dot{X}}{}_{\dot{Y}} \left(\sigma_b{}^{B\dot{Y}} v^b\right) \\
&= \mathcal{G}^A{}_B \bar{\mathcal{G}}^{\dot{X}}{}_{\dot{Y}} \sigma_b{}^{B\dot{Y}} (\delta^b_c v^c) \\
&= \mathcal{G}^A{}_B \bar{\mathcal{G}}^{\dot{X}}{}_{\dot{Y}} \sigma_b{}^{B\dot{Y}} (\Lambda_a{}^b \Lambda^a{}_c v^c) \\
&= \Lambda_a{}^b \mathcal{G}^A{}_B \bar{\mathcal{G}}^{\dot{X}}{}_{\dot{Y}} \sigma_b{}^{B\dot{Y}} (\Lambda^a{}_c v^c) \\
&= \Lambda_a{}^b \mathcal{G}^A{}_B \bar{\mathcal{G}}^{\dot{X}}{}_{\dot{Y}} \sigma_b{}^{B\dot{Y}} \hat{v}^a.
\end{aligned}$$

Thus,

$$\Lambda_a{}^b \mathcal{G}^A{}_B \bar{\mathcal{G}}^{\dot{X}}{}_{\dot{Y}} \sigma_b{}^{B\dot{Y}} \hat{v}^a = \sigma_a{}^{A\dot{X}} \hat{v}^a.$$

But v was arbitrary so we may successively select v's that give $(\hat{v}^1, \hat{v}^2, \hat{v}^3, \hat{v}^4)$ equal to $(1,0,0,0)$, $(0,1,0,0)$, $(0,0,1,0)$ and $(0,0,0,1)$ and thereby obtain (3.4.8) for $a = 1,2,3$ and 4, respectively. Since $[G^A{}_B] = -[\mathcal{G}^A{}_B]$ we again find that (3.4.8) can be written

$$\Lambda_a{}^b G^A{}_B \bar{G}^{\dot{X}}{}_{\dot{Y}} \sigma_b{}^{B\dot{Y}} = \sigma_a{}^{A\dot{X}}, \quad a = 1,2,3,4, \;\; A = 1,0, \;\; \dot{X} = \dot{1},\dot{0}. \quad (3.4.9)$$

Exercise 3.4.2. Show that

$$\mathcal{G}^A{}_B \bar{\mathcal{G}}^{\dot{X}}{}_{\dot{Y}} \sigma_a{}^{B\dot{Y}} = \Lambda^\alpha{}_a \sigma_\alpha{}^{A\dot{X}}, \quad a = 1,2,3,4, \;\; A = 1,0, \;\; \dot{X} = \dot{1},\dot{0}. \quad (3.4.10)$$

From all of this we conclude that the $\sigma_a{}^{A\dot{X}}$ behave formally like a combined world covector and spinor of valence $\left(\begin{smallmatrix}1 & 1\\ 0 & 0\end{smallmatrix}\right)$. Treating them as such we raise the index a and lower A and $\dot{X}$, i.e., we define

$$\sigma^a{}_{A\dot{X}} = \eta^{ab}\left(\sigma_b{}^{B\dot{Y}} \epsilon_{BA}\right) \bar{\epsilon}_{\dot{Y}\dot{X}}$$

for $a = 1,2,3,4$, $A = 1,0$ and $\dot{X} = \dot{1},\dot{0}$. Thus, for example, if $a = 1$, $\sigma^1{}_{A\dot{X}} = \eta^{1b}(\sigma_b{}^{B\dot{Y}}\epsilon_{BA})\bar{\epsilon}_{\dot{Y}\dot{X}} = \eta^{11}(\sigma_1{}^{B\dot{Y}}\epsilon_{BA})\bar{\epsilon}_{\dot{Y}\dot{X}} = \sigma_1{}^{B\dot{Y}}\epsilon_{BA}\bar{\epsilon}_{\dot{Y}\dot{X}}$. If $A = 1$, this becomes $\sigma^1{}_{1\dot{X}} = \sigma_1{}^{B\dot{Y}}\epsilon_{B1}\bar{\epsilon}_{\dot{Y}\dot{X}} = \sigma_1{}^{0\dot{Y}}\epsilon_{01}\bar{\epsilon}_{\dot{Y}\dot{X}} = \sigma_1{}^{0\dot{Y}}\bar{\epsilon}_{\dot{Y}\dot{X}} = \sigma_1{}^{0\dot{1}}\bar{\epsilon}_{\dot{1}\dot{X}} + \sigma_1{}^{0\dot{0}}\bar{\epsilon}_{\dot{0}\dot{X}}$. Thus, for $\dot{X} = \dot{1}$, $\sigma^1{}_{1\dot{1}} = \sigma_1{}^{0\dot{1}}\bar{\epsilon}_{\dot{1}\dot{1}} + \sigma_1{}^{0\dot{0}}\bar{\epsilon}_{\dot{0}\dot{1}} = \sigma_1{}^{0\dot{0}} = 0$ and, for $\dot{X} = \dot{0}$, $\sigma^1{}_{1\dot{0}} = \sigma_1{}^{0\dot{1}}\bar{\epsilon}_{\dot{1}\dot{0}} + \sigma_1{}^{0\dot{0}}\bar{\epsilon}_{\dot{0}\dot{0}} = -\sigma_1{}^{0\dot{1}} = -\frac{1}{\sqrt{2}}$. Similarly, $\sigma^1{}_{0\dot{1}} = -\frac{1}{\sqrt{2}}$ and $\sigma^1{}_{0\dot{0}} = 0$ so

$$\sigma^1{}_{A\dot{X}} = \begin{bmatrix} \sigma^1{}_{1\dot{1}} & \sigma^1{}_{1\dot{0}} \\ \sigma^1{}_{0\dot{1}} & \sigma^1{}_{0\dot{0}} \end{bmatrix} = -\frac{1}{\sqrt{2}}\begin{bmatrix} 0 & 1 \\ 1 & 0 \end{bmatrix} = -\sigma_1{}^{A\dot{X}}.$$

Exercise 3.4.3. Continue in this way to prove the remaining equalities in

$$\begin{aligned}
\sigma^1{}_{A\dot{X}} &= -\sigma_1{}^{A\dot{X}} &&= -\tfrac{1}{\sqrt{2}}\begin{bmatrix} 0 & 1 \\ 1 & 0 \end{bmatrix}, \\
\sigma^2{}_{A\dot{X}} &= \sigma_2{}^{A\dot{X}} &&= \tfrac{1}{\sqrt{2}}\begin{bmatrix} 0 & i \\ -i & 0 \end{bmatrix}, \\
\sigma^3{}_{A\dot{X}} &= -\sigma_3{}^{A\dot{X}} &&= -\tfrac{1}{\sqrt{2}}\begin{bmatrix} 1 & 0 \\ 0 & -1 \end{bmatrix}, \\
\sigma^4{}_{A\dot{X}} &= -\sigma_4{}^{A\dot{X}} &&= -\tfrac{1}{\sqrt{2}}\begin{bmatrix} 1 & 0 \\ 0 & 1 \end{bmatrix}.
\end{aligned} \tag{3.4.11}$$

We enumerate a number of useful properties of these so-called *Infeld-van der Waerden symbols* $\sigma_a{}^{A\dot{X}}$ and $\sigma^a{}_{A\dot{X}}$.

$$\sigma_a{}^{A\dot{X}} = \eta_{ab}\bar{\epsilon}^{\dot{X}\dot{Y}}(\epsilon^{AB}\sigma^b{}_{B\dot{Y}}), \tag{3.4.12}$$

$$\sigma_a{}^{A\dot{X}}\sigma^b{}_{A\dot{X}} = -\delta^b_a, \tag{3.4.13}$$

$$\sigma_a{}^{A\dot{X}}\sigma^a{}_{B\dot{Y}} = -\delta^A_B\delta^{\dot{X}}_{\dot{Y}}, \tag{3.4.14}$$

$$\sigma_a{}^{A\dot{X}}\sigma^a{}_{B\dot{Y}}\sigma_b{}^{B\dot{Y}} = -\sigma_b{}^{A\dot{X}}. \tag{3.4.15}$$

For the proof of (3.4.12) we insert $\sigma^b{}_{B\dot{Y}} = \eta^{bc}(\sigma_c{}^{C\dot{Z}}\epsilon_{CB})\bar{\epsilon}_{\dot{Z}\dot{Y}}$ into the right-hand side to obtain

$$\begin{aligned}
\eta_{ab}\bar{\epsilon}^{\dot{X}\dot{Y}}\Big(\epsilon^{AB}\sigma^b{}_{B\dot{Y}}\Big) &= \eta_{ab}\bar{\epsilon}^{\dot{X}\dot{Y}}\epsilon^{AB}\eta^{bc}\sigma_c{}^{C\dot{Z}}\epsilon_{CB}\bar{\epsilon}_{\dot{Z}\dot{Y}} \\
&= \big(\eta_{ab}\eta^{bc}\big)\Big(\bar{\epsilon}^{\dot{X}\dot{Y}}\bar{\epsilon}_{\dot{Z}\dot{Y}}\Big)\big(\epsilon^{AB}\epsilon_{CB}\big)\,\sigma_c{}^{C\dot{Z}}
\end{aligned}$$

$$= \delta_a^c \delta_{\dot{Z}}^{\dot{X}} \delta_C^A \sigma_c{}^{C\dot{Z}}$$
$$= \sigma_a{}^{A\dot{X}},$$

where we have used (3.3.5) and its barred and dotted equivalent.

Exercise 3.4.4. Prove (3.4.13), (3.4.14) and (3.4.15).

Similar exercises in index gymnastics yield the analogues of (3.4.8) and (3.4.10):

$$\Lambda^a{}_b G_A{}^B \bar{G}_{\dot{X}}{}^{\dot{Y}} \sigma^b{}_{B\dot{Y}} = \sigma^a{}_{A\dot{X}} \tag{3.4.16}$$

and

$$G_A{}^B \bar{G}_{\dot{X}}{}^{\dot{Y}} \sigma^a{}_{B\dot{Y}} = \Lambda_\alpha{}^a \sigma^\alpha{}_{A\dot{X}}. \tag{3.4.17}$$

Exercise 3.4.5. Prove (3.4.16) and (3.4.17) and use (3.4.16) to show that

$$\mathcal{G}^A{}_B \bar{\mathcal{G}}^{\dot{X}}{}_{\dot{Y}} \sigma^a{}_{A\dot{X}} = \Lambda^a{}_b \sigma^b{}_{B\dot{Y}}. \tag{3.4.18}$$

Given $v \in \mathcal{M}$, $\{e_a\}$ and $\{s^A\}$ we have constructed a spinor $V^{A\dot{X}} = \sigma_a{}^{A\dot{X}} v^a$. The $\sigma^a{}_{A\dot{X}}$ allow us to retrieve the v^a from the $V^{A\dot{X}}$. Indeed, multiplying on both sides of $V^{A\dot{X}} = \sigma_b{}^{A\dot{X}} v^b$ by $\sigma^a{}_{A\dot{X}}$ and summing as indicated gives

$$V^{A\dot{X}} \sigma^a{}_{A\dot{X}} = \sigma^a_{A\dot{X}} \sigma_b{}^{A\dot{X}} v^b$$
$$= -\delta_b^a v^b$$
$$= -v^a,$$

so

$$v^a = -V^{A\dot{X}} \sigma^a{}_{A\dot{X}}, \qquad a = 1, 2, 3, 4. \tag{3.4.19}$$

Note that if the $V^{A\dot{X}}$ were the components of an arbitrary spinor of valence $\left(\begin{smallmatrix} 1 & 1 \\ 0 & 0 \end{smallmatrix}\right)$, then the numbers $-V^{A\dot{X}} \sigma^a{}_{A\dot{X}}$ would, in general, be complex and so would not be the components of any world vector. However, we show next that if $V^{A\dot{X}}$ is Hermitian, then the $-V^{A\dot{X}} \sigma^a{}_{A\dot{X}}$ are real and, moreover, determine a world vector. Indeed,

$$\overline{V^{A\dot{X}} \sigma^a{}_{A\dot{X}}} = \overline{V^{1\dot{1}} \sigma^a{}_{1\dot{1}} + V^{1\dot{0}} \sigma^a{}_{1\dot{0}} + V^{0\dot{1}} \sigma^a{}_{0\dot{1}} + V^{0\dot{0}} \sigma^a{}_{0\dot{0}}}$$

$$
\begin{aligned}
&= \overline{V^{1\dot{1}}}\,\overline{\sigma^a{}_{1\dot{1}}} + \overline{V^{1\dot{0}}}\,\overline{\sigma^a{}_{1\dot{0}}} + \overline{V^{0\dot{1}}}\,\overline{\sigma^a{}_{0\dot{1}}} + \overline{V^{0\dot{0}}}\,\overline{\sigma^a{}_{0\dot{0}}} \\
&= \overline{V^{\dot{1}1}}\,\overline{\sigma^a{}_{\dot{1}1}} + \overline{V^{\dot{0}1}}\,\overline{\sigma^a{}_{\dot{0}1}} + \overline{V^{\dot{1}0}}\,\overline{\sigma^a{}_{\dot{1}0}} + \overline{V^{\dot{0}0}}\,\overline{\sigma^a{}_{\dot{0}0}} \\
&= \bar{V}^{1\dot{1}}\bar{\sigma}^a{}_{1\dot{1}} + \bar{V}^{0\dot{1}}\bar{\sigma}^a{}_{0\dot{1}} + \bar{V}^{1\dot{0}}\bar{\sigma}^a{}_{1\dot{0}} + \bar{V}^{0\dot{0}}\bar{\sigma}^a{}_{0\dot{0}} \\
&= V^{1\dot{1}}\sigma^a{}_{1\dot{1}} + V^{0\dot{1}}\sigma^a{}_{0\dot{1}} + V^{1\dot{0}}\sigma^a{}_{1\dot{0}} + V^{0\dot{0}}\sigma^a{}_{0\dot{0}} \\
&= V^{A\dot{X}}\sigma^a{}_{A\dot{X}}
\end{aligned}
$$

and so $V^{A\dot{X}}\sigma^a{}_{A\dot{X}}$ is real.

Exercise 3.4.6. Show that if $V^{A\dot{X}}$ is a spinor that satisfies $\bar{V}^{A\dot{X}} = -V^{A\dot{X}}$, then $V^{A\dot{X}}\sigma^a{}_{A\dot{X}}$ is pure imaginary so $iV^{A\dot{X}}$ is Hermitian.

Now, given a Hermitian spinor V of valence $\left(\begin{smallmatrix}1 & 1\\ 0 & 0\end{smallmatrix}\right)$, a spin frame $\{s^A\}$ and an admissible basis $\{e_a\}$, we define a vector $v \in \mathcal{M}$ by specifying its components in every admissible basis in the following way: Write $V = V^{A\dot{X}} s_A \otimes \bar{s}_{\dot{X}}$ and define the components v^a of v relative to $\{e_a\}$ by

$$v^a = -V^{A\dot{X}}\sigma^a{}_{A\dot{X}}, \qquad a = 1,2,3,4.$$

Next suppose $\{\hat{e}_a\}$ is another admissible basis for $\mathcal{M}$, related to $\{e_a\}$ by $\Lambda \in \mathcal{L}$. Let $\Lambda = \Lambda_{\pm\mathcal{G}} = \mathrm{Spin}(\pm\mathcal{G})$ and let $\{\hat{s}^A\}$ be the spin frame related to $\{s^A\}$ by $\mathcal{G}$ (or $-\mathcal{G}$). Then $V = \hat{V}^{A\dot{X}}\hat{s}_A \otimes \bar{\hat{s}}_{\dot{X}}$, where $\hat{V}^{A\dot{X}} = \mathcal{G}^A{}_B \bar{\mathcal{G}}^{\dot{X}}{}_{\dot{Y}} V^{B\dot{Y}}$ ($-\mathcal{G}$ gives the same components). We define the components of v relative to $\{\hat{e}_a\}$ by

$$\hat{v}^a = -\hat{V}^{A\dot{X}}\sigma^a{}_{A\dot{X}}, \qquad a = 1,2,3,4.$$

To justify the definition we must, as usual, verify that the v^a transform correctly, i.e., that $\Lambda^a{}_b v^b = -\hat{V}^{A\dot{X}}\sigma^a{}_{A\dot{X}}$. But

$$
\begin{aligned}
-\hat{V}^{A\dot{X}}\sigma^a{}_{A\dot{X}} &= -\mathcal{G}^A{}_B \bar{\mathcal{G}}^{\dot{X}}{}_{\dot{Y}} V^{B\dot{Y}}\sigma^a{}_{A\dot{X}} \\
&= -\left(\mathcal{G}^A{}_B \bar{\mathcal{G}}^{\dot{X}}{}_{\dot{Y}}\sigma^a{}_{A\dot{X}}\right)V^{B\dot{Y}} \\
&= -\left(\Lambda^a{}_b \sigma^b{}_{B\dot{Y}}\right)V^{B\dot{Y}} \qquad \text{by (3.4.18)} \\
&= \Lambda^a{}_b\left(-V^{B\dot{Y}}\sigma^b{}_{B\dot{Y}}\right) \\
&= \Lambda^a{}_b v^b
\end{aligned}
$$

as required. We summarize:

Theorem 3.4.1. *Let $\{e_a\}$ be an admissible basis for $\mathcal{M}$ and $\{s^A\}$ a spin frame for* ß. *The map which assigns to each vector $v \in \mathcal{M}$ $(v = v^a e_a)$ its spinor equivalent $(V = V^{A\dot{X}} s_A \otimes \bar{s}_{\dot{X}}$, where $V^{A\dot{X}} = \sigma_a{}^{A\dot{X}} v^a)$ is one-to-one and onto the set of all Hermitian spinors of valence $\left(\begin{smallmatrix}1 & 1\\ 0 & 0\end{smallmatrix}\right)$.*

Recall (Section 3.1) that every $v \in \mathcal{M}$ gives rise to a $v^* \in \mathcal{M}^*$ (the dual of $\mathcal{M}$) defined by $v^*(u) = v \cdot u$ and that every element of $\mathcal{M}^*$, i.e., every covector, arises in this way from some $v \in \mathcal{M}$. Moreover, if $\{e_a\}$ is an admissible basis for $\mathcal{M}$ and $\{e^a\}$ is its dual basis for $\mathcal{M}^*$ and if $v = v^a e_a$, then $v^* = v_a e^a$, where $v_a = \eta_{a\alpha} v^\alpha$. Now, for $A = 1, 0$ and $\dot{X} = \dot{1}, \dot{0}$, define

$$V_{A\dot{X}} = \sigma^a{}_{A\dot{X}} v_a. \tag{3.4.20}$$

Exercise 3.4.7. Show that

$$V_{A\dot{X}} = V^{B\dot{Y}} \epsilon_{BA} \bar{\epsilon}_{\dot{Y}\dot{X}}, \tag{3.4.21}$$

where $V^{B\dot{Y}} = \sigma_b{}^{B\dot{Y}} v^b$.

Since $V^{B\dot{Y}}$ are the components, relative to a spin frame $\{s^A\}$, of a spinor of valence $\left(\begin{smallmatrix}1 & 1\\ 0 & 0\end{smallmatrix}\right)$ and (3.4.21) exhibits the $V_{A\dot{X}}$ as the result of two successive contracted outer products of this spinor (with ϵ and $\bar{\epsilon}$), we conclude that the $V_{A\dot{X}}$ are the components, relative to $\{s^A\}$, of a spinor of valence $\left(\begin{smallmatrix}0 & 0\\ 1 & 1\end{smallmatrix}\right)$ which we call the *spinor equivalent* of the covector v^*.

Exercise 3.4.8. Show that, in another spin frame $\{\hat{s}^A\}$ related to $\{s^A\}$ by (3.2.1) and (3.2.6),

$$\hat{V}_{A\dot{X}} = \sigma^a{}_{A\dot{X}} \hat{v}_a, \tag{3.4.22}$$

where $\hat{v}_a = \Lambda_a{}^b v_b$, Λ being $\Lambda_{\pm\mathcal{G}}$.

Theorem 3.4.2. *Let $\{e_a\}$ be an admissible basis for $\mathcal{M}$ and $\{s^A\}$ a spin frame for* ß. *The map which assigns to each covector $v^* \in \mathcal{M}^*$ $(v^* = v_a e^a)$ its spinor equivalent $(V_{A\dot{X}} s^A \otimes \bar{s}^{\dot{X}}$, where $V_{A\dot{X}} = \sigma^a{}_{A\dot{X}} v_a)$ is one-to-one and onto the set of all Hermitian spinors of valence $\left(\begin{smallmatrix}0 & 0\\ 1 & 1\end{smallmatrix}\right)$.*

Exercise 3.4.9. Complete the proof of Theorem 3.4.2. ■

Now, let us fix an admissible basis $\{e_a\}$ and a spin frame $\{s^A\}$. Let $v = v^a e_a$ and $u = u^a e_a$ be in $\mathcal{M}$ and $V = V^{A\dot{X}} s_A \otimes \bar{s}_{\dot{X}}$ and $U = U^{A\dot{X}} s_A \otimes \bar{s}_{\dot{X}}$ the spinor equivalents of v and u. We compute $U_{A\dot{X}} V^{A\dot{X}} =$

$(\sigma^a{}_{A\dot{X}}\, u_a)\,(\sigma_b{}^{A\dot{X}}\, v^b) = (u_a\, v^b)\,(\sigma^a{}_{A\dot{X}}\, \sigma_b{}^{A\dot{X}}) = u_a\, v^b(-\delta^a_b) = -u_a\, v^a = -\eta_{ab}u^b v^a = -u \cdot v$ so

$$U_{A\dot{X}}V^{A\dot{X}} = -u \cdot v. \tag{3.4.23}$$

Observe that if we let

$$\left[V^{A\dot{X}}\right] = \begin{bmatrix} V^{1\dot{1}} & V^{1\dot{0}} \\ V^{0\dot{1}} & V^{0\dot{0}} \end{bmatrix} = \frac{1}{\sqrt{2}} \begin{bmatrix} v^3 + v^4 & v^1 + iv^2 \\ v^1 - iv^2 & -v^3 + v^4 \end{bmatrix},$$

then $\det[V^{A\dot{X}}] = -\frac{1}{2}v \cdot v$ so

$$V_{A\dot{X}}V^{A\dot{X}} = 2\det\left[V^{A\dot{X}}\right] = -v \cdot v. \tag{3.4.24}$$

Consequently, if v is null, $\det[V^{A\dot{X}}] = 0$ so, assuming $v \neq 0$, $[V^{A\dot{X}}]$ has rank 1.

Exercise 3.4.10. Show that if $\left[\begin{smallmatrix} a & b \\ c & d \end{smallmatrix}\right]$ is a 2×2 complex matrix of rank 1, then there exist pairs (ϕ^1, ϕ^0) and (ψ^1, ψ^0) of complex numbers such that

$$\begin{bmatrix} a & b \\ c & d \end{bmatrix} = \begin{bmatrix} \phi^1 \\ \phi^0 \end{bmatrix} \begin{bmatrix} \bar{\psi}^{\dot{1}} & \bar{\psi}^{\dot{0}} \end{bmatrix} = \begin{bmatrix} \phi^1\bar{\psi}^{\dot{1}} & \phi^1\bar{\psi}^{\dot{0}} \\ \phi^0\bar{\psi}^{\dot{1}} & \phi^0\bar{\psi}^{\dot{0}} \end{bmatrix}.$$

Consequently, if $v \in \mathcal{M}$ is null and nonzero we may write $V^{A\dot{X}} = \phi^A\bar{\psi}^{\dot{X}}$ for $A = 1, 0$ and $\dot{X} = \dot{1}, \dot{0}$. Observe that, in another spin frame,

$$\begin{aligned} \hat{V}^{A\dot{X}} &= \mathcal{G}^A{}_B\bar{\mathcal{G}}^{\dot{X}}{}_{\dot{Y}}V^{B\dot{Y}} \\ &= \mathcal{G}^A{}_B\bar{\mathcal{G}}^{\dot{X}}{}_{\dot{Y}}\phi^B\bar{\psi}^{\dot{Y}} \\ &= (\mathcal{G}^A{}_B\phi^B)(\bar{\mathcal{G}}^{\dot{X}}{}_{\dot{Y}}\bar{\psi}^{\dot{Y}}). \end{aligned}$$

Thus, if we define $\hat{\phi}^A = \mathcal{G}^A{}_B\phi^B$ and $\hat{\bar{\psi}}^{\dot{X}} = \bar{\mathcal{G}}^{\dot{X}}{}_{\dot{Y}}\bar{\psi}^{\dot{Y}}$, then

$$\hat{V}^{A\dot{X}} = \hat{\phi}^A\hat{\bar{\psi}}^{\dot{X}}.$$

Consequently, if we let ϕ be the spin vector whose components in $\{s^A\}$ are ϕ^A and $\bar{\psi}$ be the conjugate spin vector whose components in $\{\bar{s}^A\}$ are $\bar{\psi}^{\dot{X}}$, then V is the outer product $\phi \otimes \bar{\psi}$ of ϕ and $\bar{\psi}$. Even more can be said, however.

Exercise 3.4.11. Suppose z_1 and z_2 are two complex numbers for which $z_1\bar{z}_2$ is real. Show that one of z_1 or z_2 is a real multiple of the other.

Now, $V^{1\dot{1}}$ and $V^{0\dot{0}}$ are both real ($\pm v^3 + v^4$) so $\phi^1\bar{\psi}^{\dot{1}}$ and $\phi^0\bar{\psi}^{\dot{0}}$ are real and, since v is null, but not zero, not both can be zero. Exercise 3.4.11 gives an $r_1 \in \mathbb{R}$ such that either $\psi^1 = r_1\phi^1$ or $\phi^1 = r_1\psi^1$ and also an $r_0 \in \mathbb{R}$ such that either $\psi^0 = r_0\phi^0$ or $\phi^0 = r_0\psi^0$. Since at least one of r_1 or r_0 is nonzero we may assume without loss of generality that

$$\begin{bmatrix} \psi^1 \\ \psi^0 \end{bmatrix} = \begin{bmatrix} r_1\phi^1 \\ r_0\phi^0 \end{bmatrix}.$$

We claim that, in fact, there exists a single real number r such that

$$\begin{bmatrix} \psi^1 \\ \psi^0 \end{bmatrix} = r \begin{bmatrix} \phi^1 \\ \phi^0 \end{bmatrix}. \tag{3.4.25}$$

To prove this we first suppose $\phi^1 = 0$. Then $\psi^1 = 0$ so $\begin{bmatrix} \psi^1 \\ \psi^0 \end{bmatrix} = \begin{bmatrix} 0 \\ \psi^0 \end{bmatrix} = r_0\begin{bmatrix} 0 \\ \phi^0 \end{bmatrix} = r_0\begin{bmatrix} \phi^1 \\ \phi^0 \end{bmatrix}$. Similarly, if $\phi^0 = 0$, then $\begin{bmatrix} \psi^1 \\ \psi^0 \end{bmatrix} = r_1\begin{bmatrix} \phi^1 \\ \phi^0 \end{bmatrix}$. Now, suppose neither ϕ^1 nor ϕ^0 is zero. Then, since $\overline{V^{1\dot{0}}} = V^{0\dot{1}}$,

$$\begin{aligned} \overline{\phi^1\bar{\psi}^{\dot{0}}} &= \phi^0\bar{\psi}^{\dot{1}}, \\ \bar{\phi}^{\dot{1}}\psi^0 &= \phi^0\bar{\psi}^{\dot{1}}, \\ \frac{\bar{\phi}^{\dot{1}}}{\phi^0}\psi^0 &= \bar{\psi}^{\dot{1}}. \end{aligned} \tag{3.4.26}$$

Thus, $\bar{\psi}^{\dot{1}} = 0$ would give $\psi^0 = 0$ and $\psi^1 = 0$ so $[V^{A\dot{X}}] = \left[\begin{smallmatrix} 0 & 0 \\ 0 & 0 \end{smallmatrix}\right]$ and this gives $v^1 = v^2 = v^3 = v^4 = 0$, contrary to our assumption that $v \neq 0$. Similarly, $\psi^0 = 0$ implies $v = 0$, again a contradiction. Thus, ψ^1 and ψ^0 are nonzero so (3.4.26) gives

$$\frac{\bar{\phi}^{\dot{1}}}{\phi^0} = \frac{\bar{\psi}^{\dot{1}}}{\psi^0} = \frac{r_1\bar{\phi}^{\dot{1}}}{r_0\phi^0}$$

(since $r_1 \in \mathbb{R}$). Consequently, $r_1 = r_0$ so $\begin{bmatrix} \psi^1 \\ \psi^0 \end{bmatrix} = \begin{bmatrix} r_1\phi^1 \\ r_1\phi^0 \end{bmatrix} = r_1\begin{bmatrix} \phi^1 \\ \phi^0 \end{bmatrix}$ and (3.4.25) is proved with $r = r_1$. From this it follows that

$$\begin{aligned} V^{A\dot{X}} &= \phi^A\bar{\psi}^{\dot{X}} = \phi^A\left(r\bar{\phi}^{\dot{X}}\right) \\ &= \pm\left(|\,r\,|^{\frac{1}{2}}\,\phi^A\right)\left(|\,r\,|^{\frac{1}{2}}\,\bar{\phi}^{\dot{X}}\right) \end{aligned}$$

(+ if $r > 0$ and $-$ if $r < 0$). Now we define a spin vector ξ by $\xi^A = |\,r\,|^{\frac{1}{2}}\,\phi^A$ (relative to $\{s^A\}$). Then $\bar{\xi}^{\dot{X}} = |\,r\,|^{\frac{1}{2}}\,\bar{\phi}^{\dot{X}}$ since $|\,r\,|^{\frac{1}{2}}$ is real. Thus,

$$V^{A\dot{X}} = \pm\xi^A\bar{\xi}^{\dot{X}}.$$

Finally, observe that $v^3 + v^4 = V^{1\dot{1}} = r\phi^1\bar{\phi}^{\dot{1}}$ and $-v^3 + v^4 = V^{0\dot{0}} = r\phi^0\bar{\phi}^{\dot{0}}$ so

$$v^4 = \frac{1}{2}r\left(|\,\phi^1\,|^{\frac{1}{2}} + |\,\phi^0\,|^{\frac{1}{2}}\right)$$

and, in particular, $r > 0$ if and only if $v^4 > 0$. We have therefore proved

Theorem 3.4.3. *Let $\{e_a\}$ be an admissible basis for $\mathcal{M}$ and $\{s^A\}$ a spin frame for* ß. *Let $v \in \mathcal{M}$ be a nonzero null vector, $v = v^a e_a$, and V its spinor equivalent, $V = V^{A\dot{X}} s_A \otimes \bar{s}_{\dot{X}} = (\sigma_a{}^{A\dot{X}} v^a) s_A \otimes \bar{s}_{\dot{X}}$. Then there exists a spin vector ξ such that:*

(a) If v is future-directed, then

$$V^{A\dot{X}} = \xi^A\bar{\xi}^{\dot{X}},$$

and,

(b) If v is past-directed, then

$$V^{A\dot{X}} = -\xi^A\bar{\xi}^{\dot{X}}.$$

Notice that ξ in the theorem is certainly not unique since if $\nu^A = e^{i\theta}\xi^A$ $(\theta \in \mathbb{R})$, then $\bar{\nu}^{\dot{X}} = e^{-i\theta}\bar{\xi}^{\dot{X}}$ so $\nu^A\bar{\nu}^{\dot{X}} = \xi^A\bar{\xi}^{\dot{X}} = V^{A\dot{X}}$.

Observe that the process we have just described can be reversed as well. That is, given a nonzero spin vector ξ^A we define the spinor $V^{A\dot{X}} = \xi^A\bar{\xi}^{\dot{X}}$. Then $\det[V^{A\dot{X}}] = \xi^1\bar{\xi}^{\dot{1}}\xi^0\bar{\xi}^{\dot{0}} - \xi^1\bar{\xi}^{\dot{0}}\xi^0\bar{\xi}^{\dot{1}} = 0$ so the vector equivalent $v^a = -\sigma^a{}_{A\dot{X}}V^{A\dot{X}}$ gives a null vector $v \in \mathcal{M}$ and, moreover, $v^4 = -V^{A\dot{X}}\sigma^4{}_{A\dot{X}} = -(-\frac{1}{\sqrt{2}})\,(V^{1\dot{1}}\,\sigma^4{}_{1\dot{1}} + V^{0\dot{0}}\,\sigma^4{}_{0\dot{0}}) = \frac{1}{\sqrt{2}}(V^{1\dot{1}} + V^{0\dot{0}}) = \frac{1}{\sqrt{2}}(\xi^1\,\bar{\xi}^{\dot{1}} + \xi^0\,\bar{\xi}^{\dot{0}}) = \frac{1}{\sqrt{2}}(|\,\xi^1\,|^2 + |\,\xi^0\,|^2) > 0$ so v is future-directed. Thus, every nonzero spin vector ξ gives rise in a natural way to a future-directed null vector v which we will call the *flagpole of* ξ and which will play a prominant role in the geometrical representation of ξ that we construct in the next section.

3.5 Bivectors and Null Flags

We recall (Section 2.7) that a *bivector* on $\mathcal{M}$ is a real-valued bilinear form $\tilde{F} : \mathcal{M} \times \mathcal{M} \to \mathbb{R}$ that is skew-symmetric $[\tilde{F}(u,v) = -\tilde{F}(v,u)$ for all u and v in $\mathcal{M}]$. Thus, $\tilde{F}$ is a skew-symmetric world tensor of covariant rank 2 and contravariant rank 0. We have already seen that bivectors are useful for the description of electromagnetic fields and will return to their role in electromagnetic theory in the next section. For the present our objective is to find a "spinor equivalent" for an arbitrary bivector, show how a spin vector gives rise, in a natural way, to a bivector and construct from it a geometrical representation ("up to sign") for an arbitrary nonzero spin vector. This geometrical picture of a spin vector, called a "null flag", emphasizes what is perhaps its most fundamental characteristic, that is, an essential "two-valuedness".

Now fix an admissible basis $\{e_a\}$ for $\mathcal{M}$ and a spin frame $\{s^A\}$ for $\mathfrak{B}$. The components of $\tilde{F}$ relative to $\{e_a\}$ are given by $F_{ab} = \tilde{F}(e_a, e_b)$ and, by skew-symmetry, satisfy

$$F_{ab} = \tfrac{1}{2}(F_{ab} - F_{ba}) = F_{[ab]}. \tag{3.5.1}$$

For $A, B = 1, 0$ and $\dot{X}, \dot{Y} = \dot{1}, \dot{0}$ we define

$$F_{A\dot{X}B\dot{Y}} = F_{AB\dot{X}\dot{Y}} = \sigma^a{}_{A\dot{X}}\sigma^b{}_{B\dot{Y}}F_{ab}$$

and take these to be the components of the *spinor equivalent of* $\tilde{F}$ relative to $\{s^A\}$. Thus, in another spin frame $\{\hat{s}^A\}$, related to $\{s^A\}$ by (3.2.1) and (3.2.6),

$$\begin{aligned}
\hat{F}_{A\dot{X}B\dot{Y}} &= G_A{}^{A_1}\bar{G}_{\dot{X}}{}^{\dot{X}_1}G_B{}^{B_1}\bar{G}_{\dot{Y}}{}^{\dot{Y}_1}F_{A_1\dot{X}_1B_1\dot{Y}_1} \\
&= G_A{}^{A_1}\bar{G}_{\dot{X}}{}^{\dot{X}_1}G_B{}^{B_1}\bar{G}_{\dot{Y}}{}^{\dot{Y}_1}\left(\sigma^\alpha{}_{A_1\dot{X}_1}\sigma^\beta{}_{B_1\dot{Y}_1}F_{\alpha\beta}\right) \\
&= \left(G_A{}^{A_1}\bar{G}_{\dot{X}}{}^{\dot{X}_1}\sigma^\alpha{}_{A_1\dot{X}_1}\right)\left(G_B{}^{B_1}\bar{G}_{\dot{Y}}{}^{\dot{Y}_1}\sigma^\beta{}_{B_1\dot{Y}_1}\right)F_{\alpha\beta} \\
&= \left(\Lambda_a{}^\alpha\sigma^a{}_{A\dot{X}}\right)\left(\Lambda_b{}^\beta\sigma^b{}_{B\dot{Y}}\right)F_{\alpha\beta} \qquad \text{by (3.4.17)} \\
&= \sigma^a{}_{A\dot{X}}\sigma^b{}_{B\dot{Y}}\left(\Lambda_a{}^\alpha\Lambda_b{}^\beta F_{\alpha\beta}\right),
\end{aligned}$$

where $\Lambda = \Lambda_G$. Thus,

$$\hat{F}_{A\dot{X}B\dot{Y}} = \sigma^a{}_{A\dot{X}}\sigma^b{}_{B\dot{Y}}\hat{F}_{ab}. \tag{3.5.2}$$

We list some useful properties of the spinor equivalent of a bivector.

$$F_{ab} = \sigma_a{}^{A\dot{X}}\sigma_b{}^{B\dot{Y}}F_{A\dot{X}B\dot{Y}}, \qquad a, b = 1, 2, 3, 4, \tag{3.5.3}$$

$$\bar{F}_{A\dot{X}B\dot{Y}} = F_{A\dot{X}B\dot{Y}}, \qquad \text{i.e., } F_{A\dot{X}B\dot{Y}} \text{ is Hermitian,} \tag{3.5.4}$$

$$F_{B\dot{Y}A\dot{X}} = -F_{A\dot{X}B\dot{Y}}. \tag{3.5.5}$$

The proof of (3.5.5) proceeds as follows: $F_{B\dot{Y}A\dot{X}} = \sigma^a{}_{B\dot{Y}}\sigma^b{}_{A\dot{X}}F_{ab} = \sigma^a{}_{B\dot{Y}}\sigma^b{}_{A\dot{X}}(-F_{ba}) = -\sigma^b{}_{A\dot{X}}\sigma^a{}_{B\dot{Y}}F_{ba} = -\sigma^a{}_{A\dot{X}}\sigma^b{}_{B\dot{Y}}F_{ab} = -F_{A\dot{X}B\dot{Y}}$.

Exercise 3.5.1 Prove (3.5.3) and (3.5.4).

Now we use (3.5.5) to write

$$\begin{aligned} F_{A\dot{X}B\dot{Y}} &= \tfrac{1}{2}\left[F_{A\dot{X}B\dot{Y}} - F_{B\dot{Y}A\dot{X}}\right] \\ &= \tfrac{1}{2}\left[F_{A\dot{X}B\dot{Y}} - F_{B\dot{X}A\dot{Y}} + F_{B\dot{X}A\dot{Y}} - F_{B\dot{Y}A\dot{X}}\right] \\ &= \tfrac{1}{2}\left[F_{A\dot{X}B\dot{Y}} - F_{B\dot{X}A\dot{Y}}\right] + \tfrac{1}{2}\left[F_{B\dot{X}A\dot{Y}} - F_{B\dot{Y}A\dot{X}}\right]. \end{aligned}$$

Observe that by (3.3.14), $\epsilon_{AB}\epsilon^{CD}F_{C\dot{X}D\dot{Y}} = (\delta_A^C\delta_B^D - \delta_A^D\delta_B^C)F_{C\dot{X}D\dot{Y}} = F_{A\dot{X}B\dot{Y}} - F_{B\dot{X}A\dot{Y}}$ and, similarly, $\bar{\epsilon}_{\dot{X}\dot{Y}}\bar{\epsilon}^{\dot{U}\dot{V}}F_{B\dot{U}A\dot{V}} = F_{B\dot{X}A\dot{Y}} - F_{B\dot{Y}A\dot{X}}$ so

$$\begin{aligned} F_{A\dot{X}B\dot{Y}} &= \tfrac{1}{2}\epsilon_{AB}\epsilon^{CD}F_{C\dot{X}D\dot{Y}} + \tfrac{1}{2}\bar{\epsilon}_{\dot{X}\dot{Y}}\bar{\epsilon}^{\dot{U}\dot{V}}F_{B\dot{U}A\dot{V}} \\ &= \epsilon_{AB}\left(\tfrac{1}{2}\epsilon^{CD}F_{C\dot{X}D\dot{Y}}\right) + \bar{\epsilon}_{\dot{X}\dot{Y}}\left(\tfrac{1}{2}\epsilon^{\dot{U}\dot{V}}F_{B\dot{U}A\dot{V}}\right) \\ &= \epsilon_{AB}\left(\tfrac{1}{2}F_{C\dot{X}}{}^{C}{}_{\dot{Y}}\right) + \bar{\epsilon}_{\dot{X}\dot{Y}}\left(\tfrac{1}{2}F_{B\dot{U}A}{}^{\dot{U}}\right) \end{aligned}$$

$$F_{A\dot{X}B\dot{Y}} = \epsilon_{AB}\left(\tfrac{1}{2}F_{C\dot{X}}{}^{C}{}_{\dot{Y}}\right) + \bar{\epsilon}_{\dot{X}\dot{Y}}\left(\tfrac{1}{2}F_{\dot{U}B}{}^{\dot{U}}{}_{A}\right) \tag{3.5.6}$$

Now define ϕ_{AB} by

$$\phi_{AB} = \tfrac{1}{2}F_{\dot{U}A}{}^{\dot{U}}{}_{B}, \quad A, B = 1, 0.$$

Then we claim that

$$\phi_{BA} = \phi_{AB} \tag{3.5.7}$$

and

$$\bar{\phi}_{\dot{X}\dot{Y}} = \tfrac{1}{2}F_{C\dot{X}}{}^{C}{}_{\dot{Y}}. \tag{3.5.8}$$

To prove (3.5.7) we compute

$$\begin{aligned} \phi_{BA} &= \tfrac{1}{2}F_{\dot{U}B}{}^{\dot{U}}{}_{A} = -\tfrac{1}{2}F^{\dot{U}}{}_{A\dot{U}B} \quad \text{by (3.5.5)} \\ &= -\tfrac{1}{2}\left[\bar{\epsilon}^{\dot{U}\dot{V}}F_{\dot{V}A}{}^{\dot{W}}{}_{B}\bar{\epsilon}_{\dot{W}\dot{U}}\right] \\ &= -\tfrac{1}{2}\left[F_{\dot{V}A}{}^{\dot{W}}{}_{B}(\bar{\epsilon}^{\dot{U}\dot{V}}\bar{\epsilon}_{\dot{W}\dot{U}})\right] \\ &= -\tfrac{1}{2}\left[F_{\dot{V}A}{}^{\dot{W}}{}_{B}(-\bar{\epsilon}^{\dot{U}\dot{V}}\bar{\epsilon}_{\dot{U}\dot{W}})\right] \\ &= -\tfrac{1}{2}\left[F_{\dot{V}A}{}^{\dot{W}}{}_{B}(-\delta_{\dot{W}}^{\dot{V}})\right] \\ &= \tfrac{1}{2}F_{\dot{V}A}{}^{\dot{V}}{}_{B} = \phi_{AB}. \end{aligned}$$

Exercise 3.5.2 Prove (3.5.8).

With this we may write (3.5.6) as

$$F_{A\dot{X}B\dot{Y}} = \epsilon_{AB}\bar{\phi}_{\dot{X}\dot{Y}} + \phi_{AB}\bar{\epsilon}_{\dot{X}\dot{Y}}. \tag{3.5.9}$$

We observe next that the process which just led us from $\tilde{F}$ to $F_{A\dot{X}B\dot{Y}}$ to ϕ_{AB} can be reversed in the following sense: Given a symmetric spinor ϕ_{AB} of valence $\left(\begin{smallmatrix}0&0\\2&0\end{smallmatrix}\right)$ we can define $F_{A\dot{X}B\dot{Y}} = \epsilon_{AB}\bar{\phi}_{\dot{X}\dot{Y}} + \phi_{AB}\bar{\epsilon}_{\dot{X}\dot{Y}}$ and obtain a spinor of valence $\left(\begin{smallmatrix}0&0\\2&2\end{smallmatrix}\right)$ which satisfies (3.5.4) since $\bar{F}_{A\dot{X}B\dot{Y}} = \overline{(F_{\dot{A}X\dot{B}Y})} = \overline{(F_{X\dot{A}Y\dot{B}})} = \overline{(\epsilon_{XY}\bar{\phi}_{\dot{A}\dot{B}} + \phi_{XY}\bar{\epsilon}_{\dot{A}\dot{B}})} = \bar{\epsilon}_{\dot{X}\dot{Y}}\phi_{AB} + \bar{\phi}_{\dot{X}\dot{Y}}\epsilon_{AB} = \epsilon_{AB}\,\bar{\phi}_{\dot{X}\dot{Y}} + \phi_{AB}\bar{\epsilon}_{\dot{X}\dot{Y}} = F_{A\dot{X}B\dot{Y}}$, and (3.5.5) since $F_{B\dot{Y}A\dot{X}} = \epsilon_{BA}\bar{\phi}_{\dot{Y}\dot{X}} + \phi_{BA}\bar{\epsilon}_{\dot{Y}\dot{X}} = (-\epsilon_{AB})\,\bar{\phi}_{\dot{X}\dot{Y}} + \phi_{AB}(-\bar{\epsilon}_{\dot{X}\dot{Y}}) = -F_{A\dot{X}B\dot{Y}}$. Now define F_{ab} by (3.5.3), i.e.,

$$F_{ab} = \sigma_a{}^{A\dot{X}}\sigma_b{}^{B\dot{Y}}F_{A\dot{X}B\dot{Y}}.$$

Relative to another spin frame $\{\hat{s}^A\}$, related to $\{s^A\}$ by (3.2.1) and (3.2.6), $\hat{F}_{A\dot{X}B\dot{Y}} = G_A{}^{A_1}\bar{G}_{\dot{X}}{}^{\dot{X}_1}G_B{}^{B_1}\bar{G}_{\dot{Y}}{}^{\dot{Y}_1}F_{A_1\dot{X}_1B_1\dot{Y}_1}$.

Exercise 3.5.3. Show that $\sigma_a{}^{A\dot{X}}\sigma_b{}^{B\dot{Y}}\hat{F}_{A\dot{X}B\dot{Y}} = \Lambda_a{}^\alpha\Lambda_b{}^\beta F_{\alpha\beta}$, where $\Lambda = \Lambda_G$.

Thus, defining $\hat{F}_{ab} = \sigma_a{}^{A\dot{X}}\sigma_b{}^{B\dot{Y}}\hat{F}_{A\dot{X}B\dot{Y}}$, we find that the F_{ab} transform as the components of a bivector and we may define $\tilde{F}\colon \mathcal{M}\times\mathcal{M}\to\mathbb{R}$ by

$$\begin{aligned}\tilde{F}(u,v) &= \tilde{F}\left(u^a e_a, v^b e_b\right)\\ &= \tilde{F}\left(e_a, e_b\right)u^a v^b\\ &= F_{ab}u^a v^b\end{aligned}$$

relative to any admissible basis. Thus, every symmetric spinor ϕ of valence $\left(\begin{smallmatrix}0&0\\2&0\end{smallmatrix}\right)$ gives rise, in a natural way, to a bivector $\tilde{F}$.

Next we use the information accumulated thus far to construct a geometrical representation ("up to sign") of an arbitrary nonzero spin vector ξ. We begin, as at the end of Section 3.4, by constructing the flagpole v of ξ (the future-directed null vector equivalent of $V^{A\dot{X}} = \xi^A\bar{\xi}^{\dot{X}}$). Observe that every spin vector in the family $\{e^{i\theta}\xi : \theta\in\mathbb{R}\}$ has the same flagpole as ξ since, if ξ^A is replaced by $e^{i\theta}\xi^A$, then $\bar{\xi}^{\dot{X}}$ becomes $e^{-i\theta}\bar{\xi}^{\dot{X}}$ and $(e^{i\theta}\xi^A)(e^{-i\theta}\bar{\xi}^{\dot{X}}) = \xi^A\bar{\xi}^{\dot{X}}$. We call $e^{i\theta}$ the *phase factor* of the corresponding member of the family.

Exercise 3.5.4. Show that, conversely, if ψ is a spin vector with the same flagpole as ξ, then $\psi^A = e^{i\theta}\xi^A$ for some $\theta\in\mathbb{R}$. *Hint:* Write out v^1, v^2, v^3 and v^4 in terms of ξ^A and ψ^A, show that $\psi^1 = e^{i\theta_1}\xi^1$ and

$\psi^0 = e^{i\theta_0}\xi^0$ and then show $\theta_1 = \theta_0 + 2n\pi$ for some $n = 0, \pm 1, \ldots$.

Our geometrical representation of ξ must therefore contain more than just the flagpole if it is to distinguish spin vectors which differ only by a phase factor. To determine this additional element in the picture we now observe that ξ also determines a symmetric spinor ϕ of valence $\left(\begin{smallmatrix}0 & 0\\ 2 & 0\end{smallmatrix}\right)$ defined by

$$\phi_{AB} = \xi_A \xi_B.$$

As we saw in the discussion following (3.5.9), ϕ_{AB} gives rise to a spinor of valence $\left(\begin{smallmatrix}0 & 0\\ 2 & 2\end{smallmatrix}\right)$ defined by

$$F_{A\dot{X}B\dot{Y}} = \epsilon_{AB}\bar{\phi}_{\dot{X}\dot{Y}} + \phi_{AB}\bar{\epsilon}_{\dot{X}\dot{Y}},$$

which satisfies (3.5.4) and (3.5.5) and which, in turn, determines a bivector $\tilde{F}$ given by $F_{ab} = \sigma_a{}^{A\dot{X}}\sigma_b{}^{B\dot{Y}}F_{A\dot{X}B\dot{Y}}$, i.e.,

$$F_{ab} = \sigma_a{}^{A\dot{X}}\sigma_b{}^{B\dot{Y}}\left(\epsilon_{AB}\bar{\xi}_{\dot{X}}\bar{\xi}_{\dot{Y}} + \xi_A\xi_B\bar{\epsilon}_{\dot{X}\dot{Y}}\right). \tag{3.5.10}$$

To simplify (3.5.10) we select a spin vector η which, together with ξ, form a spin frame $\{\xi, \eta\}$ with

$$< \eta, \xi > = \xi_A\eta^A = 1 = -\xi^A\eta_A.$$

Exercise 3.5.5. Show that

$$\xi_A\eta_B - \xi_B\eta_A = \epsilon_{AB} \tag{3.5.11}$$

and

$$\bar{\xi}_{\dot{X}}\bar{\eta}_{\dot{Y}} - \bar{\xi}_{\dot{Y}}\bar{\eta}_{\dot{X}} = \bar{\epsilon}_{\dot{X}\dot{Y}}. \tag{3.5.12}$$

Substitute (3.5.11) and (3.5.12) into (3.5.10) to obtain

$$\begin{aligned}
F_{ab} &= \sigma_a{}^{A\dot{X}}\sigma_b{}^{B\dot{Y}}\left[(\xi_A\eta_B - \xi_B\eta_A)\bar{\xi}_{\dot{X}}\bar{\xi}_{\dot{Y}} + (\bar{\xi}_{\dot{X}}\bar{\eta}_{\dot{Y}} - \bar{\xi}_{\dot{Y}}\bar{\eta}_{\dot{X}})\xi_A\xi_B\right] \\
&= \sigma_a{}^{A\dot{X}}\sigma_b{}^{B\dot{Y}}\xi_A\eta_B\bar{\xi}_{\dot{X}}\bar{\xi}_{\dot{Y}} - \sigma_a{}^{A\dot{X}}\sigma_b{}^{B\dot{Y}}\xi_B\eta_A\bar{\xi}_{\dot{X}}\bar{\xi}_{\dot{Y}} \\
&\qquad + \sigma_a{}^{A\dot{X}}\sigma_b{}^{B\dot{Y}}\bar{\xi}_{\dot{X}}\bar{\eta}_{\dot{Y}}\xi_A\xi_B - \sigma_a{}^{A\dot{X}}\sigma_b{}^{B\dot{Y}}\bar{\xi}_{\dot{Y}}\bar{\eta}_{\dot{X}}\xi_A\xi_B \\
&= (\sigma_a{}^{A\dot{X}}\xi_A\bar{\xi}_{\dot{X}})(\sigma_b{}^{B\dot{Y}}\eta_B\bar{\xi}_{\dot{Y}} + \sigma_b{}^{B\dot{Y}}\xi_B\bar{\eta}_{\dot{Y}}) \\
&\qquad - (\sigma_b{}^{B\dot{Y}}\xi_B\bar{\xi}_{\dot{Y}})(\sigma_a{}^{A\dot{X}}\eta_A\bar{\xi}_{\dot{X}} + \sigma_a{}^{A\dot{X}}\xi_A\bar{\eta}_{\dot{X}}) \\
&= v_a\sigma_b{}^{B\dot{Y}}(\eta_B\bar{\xi}_{\dot{Y}} + \xi_B\bar{\eta}_{\dot{Y}}) - v_b\sigma_a{}^{A\dot{X}}(\eta_A\bar{\xi}_{\dot{X}} + \xi_A\bar{\eta}_{\dot{X}}).
\end{aligned}$$

Now define a spinor of valence $\binom{0\ 0}{1\ 1}$ by

$$W_{A\dot{X}} = \eta_A \bar{\xi}_{\dot{X}} + \xi_A \bar{\eta}_{\dot{X}}$$

and observe that $W_{A\dot{X}}$ is Hermitian since $\bar{W}_{A\dot{X}} = \overline{W_{\dot{A}X}} = \overline{(\bar{\eta}_{\dot{A}}\xi_X + \bar{\xi}_{\dot{A}}\eta_X)}$ $= \eta_A \bar{\xi}_{\dot{X}} + \xi_A \bar{\eta}_{\dot{X}} = W_{A\dot{X}}$. Consequently (Theorem 3.4.2), we may define a covector $w^* \in \mathcal{M}^*$ by

$$w_a = -\sigma_a{}^{A\dot{X}} W_{A\dot{X}} = -\sigma_a{}^{A\dot{X}} \left(\eta_A \bar{\xi}_{\dot{X}} + \xi_A \bar{\eta}_{\dot{X}}\right).$$

Thus, our expression for F_{ab} now becomes

$$F_{ab} = v_b w_a - v_a w_b. \tag{3.5.13}$$

Notice that, by (3.4.23),

$$\begin{aligned} v \cdot w &= -V^{A\dot{X}} W_{A\dot{X}} = -\xi^A \bar{\xi}^{\dot{X}} \Big(\eta_A \bar{\xi}_{\dot{X}} + \xi_A \bar{\eta}_{\dot{X}}\Big) \\ &= -\xi^A \eta_A \Big(\bar{\xi}^{\dot{X}} \bar{\xi}_{\dot{X}}\Big) - \bar{\xi}^{\dot{X}} \bar{\eta}_{\dot{X}} (\xi^A \xi_A) \\ &= -(-1)(0) - (-1)(0) \\ &= 0. \end{aligned}$$

Thus, w is orthogonal to v. Since v is null, w is spacelike.

Exercise 3.5.6. Show that, in fact, $w \cdot w = 2$.

Thus far we have found that the spin vector ξ determines a future-directed null vector v (its flagpole) and a bivector $F_{ab} = v_b w_a - v_a w_b$, where w is a spacelike vector orthogonal to v. However, w is not uniquely determined by ξ since our choice for the "spinor mate" η for ξ is not unique. We now examine the effect on w of making a different selection $\tilde{\eta}$ for η (still with $< \tilde{\eta}, \xi > = 1$). But $< \eta, \xi > = < \tilde{\eta}, \xi > = 1$ implies $< \eta - \tilde{\eta}, \xi > = < \eta, \xi > - < \tilde{\eta}, \xi > = 1 - 1 = 0$ so, by (g) of Lemma 3.2.1 and the fact that ξ is not the zero element of $\mathfrak{B}$, $\tilde{\eta} - \eta = \lambda \xi$ for some $\lambda \in \mathbb{C}$, i.e.,

$$\tilde{\eta} = \eta + \lambda \xi.$$

The new vector $\tilde{w}$ is then determined by

$$\begin{aligned} \tilde{w}_a &= -\sigma_a{}^{A\dot{X}} (\tilde{\eta}_A \bar{\xi}_{\dot{X}} + \xi_A \bar{\tilde{\eta}}_{\dot{X}}) \\ &= -\sigma_a{}^{A\dot{X}} ((\eta_A + \lambda \xi_A) \bar{\xi}_{\dot{X}} + \xi_A (\bar{\eta}_{\dot{X}} + \bar{\lambda} \bar{\xi}_{\dot{X}})) \\ &= -\sigma_a{}^{A\dot{X}} (\eta_A \bar{\xi}_{\dot{X}} + \xi_A \bar{\eta}_{\dot{X}}) - (\lambda + \bar{\lambda}) \sigma_a{}^{A\dot{X}} \xi_A \bar{\xi}_{\dot{X}} \\ &= w_a + (\lambda + \bar{\lambda}) v_a, \end{aligned}$$

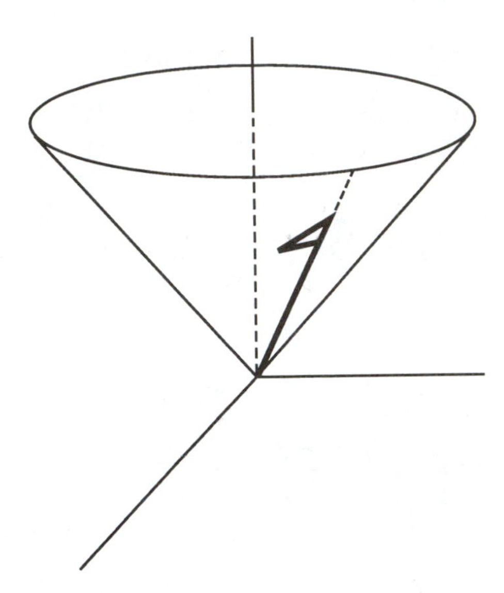

Figure 3.5.1

so

$$\tilde{w} = w + (\lambda + \bar{\lambda})v. \tag{3.5.14}$$

It follows that $\tilde{w}$ lies in the 2-dimensional plane spanned by v and w and is a spacelike vector orthogonal to v (again, $\tilde{w} \cdot \tilde{w} = 2$). Thus, ξ *uniquely* determines a future-directed null vector v and a 2-dimensional plane spanned by v and any of the spacelike vectors $w, \tilde{w}, \ldots$ determined by (3.5.14). This 2-dimensional plane lies in the 3-dimensional subspace $(\mathrm{Span}\{v\})^{\perp}$, which is tangent to the null cone along v. In a 3-dimensional picture, the null cone and $(\mathrm{Span}\{v\})^{\perp}$ appear 2-dimensional so this 2-dimensional plane is a line. However, to stress its 2-dimensionality we shall draw it as a "flag" along v as in Figure 3.5.1. The pair consisting of v and this 2-dimensional plane in $(\mathrm{Span}\{v\})^{\perp}$ is called the *null flag* of ξ and is, we claim, an accurate geometrical representation of ξ "up to sign". To see this we examine the effect of a phase change

$$\xi^A \longrightarrow e^{i\theta}\xi^A \qquad (\theta \in \mathbb{R}).$$

Of course, the flagpole v is unchanged, but $\bar{\xi}^{\dot{X}} \to e^{-i\theta}\bar{\xi}^{\dot{X}}$ so $F_{ab} \to \sigma_a{}^{A\dot{X}}\sigma_b{}^{B\dot{Y}}(e^{-2\theta i}\epsilon_{AB}\bar{\xi}_{\dot{X}}\bar{\xi}_{\dot{Y}} + e^{2\theta i}\xi_A\xi_B\bar{\epsilon}_{\dot{X}\dot{Y}})$. A spinor mate for $e^{i\theta}\xi^A$ must have the property that its $\mathfrak{B}$-inner product with $e^{i\theta}\xi^A$ is 1. Since $\dim \mathfrak{B} = 2$, it must be of the form

$$e^{-i\theta}\eta^A + k\xi^A$$

for some $k \in \mathbb{C}$. Thus,

$$\begin{aligned} w_a \longrightarrow & -\sigma_a{}^{A\dot{X}}\big[(e^{-i\theta}\eta_A + k\xi_A)(e^{-i\theta}\bar{\xi}_{\dot{X}}) + (e^{i\theta}\xi_A)(e^{i\theta}\bar{\eta}_{\dot{X}} + \bar{k}\bar{\xi}_{\dot{X}})\big] \\ = & -\sigma_a{}^{A\dot{X}}\big[e^{-2\theta i}\eta_A\bar{\xi}_{\dot{X}} + ke^{-i\theta}\xi_A\bar{\xi}_{\dot{X}} + e^{2\theta i}\xi_A\bar{\eta}_{\dot{X}} + \bar{k}e^{i\theta}\xi_A\bar{\xi}_{\dot{X}}\big] \\ = & -\sigma_a{}^{A\dot{X}}(e^{2\theta i}\xi_A\bar{\eta}_{\dot{X}} + e^{-2\theta i}\eta_A\bar{\xi}_{\dot{X}}) - (ke^{-i\theta} + \bar{k}e^{i\theta})(\sigma_a{}^{A\dot{X}}\xi_A\bar{\xi}_{\dot{X}}) \\ = & -\sigma_a{}^{A\dot{X}}\big[(\cos 2\theta + i\sin 2\theta)\xi_A\bar{\eta}_{\dot{X}} + (\cos 2\theta - i\sin 2\theta)\eta_A\bar{\xi}_{\dot{X}}\big] \\ & + rv_a \\ = & \cos 2\theta\left(-\sigma_a{}^{A\dot{X}}(\xi_A\bar{\eta}_{\dot{X}} + \eta_A\bar{\xi}_{\dot{X}})\right) \\ & + \sin 2\theta\left(-\sigma_a{}^{A\dot{X}}i(\xi_A\bar{\eta}_{\dot{X}}\eta_A\bar{\xi}_{\dot{X}})\right) + rv_a, \end{aligned}$$

where $r = ke^{-i\theta} + \bar{k}e^{i\theta} = ke^{-i\theta} + \overline{(ke^{-i\theta})} \in \mathbb{R}$. Now, $-\sigma_a{}^{A\dot{X}}(\xi_A\bar{\eta}_{\dot{X}} + \eta_A\bar{\xi}_{\dot{X}}) = w_a$. Moreover, observe that if $U_{A\dot{X}} = \xi_A\bar{\eta}_{\dot{X}} - \eta_A\bar{\xi}_{\dot{X}}$, then $\bar{U}_{A\dot{X}} = -U_{A\dot{X}}$ so, by Exercise 3.4.6, $iU_{A\dot{X}}$ is Hermitian and therefore, by Theorem 3.4.2, $u_a = -\sigma_a{}^{A\dot{X}}iU_{A\dot{X}}$ defines a covector u^* in $\mathcal{M}^*$. Thus, $w_a \to w_a\cos 2\theta + u_a\sin 2\theta + rv_a$ so the phase change $\xi^A \to e^{i\theta}\xi^A$ leaves v alone and gives a new w of

$$w \longrightarrow (\cos 2\theta)w + (\sin 2\theta)u + rv.$$

Exercise 3.5.7. Compute $w^a u_a$, $v^a u_a$ and $u^a u_a$ to show that u is orthogonal to w and v and satisfies $u \cdot u = 2$.

Thus, we picture w and u as perpendicular spacelike vectors in the 3-space $(\mathrm{Span}\{v\})^\perp$ tangent to the null cone along v. Then $(\cos 2\theta)w + (\sin 2\theta)u$ is a spacelike vector in the plane of w and u making an angle of 2θ with w. After a phase change $\xi^A \to e^{i\theta}\xi^A$ the new w is in the plane of v and $(\cos 2\theta)w + (\sin 2\theta)u$. The 2-plane containing v and this new w is the new flag. Thus, a phase change $\xi^A \to e^{i\theta}\xi^A$ leaves the flagpole v unchanged and *rotates* the flag by 2θ in the plane of w and u (in Figure 3.5.2 we have drawn the flagpole vertically even though it lies along a null line). Notice that if $\theta = \pi$, then the phase change $\xi^A \to e^{\pi i}\xi^A = -\xi^A$ carries ξ to $-\xi$, but the null flag is rotated by 2π and so returns to its original position. Thus, ξ determines a unique null flag, but the null flag representing ξ also represents $-\xi$. Hence, null flags represent spin vectors only "up to sign". This is a reflection of what might be called the "essential 2-valuedness" of spinors, which has its roots in the fact that the spinor map is two-to-one and which has been used to model some quite startling physical phenomena. We shall take up these matters in somewhat more detail in Appendix B.

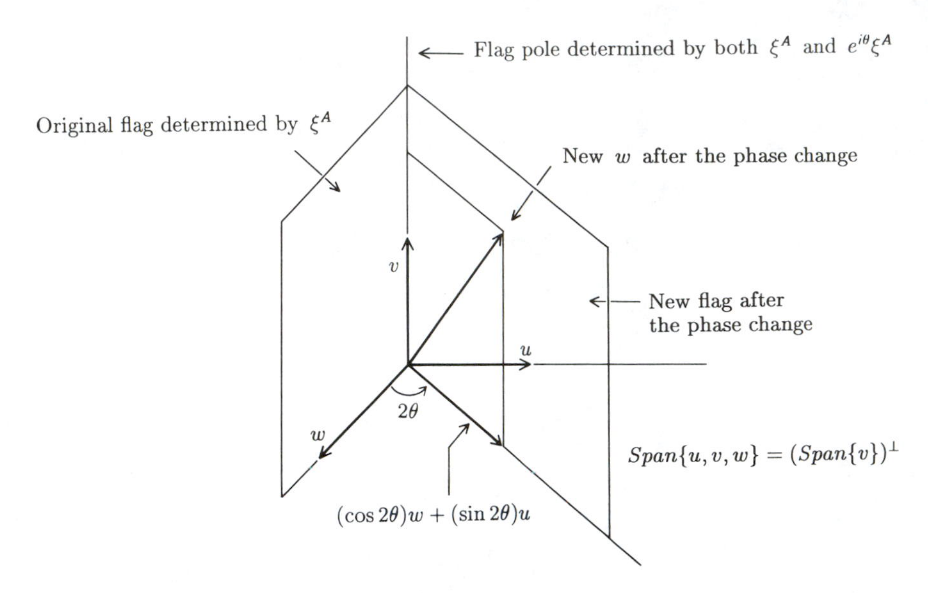

Figure 3.5.2

3.6 The Electromagnetic Field (Revisited)

In this section we shall reexamine some of our earlier results on electromagnetic fields at a point and find that, in the language of spinors, they often achieve a remarkable elegance and simplicity. We begin with a nonzero skew-symmetric linear transformation $F : \mathcal{M} \to \mathcal{M}$ (i.e., the value of an electromagnetic field at some point in $\mathcal{M}$). Select a fixed, but arbitrary admissible basis $\{e_a\}$ and spin frame $\{s^A\}$. The bivector $\tilde{F}$ associated with F is defined by (2.7.10) and has components in $\{e_a\}$ given by $F_{ab} = \tilde{F}(e_a, e_b)$. The spinor equivalent of $\tilde{F}$ is defined by $F_{A\dot{X}B\dot{Y}} = \sigma^a{}_{A\dot{X}}\sigma^b{}_{B\dot{Y}}F_{ab}$. Associated with $F_{A\dot{X}B\dot{Y}}$ is a symmetric spinor ϕ_{AB} of valence $\left(\begin{smallmatrix}0 & 0\\ 2 & 0\end{smallmatrix}\right)$ such that

$$F_{A\dot{X}B\dot{Y}} = \epsilon_{AB}\bar{\phi}_{\dot{X}\dot{Y}} + \phi_{AB}\bar{\epsilon}_{\dot{X}\dot{Y}}. \tag{3.6.1}$$

We call ϕ_{AB} the *electromagnetic spinor* associated with F.

Exercise 3.6.1. Show that if ξ is any spin vector, then $\phi_{AB}\xi^A\xi^B$ is an invariant, i.e., that, relative to another spin frame,

$$\hat{\phi}_{AB}\hat{\xi}^A\hat{\xi}^B = \phi_{AB}\xi^A\xi^B.$$

Our first objective is to obtain a canonical decomposition of ϕ_{AB} into a symmetrized outer product of spin vectors. To this end we compute the invariant in Exercise 3.6.1 for spin vectors of the form $\begin{bmatrix} \xi^1 \\ \xi^0 \end{bmatrix} = \left[\begin{smallmatrix} z \\ 1 \end{smallmatrix}\right]$, where $z \in \mathbb{C}$.

$$\begin{aligned} \phi_{AB}\xi^A\xi^B &= \phi_{11}\xi^1\xi^1 + \phi_{10}\xi^1\xi^0 + \phi_{01}\xi^0\xi^1 + \phi_{00}\xi^0\xi^0 \\ &= \phi_{11}z^2 + \phi_{10}z + \phi_{01}z + \phi_{00} \\ &= \phi_{11}z^2 + 2\phi_{10}z + \phi_{00} \end{aligned}$$

since $\phi_{01} = \phi_{10}$. Notice that this is a quadratic polynomial in the complex variable z with coefficients in $\mathbb{C}$. Consequently, it factors over $\mathbb{C}$, i.e., there exist $\alpha_1, \alpha_0, \beta_1, \beta_0 \in \mathbb{C}$ such that

$$\phi_{11}z^2 + 2\phi_{10}z + \phi_{00} = (\alpha_1 z + \alpha_0)(\beta_1 z + \beta_0) \tag{3.6.2}$$

(these are not unique, of course, since replacing α_A by α_A/γ and β_A by $\gamma\beta_A$ for any nonzero $\gamma \in \mathbb{C}$ also gives a factorization). Equating coefficients in (3.6.2) gives

$$\begin{aligned} \phi_{11} = \alpha_1\beta_1 &= \tfrac{1}{2}(\alpha_1\beta_1 + \alpha_1\beta_1) \\ \phi_{00} = \alpha_0\beta_0 &= \tfrac{1}{2}(\alpha_0\beta_0 + \alpha_0\beta_0) \\ \phi_{10} \qquad &= \tfrac{1}{2}(\alpha_1\beta_0 + \alpha_0\beta_1). \end{aligned}$$

Since $\phi_{01} = \phi_{10}$ this last equality may be written

$$\phi_{01} \qquad = \tfrac{1}{2}(\alpha_0\beta_1 + \alpha_1\beta_0).$$

Thus, for all $A, B = 1, 0$, we have

$$\phi_{AB} = \tfrac{1}{2}(\alpha_A\beta_B + \alpha_B\beta_A). \tag{3.6.3}$$

Next observe that if, in another spin frame, we define $\hat{\alpha}_A = G_A{}^{A_1}\alpha_{A_1}$ and $\hat{\beta}_B = G_B{}^{B_1}\beta_{B_1}$, i.e., if we regard α and β as spin vectors, then

$$\begin{aligned} \tfrac{1}{2}(\hat{\alpha}_A\hat{\beta}_B + \hat{\alpha}_B\hat{\beta}_A) &= \tfrac{1}{2}\Big((G_A{}^{A_1}\alpha_{A_1})(G_B{}^{B_1}\beta_{B_1}) + (G_B{}^{B_1}\alpha_{B_1})(G_A{}^{A_1}\alpha_{A_1})\Big) \\ &= \tfrac{1}{2}G_A{}^{A_1}G_B{}^{B_1}(\alpha_{A_1}\beta_{B_1} + \alpha_{B_1}\beta_{A_1}) \\ &= G_A{}^{A_1}G_B{}^{B_1}\phi_{A_1B_1} \\ &= \hat{\phi}_{AB}. \end{aligned}$$

Consequently, ϕ is the symmetrized outer product of the spin vectors α and β, i.e., in any spin frame,

$$\phi_{AB} = \tfrac{1}{2}(\alpha_A\beta_B + \alpha_B\beta_A) = \alpha_{(A}\beta_{B)}. \tag{3.6.4}$$

Although we will have no need to do so this argument, which depends only on the symmetry of ϕ, extends easily to produce analogous decompositions of higher valence symmetric spinors.

The spin vectors α and β are intimately connected with the electromagnetic field F. We will eventually show that our characterization of null and regular F's (Corollary 2.3.8) has a remarkably simple reformulation in terms of α and β (Corollary 3.6.2 asserts that F is null if and only if α and β are parallel). For the present we will content ourselves with showing that the future-directed null vectors associated with α and β (i.e., their flagpoles) are eigenvectors of the electromagnetic field F (see Section 2.4). Thus, we define future-directed null vectors v and w by

$$v^a = -\sigma^a{}_{A\dot{X}}\alpha^A\bar{\alpha}^{\dot{X}} \quad \text{and} \quad w^a = -\sigma^a{}_{A\dot{X}}\beta^A\bar{\beta}^{\dot{X}}.$$

The null directions determined by v and w are called the *principal null directions* of ϕ_{AB}. Letting $F^a{}_b = \eta^{ac}F_{cb}$ denote the entries in the matrix of F relative to $\{e_a\}$ we compute

$$\begin{aligned}
F^a{}_b v^b &= \eta^{ac}F_{cb}v^b = \eta^{ac}\sigma_c{}^{A\dot{X}}\sigma_b{}^{B\dot{Y}}F_{A\dot{X}B\dot{Y}}v^b \\
&= -\eta^{ac}\sigma_c{}^{A\dot{X}}\sigma_b{}^{B\dot{Y}}(\epsilon_{AB}\bar{\phi}_{\dot{X}\dot{Y}} + \phi_{AB}\bar{\epsilon}_{\dot{X}\dot{Y}})(\sigma^b{}_{D\dot{Z}}\alpha^D\bar{\alpha}^{\dot{Z}}) \\
&= -\eta^{ac}\sigma_c{}^{A\dot{X}}(\sigma_b{}^{B\dot{Y}}\sigma^b{}_{D\dot{Z}})(\epsilon_{AB}\bar{\phi}_{\dot{X}\dot{Y}} + \phi_{AB}\bar{\epsilon}_{\dot{X}\dot{Y}})\alpha^D\bar{\alpha}^{\dot{Z}} \\
&= -\eta^{ac}\sigma_c{}^{A\dot{X}}(-\delta^B_D\delta^{\dot{Y}}_{\dot{Z}})(\epsilon_{AB}\bar{\phi}_{\dot{X}\dot{Y}} + \phi_{AB}\bar{\epsilon}_{\dot{X}\dot{Y}})\alpha^D\bar{\alpha}^{\dot{Z}} \\
&= \eta^{ac}\sigma_c{}^{A\dot{X}}(\epsilon_{AB}\bar{\phi}_{\dot{X}\dot{Y}} + \phi_{AB}\bar{\epsilon}_{\dot{X}\dot{Y}})\alpha^B\bar{\alpha}^{\dot{Y}} \\
&= \eta^{ac}\sigma_c{}^{A\dot{X}}\left[(\epsilon_{AB}\alpha^B)(\bar{\phi}_{\dot{X}\dot{Y}}\bar{\alpha}^{\dot{Y}}) + (\phi_{AB}\alpha^B)(\bar{\epsilon}_{\dot{X}\dot{Y}}\bar{\alpha}^{\dot{Y}})\right]
\end{aligned}$$

$$F^a{}_b v^b = \eta^{ac}\sigma_c{}^{A\dot{X}}\left[(-\alpha_A)(\bar{\phi}_{\dot{X}\dot{Y}}\bar{\alpha}^{\dot{Y}}) + (\phi_{AB}\alpha^B)(-\bar{\alpha}_{\dot{X}})\right]. \tag{3.6.5}$$

Exercise 3.6.2. Show that $\phi_{AB}\alpha^B = \frac{1}{2}(\alpha_1\beta_0 - \alpha_0\beta_1)\alpha_A$ and $\bar{\phi}_{\dot{X}\dot{Y}}\bar{\alpha}^{\dot{Y}} = \overline{\frac{1}{2}(\alpha_1\beta_0 - \alpha_0\beta_1)}\bar{\alpha}_{\dot{X}}$.

Letting $\mu = \frac{1}{2}(\alpha_1\beta_0 - \alpha_0\beta_1)$ we obtain, from Exercise 3.6.2,

$$\phi_{AB}\alpha^B = \mu\alpha_A \tag{3.6.6}$$

and

$$\bar{\phi}_{\dot{X}\dot{Y}}\bar{\alpha}^{\dot{Y}} = \bar{\mu}\bar{\alpha}_{\dot{X}}, \tag{3.6.7}$$

which we now substitute into (3.6.5).

$$
\begin{aligned}
F^a{}_b v^b &= -\eta^{ac}\sigma_c{}^{A\dot{X}}\left(\alpha_A(\bar{\mu}\bar{\alpha}_{\dot{X}}) + (\mu\alpha_A)\bar{\alpha}_{\dot{X}}\right) \\
&= -\eta^{ac}\sigma_c{}^{A\dot{X}}(\mu+\bar{\mu})(\alpha_A\bar{\alpha}_{\dot{X}}) \\
&= -(\mu+\bar{\mu})\eta^{ac}(\sigma_c{}^{A\dot{X}}\alpha_A\bar{\alpha}_{\dot{X}}) \\
&= -(\mu+\bar{\mu})\eta^{ac}v_c \\
&= -(\mu+\bar{\mu})v^a.
\end{aligned}
$$

If we let

$$
\begin{aligned}
\lambda &= -(\mu+\bar{\mu}) = -2\mathrm{Re}(\mu) = -2\mathrm{Re}\left(\tfrac{1}{2}(\alpha_1\beta_0 - \alpha_0\beta_1)\right) \\
&= -\mathrm{Re}(\alpha_1\beta_0 - \alpha_0\beta_1) \\
&= -\mathrm{Re} < \alpha, \beta >,
\end{aligned}
$$

we obtain

$$
F^a{}_b v^b = \lambda v^a = -\mathrm{Re} < \alpha, \beta > v^a, \tag{3.6.8}
$$

or, equivalently,

$$
Fv = \lambda v = -\mathrm{Re} < \alpha, \beta > v, \tag{3.6.9}
$$

so v is an eigenvector of F with eigenvalue $\lambda = -\mathrm{Re} < \alpha, \beta >$.

Exercise 3.6.3. Show in the same way that

$$
Fw = -\lambda w = \mathrm{Re} < \alpha, \beta > w. \tag{3.6.10}
$$

We conclude that the flagpoles of α and β are two (possibly coincident) future-directed null eigenvectors of F with eigenvalues $-\mathrm{Re} < \alpha, \beta >$ and $\mathrm{Re} < \alpha, \beta >$ respectively.

Let us rearrange (3.6.6) a bit.

$$
\begin{aligned}
\phi_{AC}\alpha^C &= \mu\alpha_A, \\
\phi_{AC}(\epsilon^{CB}\alpha_B) &= \mu\alpha_A, \\
(\phi_{AC}\epsilon^{CB})\alpha_B &= \mu\alpha_A, \\
(-\epsilon^{BC}\phi_{AC})\alpha_B &= \mu\alpha_A,
\end{aligned}
$$

$$
\phi_A{}^B\alpha_B = -\mu\alpha_A. \tag{3.6.11}
$$

Thinking of $[\phi_A{}^B]$ as the matrix, relative to $\{s^A\}$, of a linear transformation $\phi\colon \mathfrak{S} \to \mathfrak{S}$ on spin space motivates the following definitions. A complex number λ is an *eigenvalue* of ϕ_{AB} if there exists a nonzero spin vector $\alpha \in \mathfrak{B}$, called an *eigenspinor* of ϕ_{AB}, such that $\phi_A{}^B \alpha_B = \lambda \alpha_A$. Such an α will exist if and only if λ satisfies

$$\det\left[\begin{bmatrix} \phi_1{}^1 & \phi_1{}^0 \\ \phi_0{}^1 & \phi_0{}^0 \end{bmatrix} - \lambda \begin{bmatrix} 1 & 0 \\ 0 & 1 \end{bmatrix}\right] = 0,$$

which, when expanded, gives

$$\lambda^2 - \left(\phi_1{}^1 + \phi_0{}^0\right) + \det[\phi_A{}^B] \;=\; 0. \tag{3.6.12}$$

However, #1 of Exercise 3.3.6 and the symmetry of ϕ_{AB} gives $\phi_1{}^1 + \phi_0{}^0 = 0$, whereas #3 of that same Exercise gives $\det[\phi_A{}^B] = \det[\phi_{AB}] = \frac{1}{2}\phi_{AB}\phi^{AB}$, so the solutions to (3.6.12) are

$$\lambda = \pm\left(-\det[\phi_{AB}]\right)^{\frac{1}{2}} \;=\; \pm\left(-\tfrac{1}{2}\phi_{AB}\phi^{AB}\right)^{\frac{1}{2}}. \tag{3.6.13}$$

The physical significance of these eigenvalues of ϕ_{AB} will emerge when we compute $\det[\phi_{AB}]$ in terms of the 3-vectors $\vec{E}$ and $\vec{B}$, which we accomplish by means of (2.7.14). First observe that

$$\begin{aligned}
\phi_{AB} &= \tfrac{1}{2}F_{\dot{U}A}{}^{\dot{U}}{}_B = \tfrac{1}{2}\left[F_{\dot{1}A}{}^{\dot{1}}{}_B + F_{\dot{0}A}{}^{\dot{0}}{}_B\right] \\
&= \tfrac{1}{2}\left[\bar{\epsilon}^{\dot{1}\dot{X}}F_{\dot{1}A\dot{X}B} + \bar{\epsilon}^{\dot{0}\dot{X}}F_{\dot{0}A\dot{X}B}\right] \\
&= \tfrac{1}{2}\left[\bar{\epsilon}^{\dot{1}\dot{0}}F_{\dot{1}A\dot{0}B} + \bar{\epsilon}^{\dot{0}\dot{1}}F_{\dot{0}A\dot{1}B}\right] \\
&= \tfrac{1}{2}\left[-F_{\dot{1}A\dot{0}B} + F_{\dot{0}A\dot{1}B}\right] = \tfrac{1}{2}\left[F_{A\dot{0}B\dot{1}} - F_{A\dot{1}B\dot{0}}\right].
\end{aligned}$$

Thus, for example,

$$\begin{aligned}
\phi_{11} &= \tfrac{1}{2}\left[F_{1\dot{0}1\dot{1}} - F_{1\dot{1}1\dot{0}}\right] = \tfrac{1}{2}\left[F_{1\dot{0}1\dot{1}} - (-F_{1\dot{0}1\dot{1}})\right] \\
&= F_{1\dot{0}1\dot{1}} = \sigma^a{}_{1\dot{0}}\sigma^b{}_{1\dot{1}}F_{ab}.
\end{aligned}$$

Now, if $a = b$ the corresponding term in this sum is zero since $F_{aa} = 0$. The $a = 3,4$ and $b = 1,2$ terms vanish by the definitions of the $\sigma^a{}_{A\dot{X}}$. Thus, only the $ab = 13, 14, 23, 24$ terms survive so

$$\begin{aligned}
\phi_{11} &= \sigma^1{}_{1\dot{0}}\sigma^3{}_{1\dot{1}}F_{13} + \sigma^1{}_{1\dot{0}}\sigma^4{}_{1\dot{1}}F_{14} + \sigma^2{}_{1\dot{0}}\sigma^3{}_{1\dot{1}}F_{23} + \sigma^2{}_{1\dot{0}}\sigma^4{}_{1\dot{1}}F_{24} \\
&= \left(-\tfrac{1}{\sqrt{2}}\right)\left(-\tfrac{1}{\sqrt{2}}\right)F_{13} + \left(-\tfrac{1}{\sqrt{2}}\right)\left(-\tfrac{1}{\sqrt{2}}\right)F_{14} \\
&\quad + \left(\tfrac{i}{\sqrt{2}}\right)\left(-\tfrac{1}{\sqrt{2}}\right)F_{23} + \left(\tfrac{i}{\sqrt{2}}\right)\left(-\tfrac{1}{\sqrt{2}}\right)F_{24}
\end{aligned}$$

$$= \tfrac{1}{2}(-B^2) + \tfrac{1}{2}(E^1) - \tfrac{1}{2}i(B^1) - \tfrac{1}{2}i(E^2)$$

$$\phi_{11} = \tfrac{1}{2}\left[(E^1 - B^2) - i(E^2 + B^1)\right]. \tag{3.6.14}$$

Exercise 3.6.4. Continue in the same way to show

$$\phi_{10} = \phi_{01} = \tfrac{1}{2}(-E^3 + iB^3) \tag{3.6.15}$$

and

$$\phi_{00} = \tfrac{1}{2}\left[-(E^1 + B^2) + i(-E^2 + B^1)\right]. \tag{3.6.16}$$

Exercise 3.6.5. Compute $\phi_{11}\phi_{00} - \phi_{10}\phi_{01}$ from (3.6.14)-(3.6.16) to show that

$$\det[\phi_{AB}] = \tfrac{1}{4}\left(|\vec{B}|^2 - |\vec{E}|^2\right) + \tfrac{1}{2}(\vec{E} \cdot \vec{B})i. \tag{3.6.17}$$

Returning now to (3.6.13) we find that the eigenvalues of the electromagnetic spinor ϕ_{AB} are given by

$$\lambda = \pm\left[-\tfrac{1}{4}(|\vec{B}|^2 - |\vec{E}|^2) - \tfrac{1}{2}(\vec{E} \cdot \vec{B})i\right]^{\frac{1}{2}}. \tag{3.6.18}$$

But then $\lambda = 0$ if and only if $|\vec{B}|^2 - |\vec{E}|^2 = \vec{E} \cdot \vec{B} = 0$ so F is null if and only if the only eigenvalue of ϕ_{AB} is 0 and we have proved:

Theorem 3.6.1. *Let $F: \mathcal{M} \to \mathcal{M}$ be a nonzero, skew-symmetric linear transformation, $\tilde{F}$ its associated bivector, $F_{A\dot{X}B\dot{Y}}$ the spinor equivalent of $\tilde{F}$ and ϕ_{AB} the symmetric spinor for which $F_{A\dot{X}B\dot{Y}} = \epsilon_{AB}\bar{\phi}_{\dot{X}\dot{Y}} + \phi_{AB}\bar{\epsilon}_{\dot{X}\dot{Y}}$. Then F is null if and only if $\lambda = 0$ is the only eigenvalue of ϕ_{AB}.*

Another equally elegant form of this characterization theorem is:

Corollary 3.6.2. *Let $F: \mathcal{M} \to \mathcal{M}$ be a nonzero, skew-symmetric linear transformation, $\tilde{F}$ its associated bivector, $F_{A\dot{X}B\dot{Y}}$ the spinor equivalent of $\tilde{F}$, ϕ_{AB} the symmetric spinor for which $F_{A\dot{X}B\dot{Y}} = \epsilon_{AB}\bar{\phi}_{\dot{X}\dot{Y}} + \phi_{AB}\bar{\epsilon}_{\dot{X}\dot{Y}}$, and α and β spin vectors for which $\phi_{AB} = \alpha_{(A}\beta_{B)}$. Then F is null if and only if α and β are linearly dependent.*

Proof: First we compute

$$\begin{aligned}\phi_{AB}\phi^{AB} &= \tfrac{1}{2}(\alpha_A\beta_B + \alpha_B\beta_A)\tfrac{1}{2}(\alpha^A\beta^B + \alpha^B\beta^A)\\ &= \tfrac{1}{4}[(\alpha_A\alpha^A)(\beta_B\beta^B) + (\alpha_A\beta^A)(\beta_B\alpha^B)\\ &\quad + (\alpha_B\beta^B)(\beta_A\alpha^A) + (\alpha_B\alpha^B)(\beta_A\beta^A)]\\ &= \tfrac{1}{4}[(0)(0) + <\beta,\alpha><\alpha,\beta> + <\beta,\alpha><\alpha,\beta> + (0)(0)]\\ &= \tfrac{1}{4}[-<\alpha,\beta><\alpha,\beta> - <\alpha,\beta><\alpha,\beta>]\\ &= -\tfrac{1}{2}<\alpha,\beta>^2 .\end{aligned}$$

Thus, (3.6.13) gives

$$\lambda = \pm\left(\tfrac{1}{4}<\alpha,\beta>^2\right)^{\frac{1}{2}} = \pm\tfrac{1}{2}<\alpha,\beta> .$$

Theorem 3.6.1 therefore implies that F is null if and only if $<\alpha,\beta> = 0$ which, by Lemma 3.2.1 (g), is the case if and only if α and β are linearly dependent. ∎

We have defined the spinor equivalent of a bivector in Section 3.5, but the same definition yields a spinor equivalent of any bilinear form on $\mathcal{M}$. Specifically, if we fix an admissible basis $\{e_a\}$ and a spin frame $\{s^A\}$ and let $H\colon \mathcal{M}\times\mathcal{M}\to\mathbb{R}$ be a bilinear form on $\mathcal{M}$, then the *spinor equivalent* of H is the spinor of valence $\left(\begin{smallmatrix}0&0\\2&2\end{smallmatrix}\right)$ whose components in $\{s^A\}$ are given by

$$H_{A\dot{X}B\dot{Y}} = \sigma^a{}_{A\dot{X}}\sigma^b{}_{B\dot{Y}}H_{ab},$$

where $H_{ab} = H(e_a, e_b)$.

Exercise 3.6.6. Show that, in another spin frame $\{\hat{s}^A\}$, related to $\{s^A\}$ by (3.2.1) and (3.2.6), $\hat{H}_{A\dot{X}B\dot{Y}} = \sigma^a{}_{A\dot{X}}\sigma^b{}_{B\dot{Y}}\hat{H}_{ab}$, where $\hat{H}_{ab} = \Lambda_a{}^\alpha\Lambda_b{}^\beta H_{\alpha\beta}$, Λ being $\Lambda_{\mathcal{G}}$.

A particularly important example of a bilinear form is the Lorentz inner product itself: $g\colon \mathcal{M}\times\mathcal{M}\to\mathbb{R}$, defined by $g(u,v) = u\cdot v$. Relative to any $\{e_a\}$, the components of g are

$$g(e_a, e_b) = e_a \cdot e_b = \eta_{ab}.$$

The spinor equivalent of g is defined by

$$g_{A\dot{X}B\dot{Y}} = \sigma^a{}_{A\dot{X}}\sigma^b{}_{B\dot{Y}}\eta_{ab}$$

$$g_{A\dot{X}B\dot{Y}} = \sigma^1{}_{A\dot{X}}\sigma^1{}_{B\dot{Y}} + \sigma^2{}_{A\dot{X}}\sigma^2{}_{B\dot{Y}} + \sigma^3{}_{A\dot{X}}\sigma^3{}_{B\dot{Y}} - \sigma^4{}_{A\dot{X}}\sigma^4{}_{B\dot{Y}} . \quad (3.6.19)$$

We claim that

$$g_{A\dot{X}B\dot{Y}} = -\epsilon_{AB}\bar{\epsilon}_{\dot{X}\dot{Y}}. \tag{3.6.20}$$

One verifies (3.6.20) by simply considering all possible choices for $A, B, \dot{X}$ and $\dot{Y}$. For example, if either (i) A and B are the same, but $\dot{X}$ and $\dot{Y}$ are different, or (ii) $\dot{X}$ and $\dot{Y}$ are the same, but A and B are different, then both sides of (3.6.20) are zero [every ${\sigma^a}_{A\dot{X}}$ has either ${\sigma^a}_{1\dot{1}} = {\sigma^a}_{0\dot{0}} = 0$ or ${\sigma^a}_{1\dot{0}} = {\sigma^a}_{0\dot{1}} = 0$, so all of the ${\sigma^a}_{A\dot{X}}{\sigma^a}_{B\dot{Y}}$ in (3.6.19) are zero]. All that remain then are the cases in which (iii) $A = B$ and $\dot{X} = \dot{Y}$, or (iv) $A \neq B$ and $\dot{X} \neq \dot{Y}$, i.e., $A\dot{X}B\dot{Y} = 1\dot{1}1\dot{1}, 1\dot{0}1\dot{0}, 0\dot{1}0\dot{1}, 0\dot{0}0\dot{0}, 1\dot{1}0\dot{0}, 1\dot{0}0\dot{1}, 0\dot{1}1\dot{0}, 0\dot{0}1\dot{1}$. For example,

$$\begin{aligned}
g_{1\dot{0}0\dot{1}} &= {\sigma^1}_{1\dot{0}}{\sigma^1}_{0\dot{1}} + {\sigma^2}_{1\dot{0}}{\sigma^2}_{0\dot{1}} + {\sigma^3}_{1\dot{0}}{\sigma^3}_{0\dot{1}} - {\sigma^4}_{1\dot{0}}{\sigma^4}_{0\dot{1}} \\
&= {\sigma^1}_{1\dot{0}}{\sigma^1}_{0\dot{1}} + {\sigma^2}_{1\dot{0}}{\sigma^2}_{0\dot{1}} = (-\tfrac{1}{\sqrt{2}})(-\tfrac{1}{\sqrt{2}}) + (\tfrac{i}{\sqrt{2}})(-\tfrac{i}{\sqrt{2}}) \\
&= \tfrac{1}{2} - \tfrac{1}{2}i^2 = \tfrac{1}{2} + \tfrac{1}{2} \\
&= 1 \\
&= -\epsilon_{10}\bar{\epsilon}_{\dot{0}\dot{1}}\,.
\end{aligned}$$

Exercise 3.6.7. Verify the remaining cases.

The energy-momentum transformation $T\colon \mathcal{M} \to \mathcal{M}$ of an electromagnetic field $F\colon \mathcal{M} \to \mathcal{M}$ also has an associated (symmetric) bilinear form $\tilde{T} : \mathcal{M} \times \mathcal{M} \to \mathbb{R}$ defined by $\tilde{T}(u,v) = u \cdot Tv$ and with components $T_{ab} = T(e_a, e_b)$ given, according to Exercise 2.7.8, by

$$T_{ab} = \tfrac{1}{4\pi}\big[F_{a\alpha}{F_b}^{\alpha} - \tfrac{1}{4}\eta_{ab}F_{\alpha\beta}F^{\alpha\beta}\big]\,. \tag{3.6.21}$$

We show next that the spinor equivalent of $\tilde{T}$ takes the following particularly simple form:

$$T_{A\dot{X}B\dot{Y}} = \tfrac{1}{2\pi}\phi_{AB}\bar{\phi}_{\dot{X}\dot{Y}}\,, \tag{3.6.22}$$

where ϕ_{AB} is the electromagnetic spinor associated with F. By definition, the spinor equivalent of $\tilde{T}$ is given by

$$\begin{aligned}
T_{A\dot{X}B\dot{Y}} &= {\sigma^a}_{A\dot{X}}{\sigma^b}_{B\dot{Y}}T_{ab} = \tfrac{1}{4\pi}{\sigma^a}_{A\dot{X}}{\sigma^b}_{B\dot{Y}}\left[F_{a\alpha}{F_b}^{\alpha} - \tfrac{1}{4}\eta_{ab}F_{\alpha\beta}F^{\alpha\beta}\right] \\
&= \tfrac{1}{4\pi}\Big[{\sigma^a}_{A\dot{X}}{\sigma^b}_{B\dot{Y}}F_{a\alpha}{F_b}^{\alpha} - \tfrac{1}{4}({\sigma^a}_{A\dot{X}}{\sigma^b}_{B\dot{Y}}\eta_{ab})F_{\alpha\beta}F^{\alpha\beta}\Big]
\end{aligned}$$

$$T_{A\dot{X}B\dot{Y}} = \tfrac{1}{4\pi}\Big[{\sigma^a}_{A\dot{X}}{\sigma^b}_{B\dot{Y}}F_{a\alpha}{F_b}^{\alpha} + \tfrac{1}{4}\epsilon_{AB}\bar{\epsilon}_{\dot{X}\dot{Y}}(F_{\alpha\beta}F^{\alpha\beta})\Big] \tag{3.6.23}$$

by (3.6.20). We begin simplifying (3.6.23) with two observations:

$$F_{\alpha\beta}F^{\alpha\beta} = F_{C\dot{Z}D\dot{W}}F^{C\dot{Z}D\dot{W}} \tag{3.6.24}$$

and

$$\sigma^a{}_{A\dot{X}}\sigma^b{}_{B\dot{Y}}F_{a\alpha}F_b{}^\alpha = -F_{A\dot{X}C\dot{Z}}F_{B\dot{Y}}{}^{C\dot{Z}}. \tag{3.6.25}$$

For the proof of (3.6.24) we compute

$$\begin{aligned}
F_{C\dot{Z}D\dot{W}}F^{C\dot{Z}D\dot{W}} &= F_{C\dot{Z}D\dot{W}}\epsilon^{CC_1}\bar{\epsilon}^{\dot{Z}\dot{Z}_1}\epsilon^{DD_1}\bar{\epsilon}^{\dot{W}\dot{W}_1}F_{C_1\dot{Z}_1D_1\dot{W}_1}\\
&= (\sigma^a{}_{C\dot{Z}}\sigma^b{}_{D\dot{W}}F_{ab})\epsilon^{CC_1}\bar{\epsilon}^{\dot{Z}\dot{Z}_1}\epsilon^{DD_1}\bar{\epsilon}^{\dot{W}\dot{W}_1}\left(\sigma^\alpha{}_{C_1\dot{Z}_1}\sigma^\beta{}_{D_1\dot{W}_1}(\eta_{\alpha\mu}\eta_{\beta\nu}F^{\mu\nu})\right)\\
&= (\epsilon^{CC_1}\bar{\epsilon}^{\dot{Z}\dot{Z}_1}\eta_{\mu\alpha}\sigma^\alpha{}_{C_1\dot{Z}_1})(\epsilon^{DD_1}\bar{\epsilon}^{\dot{W}\dot{W}_1}\eta_{\nu\beta}\sigma^\beta{}_{D_1\dot{W}_1})\sigma^a{}_{C\dot{Z}}\sigma^b{}_{D\dot{W}}F_{ab}F^{\mu\nu}\\
&= \sigma_\mu{}^{C\dot{Z}}\sigma_\nu{}^{D\dot{W}}\sigma^a{}_{C\dot{Z}}\sigma^b{}_{D\dot{W}}F_{ab}F^{\mu\nu} = (\sigma_\mu{}^{C\dot{Z}}\sigma^a{}_{C\dot{Z}})(\sigma_\nu{}^{D\dot{W}}\sigma^b{}_{D\dot{W}})F_{ab}F^{\mu\nu}\\
&= (-\delta^a_\mu)(-\delta^b_\nu)F_{ab}F^{\mu\nu} = F_{ab}F^{ab}.
\end{aligned}$$

Exercise 3.6.8. Prove (3.6.25).

Substituting (3.6.24) and (3.6.25) into (3.6.23) gives

$$T_{A\dot{X}B\dot{Y}} = \tfrac{1}{4\pi}\Big[-F_{A\dot{X}C\dot{Z}}F_{B\dot{Y}}{}^{C\dot{Z}} + \tfrac{1}{4}\epsilon_{AB}\bar{\epsilon}_{\dot{X}\dot{Y}}F_{C\dot{Z}D\dot{W}}F^{C\dot{Z}D\dot{W}}\Big]. \tag{3.6.26}$$

Now we claim that if $F_{A\dot{X}B\dot{Y}} = \epsilon_{AB}\bar{\phi}_{\dot{X}\dot{Y}} + \phi_{AB}\bar{\epsilon}_{\dot{X}\dot{Y}}$, then

$$F_{C\dot{Z}D\dot{W}}F^{C\dot{Z}D\dot{W}} = 2(\phi_{CD}\phi^{CD} + \bar{\phi}_{\dot{Z}\dot{W}}\bar{\phi}^{\dot{Z}\dot{W}}) \tag{3.6.27}$$

and

$$F_{A\dot{X}C\dot{Z}}F_{B\dot{Y}}{}^{C\dot{Z}} = -2\phi_{AB}\bar{\phi}_{\dot{X}\dot{Y}} + \epsilon_{AB}\bar{\phi}_{\dot{X}\dot{Z}}\bar{\phi}_{\dot{Y}}{}^{\dot{Z}} + \bar{\epsilon}_{\dot{X}\dot{Y}}\phi_{AC}\phi_B{}^C. \tag{3.6.28}$$

We prove (3.6.27) as follows:

$$\begin{aligned}
F_{C\dot{Z}D\dot{W}}F^{C\dot{Z}D\dot{W}} &= (\epsilon_{CD}\bar{\phi}_{\dot{Z}\dot{W}} + \phi_{CD}\bar{\epsilon}_{\dot{Z}\dot{W}})(\epsilon^{CD}\bar{\phi}^{\dot{Z}\dot{W}} + \phi^{CD}\bar{\epsilon}^{\dot{Z}\dot{W}})\\
&= (\epsilon_{CD}\epsilon^{CD})(\bar{\phi}_{\dot{Z}\dot{W}}\bar{\phi}^{\dot{Z}\dot{W}}) + (\epsilon_{CD}\phi^{CD})(\bar{\phi}_{\dot{Z}\dot{W}}\bar{\epsilon}^{\dot{Z}\dot{W}})\\
&\quad + (\phi_{CD}\epsilon^{CD})(\bar{\epsilon}_{\dot{Z}\dot{W}}\bar{\phi}^{\dot{Z}\dot{W}}) + (\phi_{CD}\phi^{CD})(\bar{\epsilon}_{\dot{Z}\dot{W}}\bar{\epsilon}^{\dot{Z}\dot{W}}).
\end{aligned}$$

But observe that, by symmetry of ϕ, $\epsilon_{CD}\phi^{CD} = \epsilon_{10}\phi^{10} + \epsilon_{01}\phi^{01} = -\phi^{10} + \phi^{01} = 0$ and, similarly, $\bar{\epsilon}_{\dot{Z}\dot{W}}\bar{\phi}^{\dot{Z}\dot{W}} = 0$. Moreover, by (3.3.7), $\epsilon_{CD}\epsilon^{CD} = \bar{\epsilon}_{\dot{Z}\dot{W}}\bar{\epsilon}^{\dot{Z}\dot{W}} = 2$ so

$$F_{C\dot{Z}D\dot{W}}F^{C\dot{Z}D\dot{W}} = 2\bar{\phi}_{\dot{Z}\dot{W}}\bar{\phi}^{\dot{Z}\dot{W}} + 0 + 0 + 2\phi_{CD}\phi^{CD}$$

which gives (3.6.27).

Exercise 3.6.9. Prove (3.6.28).

With (3.6.27) and (3.6.28), (3.6.26) becomes

$$\begin{aligned} T_{A\dot{X}B\dot{Y}} &= \tfrac{1}{4\pi}\Big[2\phi_{AB}\bar{\phi}_{\dot{X}\dot{Y}} - \epsilon_{AB}\bar{\phi}_{\dot{X}\dot{Z}}\bar{\phi}_{\dot{Y}}{}^{\dot{Z}} - \bar{\epsilon}_{\dot{X}\dot{Y}}\phi_{AC}\phi_{B}{}^{C} \\ &\quad + \tfrac{1}{2}\epsilon_{AB}\bar{\epsilon}_{\dot{X}\dot{Y}}(\phi_{CD}\phi^{CD} + \bar{\phi}_{\dot{Z}\dot{W}}\bar{\phi}^{\dot{Z}\dot{W}})\Big]. \end{aligned}$$

$$\begin{aligned} T_{A\dot{X}B\dot{Y}} &= \tfrac{1}{4\pi}\Big[2\phi_{AB}\bar{\phi}_{\dot{X}\dot{Y}} - \epsilon_{AB}\bar{\phi}_{\dot{X}\dot{Z}}\bar{\phi}_{\dot{Y}}{}^{\dot{Z}} - \bar{\epsilon}_{\dot{X}\dot{Y}}\phi_{AC}\phi_{B}{}^{C} \\ &\quad + (\det[\phi_{AB}])\epsilon_{AB}\bar{\epsilon}_{\dot{X}\dot{Y}} + (\det[\bar{\phi}_{\dot{X}\dot{Y}}])\epsilon_{AB}\bar{\epsilon}_{\dot{X}\dot{Y}}\Big], \end{aligned} \tag{3.6.29}$$

where we have appealed to part (3) of Exercise 3.3.6 and its conjugated version. For the remaining simplifications we use part (4) of this same exercise. If either $A = B$ or $\dot{X} = \dot{Y}$ all the terms on the right-hand side of (3.6.29) except the first are zero so $T_{A\dot{X}B\dot{Y}} = \frac{1}{2\pi}\phi_{AB}\bar{\phi}_{\dot{X}\dot{Y}}$ and (3.6.22) is proved. The remaining cases are $A\dot{X}B\dot{Y} = 1\dot{1}0\dot{0}, 1\dot{0}0\dot{1}, 0\dot{1}1\dot{0}$ and $0\dot{0}1\dot{1}$ and all are treated in the same way, e.g.,

$$\begin{aligned} T_{1\dot{1}0\dot{0}} &= \tfrac{1}{4\pi}\Big[2\phi_{10}\bar{\phi}_{\dot{1}\dot{0}} - \epsilon_{10}\bar{\phi}_{\dot{1}\dot{Z}}\bar{\phi}_{\dot{0}}{}^{\dot{Z}} - \bar{\epsilon}_{\dot{1}\dot{0}}\phi_{1C}\phi_{0}{}^{C} \\ &\quad + (\det[\phi_{AB}])\epsilon_{10}\bar{\epsilon}_{\dot{1}\dot{0}} + (\det[\bar{\phi}_{\dot{X}\dot{Y}}])\epsilon_{10}\bar{\epsilon}_{\dot{1}\dot{0}}\Big] \\ &= \tfrac{1}{4\pi}[2\phi_{10}\bar{\phi}_{\dot{1}\dot{0}} - (-1)(-\det[\bar{\phi}_{\dot{X}\dot{Y}}]) - (-1)(-\det[\phi_{AB}]) \\ &\quad + (\det[\phi_{AB}])(-1)(-1) + (\det[\bar{\phi}_{\dot{X}\dot{Y}}])(-1)(-1)] \\ &= \tfrac{1}{4\pi}[2\phi_{10}\bar{\phi}_{\dot{1}\dot{0}}] \\ &= \tfrac{1}{2\pi}\phi_{10}\bar{\phi}_{\dot{1}\dot{0}}. \end{aligned}$$

Exercise 3.6.10. Check the remaining cases to complete the proof of (3.6.22).

We use the spinor equivalent $T_{A\dot{X}B\dot{Y}} = \frac{1}{2\pi}\phi_{AB}\bar{\phi}_{\dot{X}\dot{Y}}$ of the energy-momentum T of F to give another proof of the dominant energy condition (Exercise 2.5.6) that does not depend on the canonical forms of F. Begin with two future-directed null vectors $u = u^a e_a$ and $v = v^b e_b$ in $\mathcal{M}$. By Theorem 3.4.3, the spinor equivalents of u and v can be written

$U^{A\dot{X}} = \mu^A \bar{\mu}^{\dot{X}}$ and $V^{A\dot{X}} = \nu^A \bar{\nu}^{\dot{X}}$, where μ and ν are two spin vectors. Thus, we may write $u^a = -\sigma^a{}_{A\dot{X}} \mu^A \bar{\mu}^{\dot{X}}$ and $v^b = -\sigma^b{}_{B\dot{Y}} \nu^B \bar{\nu}^{\dot{Y}}$ so that

$$\begin{aligned} Tu \cdot v &= T_{ab} u^a v^b = T_{ab}(-\sigma^a{}_{A\dot{X}} \mu^A \bar{\mu}^{\dot{X}})(-\sigma^b{}_{B\dot{Y}} \nu^B \bar{\nu}^{\dot{Y}}) \\ &= (\sigma^a{}_{A\dot{X}} \sigma^b{}_{B\dot{Y}} T_{ab}) \mu^A \bar{\mu}^{\dot{X}} \nu^B \bar{\nu}^{\dot{Y}} \\ &= T_{A\dot{X}B\dot{Y}} \mu^A \bar{\mu}^{\dot{X}} \nu^B \bar{\nu}^{\dot{Y}} \\ &= \tfrac{1}{2\pi} \phi_{AB} \bar{\phi}_{\dot{X}\dot{Y}} \mu^A \bar{\mu}^{\dot{X}} \nu^B \bar{\nu}^{\dot{Y}} \\ &= \tfrac{1}{2\pi} (\phi_{AB} \mu^A \mu^B)(\bar{\phi}_{\dot{X}\dot{Y}} \bar{\mu}^{\dot{X}} \bar{\nu}^{\dot{Y}}), \end{aligned}$$

so

$$Tu \cdot v = \tfrac{1}{2\pi} |\ \phi_{AB} \mu^A \nu^B\ |^2 . \tag{3.6.30}$$

In particular, $Tu \cdot v \geq 0$.

Exercise 3.6.11. Show that $Tu \cdot v \geq 0$ whenever u and v are timelike or null and both are future-directed. *Hint:* Any future-directed timelike vector can be written as a sum of two future-directed null vectors.

We recall from Section 2.5 that, for any future-directed unit timelike vector U, $TU \cdot U = \frac{1}{8\pi}[\,|\vec{E}|^2 + |\vec{B}|^2\,]$ is the energy density in any admissible frame with $e_4 = U$. Consequently, if F is nonzero, $Tu \cdot u \neq 0$ for any timelike vector u. We now investigate the circumstances under which $Tv \cdot v = 0$ for some future-directed null vector v. Suppose then that v is null and future-directed and $Tv \cdot v = 0$. Write $v^a = -\sigma^a{}_{A\dot{X}} \nu^A \bar{\nu}^{\dot{X}}$ for some spin vector ν (Theorem 3.4.3). Then, by (3.6.30), $\phi_{AB} \nu^A \nu^B = 0$. Using the decomposition (3.6.4) of ϕ this is equivalent to

$$(\alpha_A \beta_B + \alpha_B \beta_A)(\nu^A \nu^B) = 0,$$

$$(\alpha_A \nu^A)(\beta_B \nu^B) + (\alpha_B \nu^B)(\beta_A \nu^A) = 0,$$

$$2 < \nu, \alpha > < \nu, \beta > = 0$$

which is the case if and only if either $< \nu, \alpha > = 0$ or $< \nu, \beta > = 0$. But $< \nu, \alpha > = 0$ if and only if ν is a multiple of α [Lemma 3.2.1(g)] and similarly for $< \nu, \beta > = 0$. But if ν is a multiple of either α or β, then v is a multiple of one of the two null vectors determined by α or β, i.e., v is along a principal null direction of ϕ_{AB}. Thus, a future-directed null vector v for which $Tv \cdot v = 0$ must lie along a principal null direction of ϕ_{AB}. Moreover, by reversing the steps above, one finds that the converse is also true so we have proved that a nonzero null vector v satisfies $Tv \cdot v = 0$ if and only if v lies along a principal null direction of ϕ_{AB}.

Exercise 3.6.12. Let $F\colon \mathcal{M} \to \mathcal{M}$ be a nonzero, skew-symmetric linear transformation, $\tilde{F}$ the associated bivector and ${}^*\tilde{F}$ the dual of $\tilde{F}$ (Section 2.7). By (2.7.16) (with $M = \Lambda$), the Levi-Civita symbol ϵ_{abcd} defines a (constant) covariant world tensor of rank 4. We define its *spinor equivalent* by $\epsilon_{A\dot{X}B\dot{Y}C\dot{Z}D\dot{W}} = \sigma^a{}_{A\dot{X}}\sigma^b{}_{B\dot{Y}}\sigma^c{}_{C\dot{Z}}\sigma^d{}_{D\dot{W}}\epsilon_{abcd}$. Show that

$$\epsilon_{A\dot{X}B\dot{Y}C\dot{Z}D\dot{W}} = i(\epsilon_{AC}\epsilon_{BD}\bar{\epsilon}_{\dot{X}\dot{W}}\bar{\epsilon}_{\dot{Y}\dot{Z}} - \epsilon_{AD}\epsilon_{BC}\bar{\epsilon}_{\dot{X}\dot{Z}}\bar{\epsilon}_{\dot{Y}\dot{W}})$$

and then raise indices to obtain

$$\epsilon_{A\dot{X}B\dot{Y}}{}^{C\dot{Z}D\dot{W}} = i\left(\delta_A^C\,\delta_B^D\,\delta_{\dot{X}}^{\dot{W}}\,\delta_{\dot{Y}}^{\dot{Z}} - \delta_A^D\,\delta_B^C\,\delta_{\dot{X}}^{\dot{Z}}\,\delta_{\dot{Y}}^{\dot{W}}\right).$$

Now show that the spinor equivalent of ${}^*\tilde{F}$ is given by

$${}^*F_{A\dot{X}B\dot{Y}} = i\left(\epsilon_{AB}\bar{\phi}_{\dot{X}\dot{Y}} - \phi_{AB}\bar{\epsilon}_{\dot{X}\dot{Y}}\right).$$

Exercise 3.6.13. Define the *spinor equivalent* of an arbitrary world tensor (Section 3.1) of contravariant rank r and covariant rank s, being sure to verify the appropriate transformation law, and show that any such spinor equivalent is Hermitian. Find an "inversion formula" analogous to (3.4.19) and (3.5.3) that retrieves the world tensor from its spinor equivalent. For what type of spinor can this process be reversed to yield a world tensor equivalent?

We conclude our discussion of electromagnetic theory by deriving the elegant spinor form of the source-free Maxwell equations. For this we fix an admissible basis $\{e_a\}$ and a spin frame $\{s^A\}$ and let F denote an electromagnetic field on (a region in) $\mathcal{M}$. As usual, we denote by $F_{ab} = \eta_{a\alpha}F^\alpha{}_b$ the components of the corresponding bivector $\tilde{F}$, all of which are functions of (x^1, x^2, x^3, x^4). Then the spinor equivalent of $\tilde{F}$ is $F_{A\dot{X}B\dot{Y}} = \sigma^a{}_{A\dot{X}}\sigma^b{}_{B\dot{Y}}F_{ab}$ and the electromagnetic spinor ϕ_{AB} is given by

$$\phi_{AB} = \tfrac{1}{2}F_{\dot{U}A}{}^{\dot{U}}{}_B = \tfrac{1}{2}[F_{A\dot{0}B\dot{1}} - F_{A\dot{1}B\dot{0}}].$$

Next we introduce "spinor equivalents" for the differential operators $\partial_a = \frac{\partial}{\partial x^a}$, $a = 1, 2, 3, 4$. Specifically, we define, for each $A = 1, 0$ and $\dot{X} = \dot{1}, \dot{0}$, an operator $\nabla^{A\dot{X}}$ by

$$\nabla^{A\dot{X}} = \sigma_a{}^{A\dot{X}}\partial^a = \sigma_a{}^{A\dot{X}}(\eta^{a\alpha}\partial_\alpha).$$

Thus, for example,

$$\begin{aligned}
\nabla^{1\dot{1}} &= \sigma_a{}^{1\dot{1}}\partial^a = \sigma_1{}^{1\dot{1}}\partial^1 + \sigma_2{}^{1\dot{1}}\partial^2 + \sigma_3{}^{1\dot{1}}\partial^3 + \sigma_4{}^{1\dot{1}}\partial^4 \\
&= \sigma_3{}^{1\dot{1}}\partial^3 + \sigma_4{}^{1\dot{1}}\partial^4 = \tfrac{1}{\sqrt{2}}\partial^3 + \tfrac{1}{\sqrt{2}}\partial^4 \\
&= \tfrac{1}{\sqrt{2}}(\partial_3 - \partial_4).
\end{aligned}$$

Exercise 3.6.14. Prove the remaining identities in (3.6.31):

$$\begin{aligned} \nabla^{1\dot{1}} &= \tfrac{1}{\sqrt{2}}(\partial_3 - \partial_4), & \nabla^{1\dot{0}} &= \tfrac{1}{\sqrt{2}}(\partial_1 + i\partial_2), \\ \nabla^{0\dot{1}} &= \tfrac{1}{\sqrt{2}}(\partial_1 - i\partial_2), & \nabla^{0\dot{0}} &= -\tfrac{1}{\sqrt{2}}(\partial_3 + \partial_4). \end{aligned} \tag{3.6.31}$$

With this notation we claim that all of the information contained in the source-free Maxwell equations (2.7.15) and (2.7.21) can be written concisely as

$$\nabla^{A\dot{X}}\phi_{AB} = 0, \quad A = 1, 0, \ \dot{X} = \dot{1}, \dot{0}. \tag{3.6.32}$$

Equations (3.6.32) are the *spinor form* of the source-free Maxwell equations. To verify the claim we write

$$\begin{aligned} \phi_{11} &= \tfrac{1}{2}[(F_{13} + F_{14}) + i(F_{32} + F_{42})], \\ \phi_{10} &= \phi_{01} = \tfrac{1}{2}[F_{43} + iF_{12}], \\ \phi_{00} &= \tfrac{1}{2}[(F_{41} + F_{13}) + i(F_{42} + F_{23})] \end{aligned}$$

[see (3.6.14)-(3.6.16)]. Now compute, for example,

$$\begin{aligned} \nabla^{A\dot{1}}\phi_{A0} &= \nabla^{1\dot{1}}\phi_{10} + \nabla^{0\dot{1}}\phi_{00} = \tfrac{1}{2\sqrt{2}}\{(\partial_3 - \partial_4)(F_{43} + iF_{12}) \\ &\quad + (\partial_1 - i\partial_2)\left((F_{41} + F_{13}) + i(F_{42} + F_{23})\right)\} \\ &= \tfrac{1}{2\sqrt{2}}\{[-(F_{14,1} + F_{24,2} + F_{34,3}) + (F_{13,1} + F_{23,2} - F_{43,4})] \\ &\quad + i\,[(F_{12,3} + F_{31,2} + F_{23,1}) - (F_{12,4} + F_{41,2} + F_{24,1})]\} \\ &= \tfrac{1}{2\sqrt{2}}\left\{\left[-\text{div}\,\vec{E} - [(\text{curl}\,\vec{B}) \cdot e_3 - \tfrac{\partial E^3}{\partial x^4}]\right]\right. \\ &\quad \left. + i\left[\text{div}\,\vec{B} - [(\text{curl}\,\vec{E}) \cdot e_3 + \tfrac{\partial B^3}{\partial x^4}]\right]\right\}. \end{aligned}$$

Exercise 3.6.15. Calculate, in the same way, $\nabla^{A\dot{1}}\phi_{A1}$, $\nabla^{A\dot{0}}\phi_{A1}$, and $\nabla^{A\dot{0}}\phi_{A0}$, and show that (3.6.32) is equivalent to Maxwell's equations.

Generalizations of (3.6.32) are used in relativistic quantum mechanics as field equations for various types of massless particles. Specifically, if n is a positive integer and $\phi_{A_1 A_2 \cdots A_n}$ is a symmetric spinor of valence $\left(\begin{smallmatrix} 0 & 0 \\ n & 0 \end{smallmatrix}\right)$, then

$$\nabla^{A\dot{X}}\phi_{AA_2\cdots A_n} = 0, \qquad A_2, \ldots, A_n = 1, 0, \ \dot{X} = \dot{1}, \dot{0},$$

is taken to be the massless free-field equation for arbitrary spin $\frac{1}{2}n$ particles (see 5.7 of [**PR**]). In particular, if $n = 1$, then ϕ_A is a spin vector and one obtains the *Weyl neutrino equation*

$$\nabla^{A\dot{X}}\phi_A = 0, \quad \dot{X} = \dot{1}, \dot{0},$$

which suggested the possibility of parity nonconservation in weak interactions years before the phenomenon itself was observed (see [**LY**]).

Appendix A

Topologies For $\mathcal{M}$

A.1 The Euclidean Topology

In this appendix we wish to lay before the reader certain material which requires a bit more in the way of background than the text itself and which admittedly has not had a profound impact on subsequent research in relativity, but which is nonetheless remarkable from both the physical and the mathematical points of view. We will assume a very basic familiarity with elementary point-set topology and adopt [**Wi**] as our canonical reference.

The subject we wish to address had its origins in the extraordinary paper [**Z2**] of Zeeman in 1967. Zeeman observed that the ordinary Euclidean topology for $\mathcal{M}$ (defined below) has, from the relativistic viewpoint, no physical significance or justification and proposed an alternative he called the "fine" topology. This topology was easy to describe, physically well motivated and had the remarkable property that its homeomorphism group (also defined below) was essentially just the Lorentz group (together with translations and nonzero scalar multiplications). Thus, perhaps the most important group in all of physics is seen to emerge at the very primitive level of topology, i.e., from just an appropriate definition of "nearby" events. The fine topology is, however, from the technical point of view, rather difficult to work with and the arguments in [**Z2**] are by no means simple. In 1976, Hawking, King and McCarthy [**HKM**] described another topology on $\mathcal{M}$ which seemed physically even more natural, had precisely the same homeomorphism group as Zeeman's fine topology and required for the proof of this nothing beyond the most rudimentary point-set topology and Zeeman's Theorem 1.6.2. This so-called "path topology" for $\mathcal{M}$ is the object of our investigations in this appendix.

We begin by transferring to $\mathcal{M}$ the standard Euclidean topology of $\mathbb{R}^4$ via a linear isomorphism. Specifically, we select some fixed admissible basis $\{e_a\}_{a=1}^4$ for $\mathcal{M}$ (this determines an obvious linear isomorphism of $\mathcal{M}$ onto $\mathbb{R}^4$). If $x = x^a e_a$ and $x_0 = x_0^a e_a$ are two points in $\mathcal{M}$ we define the *E-distance* from x_0 to x by $d_E(x_0, x) = ((x^1 - x_0^1)^2 + (x^2 - x_0^2)^2 + (x^3 - x_0^3)^2 + (x^4 - x_0^4)^2)^{1/2}$. Then d_E is a metric on $\mathcal{M}$, i.e., satisfies (1) $d_E(x, x_0) = d_E(x_0, x)$, (2) $d_E(x_0, x) \geq 0$ and $d_E(x_0, x) = 0$ if and only if $x = x_0$, and (3) $d_E(x_0, x) \leq d_E(x_0, y) + d_E(y, x)$ for all x_0, x and y in $\mathcal{M}$. Consequently, d_E determines, in the usual way (3.2 of [**Wi**]) a topology E for $\mathcal{M}$ called the *Euclidean* (or E-) *topology.* Specifically, if x_0 is in $\mathcal{M}$ and $\varepsilon > 0$ we define the *E-open ball of radius* ε *about* x_0 by

$$N_\varepsilon^E(x_0) = \{x \in \mathcal{M} : d_E(x_0, x) < \varepsilon\} .$$

A subset V of $\mathcal{M}$ is then said to be *E-open* if for every x_0 in V there exists an $\varepsilon > 0$ such that $N_\varepsilon^E(x_0) \subseteq V$. The collection of all E-open sets in $\mathcal{M}$ constitutes the E-topology for $\mathcal{M}$. When thinking of $\mathcal{M}$ as being endowed with the Euclidean topology we will denote it $\mathcal{M}^E$. E's will likewise be appended to various other terms and symbols to emphasize that we are operating in the Euclidean topology, e.g., maps will be referred to as "E-continuous", "$\mathrm{Cl}_E\, A$" and "$\mathrm{bdy}_E\, A$" will designate the E-closure and E-boundary of A and so on. $\mathcal{M}^E$ is, of course, homeomorphic to $\mathbb{R}^4$ with its customary Euclidean topology so that its basic topological properties are well-known,[1] e.g., it is first countable, separable, locally compact, but not compact, pathwise connected, etc.

Notice that the definition of the E-metric d_E on $\mathcal{M}$ is *not* invariant under Lorentz transformations. That is, if $d_E(x_0, x)$ is computed by the defining formula from the coordinates of x_0 and x relative to another admissible basis $\{\hat{e}_a\}$ for $\mathcal{M}$ the result will, in general, be different. The reason for this is clear since the two bases are related by an element of $\mathcal{L}$ and elements of $\mathcal{L}$ preserve the Lorentz inner product and not the Euclidean inner product (i.e., they satisfy $\Lambda^{-1} = \eta\,\Lambda^T\eta$ rather than $\Lambda^{-1} = \Lambda^T$). Nevertheless, two such metrics, while not equal, are *equivalent* in the sense that they determine the same topology for $\mathcal{M}$ (because an element of $\mathcal{L}$ is a one-to-one linear map of $\mathcal{M}$ onto $\mathcal{M}$ and so an E-homeomorphism).

A.2 E-Continuous Timelike Curves

In Section 1.4 we defined what it meant for a *smooth* curve in $\mathcal{M}$ to be "timelike" and "future- (or past-) directed". For the definition of the topology we propose to describe in the next section it is essential to extend these notions to the class of curves in $\mathcal{M}$ that are E-continuous, but need not have a velocity vector at each point. Thus, we let I denote a (nondegenerate) interval in $\mathbb{R}$ (open, closed, or half-open) and consider a curve $\alpha : I \to \mathcal{M}$ that is E-continuous [i.e., $\alpha^{-1}(V)$ is open in I for every E-open set V in $\mathcal{M}$]. Fix a t_0 in I. We say that α is *future-timelike at* t_0 if there exists a connected, relatively open subset U of I containing t_0

[1] Its not-so-basic topological properties are quite another matter, however. Indeed, in many topological ways, $\mathbb{R}^4$ is unique among the Euclidean spaces $\mathbb{R}^n$ (see, for example, [**FL**]).

such that

$$t \in U \text{ and } t < t_0 \implies \alpha(t) \ll \alpha(t_0)$$

and

$$t \in U \text{ and } t_0 < t \implies \alpha(t_0) \ll \alpha(t).$$

(U is an interval which may contain one or both of the endpoints of I, if I happens to have endpoints). *Past-timelike at* t_0 is defined similarly. α is said to be *future-timelike* (resp., *past-timelike*) if it is future-timelike (resp., past-timelike) at every t_0 in I. Finally, α is *timelike* if it is either future-timelike or past-timelike.

Any curve $\alpha : I \to \mathcal{M}$ that is smooth has component functions relative to any admissible basis that are continuous as maps from I into $\mathbb{R}$. Since $\mathcal{M}^E$ is homeomorphic to $\mathbb{R}^4$ with its product topology, such an α is E-continuous (8.8 of [**Wi**]). According to Lemma 1.4.7, a smooth curve that is timelike and future-directed in the sense of Section 1.4 is therefore also future-timelike in our new sense. Of course, the same is true of smooth, timelike and past-directed curves. However, any timelike polygon (which has no velocity vector at its "joints") can obviously be parametrized so as to become either future-timelike or past-timelike, but is not "smooth-timelike". Oddly enough, an E-continuous curve can be timelike and smooth without being smooth-timelike in the sense of Section 1.4. For example, if $\{e_a\}$ is an admissible basis and if one defines $\alpha : \mathbb{R} \to \mathcal{M}$ by $\alpha(t) = (\sin t)e_1 + te_4$, then α is future-timelike and smooth, but $\alpha'(t) = (\cos t)e_1 + e_4$ which is null at $t = n\pi$, $n = 0, \pm 1, \pm 2, \ldots$ (see Figure A.2.1). This is unfortunate since it complicates the physical interpretation of "E-continuous future-timelike" somewhat. One would like to regard such a curve as the worldline of a material particle which may be undergoing abrupt changes in speed and direction (due, say, to collisions). Of course, having a null velocity vector at some point would tend to indicate a particle momentarily attaining the speed of light and this we prefer not to admit as a realistic possibility. One would seem forced to accept a curve of the type just described as an acceptable model for the worldline of a material particle only on the intervals between points at which the tangent is null (notice that the situation cannot get much worse, i.e., the velocity vector of a smooth future-timelike curve cannot be null on an interval, nor can it ever be spacelike).

We proceed now to derive a sequence of results that will be needed in the next section.

Lemma A.2.1. *Let* $\{e_a\}_{a=1}^4$ *be an admissible basis for* $\mathcal{M}$ *and* $\alpha : I \to \mathcal{M}$ *an* E*-continuous timelike curve. If* α *is future-timelike, then* $x^4(\alpha(t))$ *is increasing on* I*. If* α *is past-timelike, then* $x^4(\alpha(t))$ *is decreasing on* I*. In particular, if* α *is timelike, it is one-to-one.*

Proof: Suppose α is future-timelike (the argument for α past-timelike is

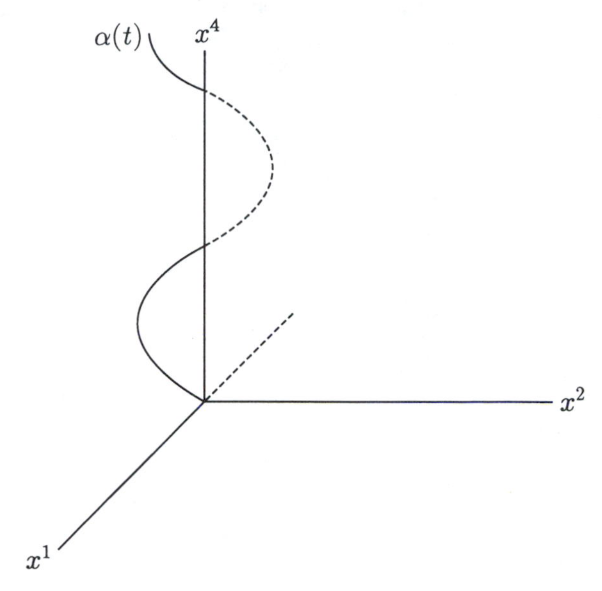

Figure A.2.1

similar). Let $t_0,\ t_1 \in I$ with $t_0 < t_1$. We show that $x^4(\alpha(t_0)) < x^4(\alpha(t_1))$. Suppose, to the contrary, that $x^4(\alpha(t_0)) \geq x^4(\alpha(t_1))$. $x^4(\alpha(t))$ is a real-valued continuous (8.8 of [**Wi**]) function on the closed bounded interval $[t_0,\ t_1]$ and so achieves a maximum value at some $t_2 \in [t_0,\ t_1]$. Since α is future-timelike at t_0 and $p \ll q$ implies $x^4(p) < x^4(q)$, $x^4(\alpha(t))$ must increase immediately to the right of t_0 so $t_2 > t_0$. But $x^4(\alpha(t_2)) > x^4(\alpha(t_0)) \geq x^4(\alpha(t_1))$ implies $t_2 < t_1$ so $t_2 \in (t_0,\ t_1)$. But α is future-timelike at t_2 and so $x^4(\alpha(t))$ must increase immediately to the right of t_2 and this contradicts the fact that, on $[t_0,\ t_1]$, $x^4(\alpha(t))$ has a maximum at t_2. ∎

Next we show that Theorem 1.4.6 remains true if "smooth future-directed timelike" is replaced with "*E*-continuous future-timelike".

Theorem A.2.2. *Let p and q be two points in $\mathcal{M}$. Then $p \ll q$ if and only if there exists an E-continuous future-timelike curve $\alpha : [a,b] \to \mathcal{M}$ such that $\alpha(a) = p$ and $\alpha(b) = q$.*

Proof: The necessity is clear from Theorem 1.4.6. For the sufficiency we assume $\alpha : [a,b] \to \mathcal{M}$ is E-continuous future-timelike with $\alpha(a) = p$ and $\alpha(b) = q$. For each t in $[a,b]$ we select a connected, relatively open subset U_t of $[a,b]$ containing t as in the definition of future-timelike at t. Then $\{U_t : t \in [a,b]\}$ is an open cover of $[a,b]$ so, by compactness (17.9 of [**Wi**]), we may select a finite subcover $\mathcal{U} = \{U_a, U_{t_1}, \ldots, U_{t_n}\}$.

By definition, $a \in U_a$. Moreover, if $b \in U_a$, then $\alpha(a) \ll \alpha(b)$ and we are done. If $b \notin U_a$, then the right-hand endpoint s_0 of U_a is less than b and not in U_a.

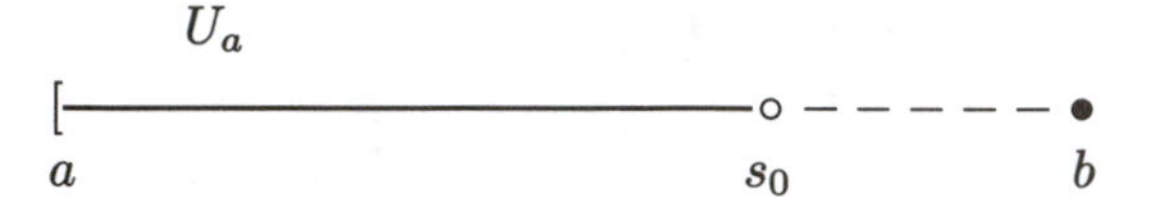

Select a U_{t_i} in $\mathcal{U}$ such that $s_0 \in U_{t_i}$. Then $U_{t_i} \neq U_a$, but $U_a \cap U_{t_i} \neq \emptyset$. Select a $T_0 \in U_a \cap U_{t_i}$ such that $a < T_0 < t_i$. Now, if $b \in U_{t_i}$, then $\alpha(a) \ll \alpha(T_0) \ll \alpha(t_i) \ll \alpha(b)$ and we are done. Otherwise, the right-hand endpoint s_1 of U_{t_i} is less than b and not in U_{t_i}. Repeat the process, beginning at T_0 rather than a. Select a U_{t_j} in $\mathcal{U}$ with $s_1 \in U_{t_j}$. Observe that $U_{t_j} \neq U_a$ and $U_{t_j} \neq U_{t_i}$ since s_1 is in neither U_a nor U_{t_i}. However, $U_{t_i} \cap U_{t_j} \neq \emptyset$. Select T_1 as above and continue to repeat the process. Since $\mathcal{U}$ is finite and covers $[a, b]$ the procedure must terminate in a finite number of steps with $\alpha(a) \ll \alpha(b)$ as required. ■

Next we prove that an E-continuous curve that is timelike at each point in an interval must have the same causal character (future-timelike or past-timelike) at each point. In fact, we prove more.

Lemma A.2.3. *Let $\alpha : I \to \mathcal{M}$ be an E-continuous curve. If α is timelike at each t_0 in the interior* $\operatorname{Int} I$ *of I, then α is timelike.*

Proof: We first show that α is either future-timelike at each $t_0 \in \operatorname{Int} I$ or past-timelike at each $t_0 \in \operatorname{Int} I$. The procedure will be to show that the set $S = \{t_0 \in \operatorname{Int} I : \alpha$ is future-timelike at $t_0\}$ is both open and closed in $\operatorname{Int} I$ and so, since $\operatorname{Int} I$ is connected, is either $\emptyset$ or all of $\operatorname{Int} I$ (26.1 of [**Wi**]). Suppose then that $S \neq \emptyset$. Let $t_0 \in S$ and select some $U \subseteq \operatorname{Int} I$ as in the definition of "future-timelike at t_0". We show that α is future-timelike at each t in U so $t_0 \in U \subseteq S$ and, since $t_0 \in S$ was arbitrary, conclude that S is open. First suppose there were a $t_1 > t_0$ in U at which α is past-timelike.

Exercise A.2.1. Relative to an admissible basis consider $x^4(\alpha(t))$ on $[t_0, t_1]$ and argue as in the proof of Lemma A.2.1 to derive a contradiction.

A similar argument shows that there can be no $t_1 < t_0$ in U at which α is past-timelike. Thus, $U \subseteq S$ as required so S is open. The same argument shows that $\{t_0 \in \operatorname{Int} I : \alpha$ is past-timelike at $t_0\}$, which is the complement of S in $\operatorname{Int} I$, is open in $\operatorname{Int} I$ so S is open and closed in $\operatorname{Int} I$ as required. Thus, either $S = \emptyset$ or $S = \operatorname{Int} I$ so α is either past-timelike at every $t_0 \in \operatorname{Int} I$ or future-timelike at every $t_0 \in \operatorname{Int} I$.

Now we show that if I has endpoints then α must be timelike and have the same causal character at these points that it has on $\operatorname{Int} I$. The arguments are similar in all cases so we suppose α is future-timelike on $\operatorname{Int} I$ and that $t = a$ is the left-hand endpoint of I. We show that α is future-timelike at a. Let U be a connected, relatively open subset of I containing a, but not containing the right-hand endpoint of I (should I happen to have a right-hand endpoint). Let $t_1 > a$ be in U and set $q = \alpha(a)$ and $r = \alpha(t_1)$. We show that $q \ll r$. Since $(a, t_1) \subseteq \operatorname{Int} I$, it follows from Theorem A.2.2 that $\alpha(a, t_1) \subseteq \mathcal{C}_T^-(r)$. Since a is in the closure of (a, t_1) in I and α is E-continuous, $q = \alpha(a)$ is in $\operatorname{Cl}_E \mathcal{C}_T^-(r)$ (7.2 of [**Wi**]). But $\operatorname{Cl}_E \mathcal{C}_T^-(r) = \mathcal{C}_T^-(r) \cup \mathcal{C}_N^-(r) \cup \{r\}$ so q must be in one of these sets. $q = r$ is impossible since, for every t in $\operatorname{Int} I$ with $t < t_1$, $x^4(\alpha(t)) < x^4(r)$ so $x^4(q) \leq x^4(\alpha(t)) < x^4(r)$. We show now that q must be in $\mathcal{C}_T^-(r)$. Select a $t_2 \in (a, t_1)$ and set $s = \alpha(t_2)$. Then $s \in \mathcal{C}_T^-(r)$ and, as above, $q \in \operatorname{Cl}_E \mathcal{C}_T^-(s)$ and $q = s$ is impossible so either $q \in \mathcal{C}_T^-(s)$ or $q \in \mathcal{C}_N^-(s)$. But then $r - s$ is timelike and future-directed and $s - q$ is either timelike or null and future-directed. Lemma 1.4.3 then implies that $r - q = (r - s) + (s - q)$ is timelike and future-directed, i.e., $q \in \mathcal{C}_T^-(r)$, so $q \ll r$. Since there are no points in U less than a, α is future-timelike at a. ∎

A.3 The Path Topology

The E-topology on $\mathcal{M}$ has the following property: For any E-continuous timelike curve $\alpha : I \to \mathcal{M}$, the image $\alpha(I)$ inherits, as a subspace of $\mathcal{M}^E$, the ordinary Euclidean topology. The *path topology* (or *P-topology*) is the finest topology on $\mathcal{M}$ that has this property (i.e., which gives the familiar notion of "nearby" to events on a continuous timelike worldline). Specifically, a subset V of $\mathcal{M}$ is *P-open* if and only if for every E-continuous timelike curve $\alpha : I \to \mathcal{M}$ there exists an E-open subset U of $\mathcal{M}$ such that

$$\alpha(I) \cap V = \alpha(I) \cap U\,,$$

which we henceforth abbreviate $\alpha \cap V = \alpha \cap U$.

Exercise A.3.1. Show that the collection of all such sets V does, indeed, form a topology for $\mathcal{M}$ (3.1 of [**Wi**]).

Obviously, any E-open set is P-open so that the P-topology is finer than (3.1 of [**Wi**]) the E-topology. It is strictly finer by virtue of:

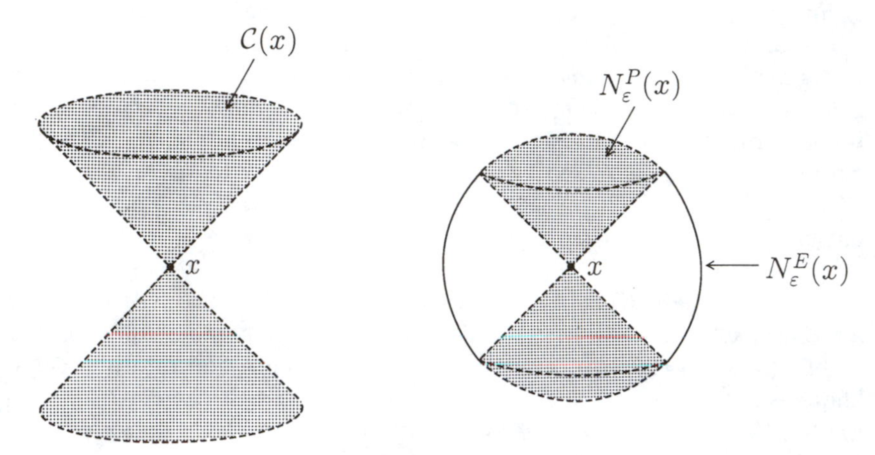

Figure A.3.1

Lemma A.3.1. *For each x in $\mathcal{M}$ and $\varepsilon > 0$ let*

$$\mathcal{C}(x) = \mathcal{C}_T^-(x) \cup \mathcal{C}_T^+(x) \cup \{x\}$$

and

$$N_\varepsilon^P(x) = \mathcal{C}(x) \cap N_\varepsilon^E(x)\,.$$

Then $\mathcal{C}(x)$ and $N_\varepsilon^P(x)$ are P-open, but not E-open (see Figure A.3.1).

Proof: Neither set contains an $N_\delta^E(x)$ so they both fail to be E-open. Now, let $\alpha : I \to \mathcal{M}$ be an E-continuous timelike curve. If α goes through x, then $\alpha(I)$ is entirely contained in $\mathcal{C}(x)$ by Theorem A.2.2 so $\alpha \cap \mathcal{C}(x) = \alpha \cap \mathcal{M}$. If α does not go through x, then $\alpha \cap \mathcal{C}(x) = \alpha \cap (\mathcal{C}_T^-(x) \cup \mathcal{C}_T^+(x))$. In either case $\alpha \cap \mathcal{C}(x) = \alpha \cap U$ for some E-open set U in $\mathcal{M}$ so $\mathcal{C}(x)$ is P-open. But then $N_\varepsilon^P(x)$ is the intersection of two P-open sets and so is P-open. ∎

$\mathcal{M}$ endowed with the P-topology is denoted $\mathcal{M}^P$ and we now show that the sets $N_\varepsilon^P(x)$ form a base (5.1 of [**Wi**]) for $\mathcal{M}^P$.

Theorem A.3.2. *The sets $N_\varepsilon^P(x)$ for $x \in \mathcal{M}$ and $\varepsilon > 0$ form a base for the open sets in $\mathcal{M}^P$.*

Proof: Let $V \subseteq \mathcal{M}$ be P-open and $x \in V$. We must show that there exists an $\varepsilon > 0$ such that $N_\varepsilon^P(x) \subseteq V$. We assume that no such ε exists

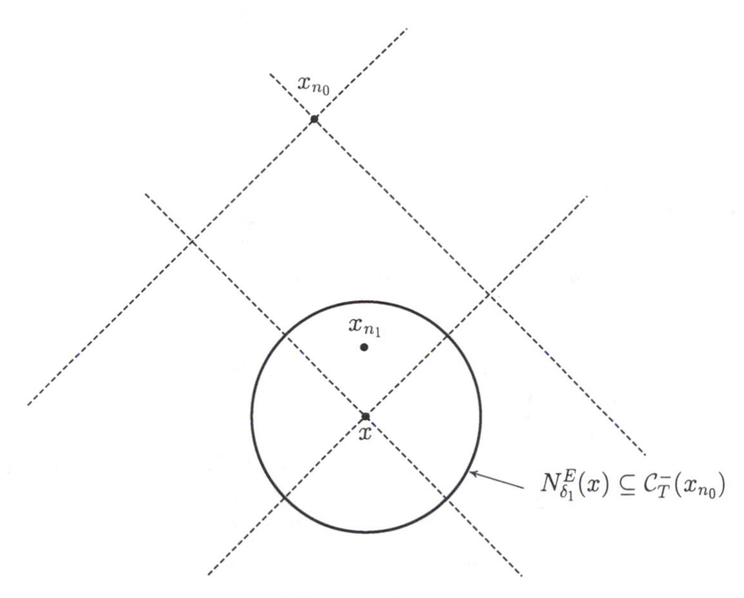

Figure A.3.2

and produce an E-continuous timelike curve α such that $\alpha \cap V$ cannot be written as $\alpha \cap U$ for any E-open set U and this is, of course, a contradiction.

We begin with $N^P_1(x)$ which, by assumption, is not contained in V. Since no $N^P_\varepsilon(x)$ is contained in V one or the other of $\mathcal{C}^+_T(x) \cap N^P_1(x)$ or $\mathcal{C}^-_T(x) \cap N^P_1(x)$ (or both) must contain an infinite sequence $\{x_1, x_2, \ldots\}$ of points not in V which E-converges to x. Since the proof is the same in both cases we assume that this sequence is in $\mathcal{C}^+_T(x) \cap N^P_1(x)$. We select a subsequence $\{x_{n_i}\}_{i=0}^\infty$ as follows: Let $x_{n_0} = x_1$. Since $x \in \mathcal{C}^-_T(x_{n_0})$ we may select a $\delta_1 > 0$ such that $N^E_{\delta_1}(x) \subseteq \mathcal{C}^-_T(x_{n_0})$ (see Figure A.3.2). Let $\varepsilon_1 = \min\{\delta_1, 1/2\}$. Select an x_{n_1} in the sequence which lies in $N^P_{\varepsilon_1}(x)$. Then $x \ll x_{n_1} \ll x_{n_0}$. Repeat the procedure. Since $x \in \mathcal{C}^-_T(x_{n_1})$ there exists a $\delta_2 > 0$ such that $N^E_{\delta_2}(x) \subseteq \mathcal{C}^-_T(x_{n_1})$. Let $\varepsilon_2 = \min\{\delta_2, 1/2^2\}$ and select an x_{n_2} in the sequence which lies in $N^E_{\varepsilon_2}(x)$. Then $x \ll x_{n_2} \ll x_{n_1} \ll x_{n_0}$. Continuing inductively we construct a subsequence $\{x_{n_0}, x_{n_1}, x_{n_2}, \ldots\}$ of $\{x_n\}$ such that

$$x \ll \cdots \ll x_{n_i} \ll \cdots \ll x_{n_2} \ll x_{n_1} \ll x_{n_0}$$

and $\{x_{n_i}\}_{i=0}^\infty$ E-converges to x. Now define $\hat{\alpha}: (0,1] \to \mathcal{M}$ as follows: On $[\frac{1}{2}, 1]$, $\hat{\alpha}$ is a linear parametrization of the future-timelike segment from x_{n_1} to x_{n_0}. On $[\frac{1}{3}, \frac{1}{2}]$, $\hat{\alpha}$ is a linear parametrization of the future-timelike segment from x_{n_2} to x_{n_1}, and so on. Then $\hat{\alpha}$ is obviously E-continuous and future-timelike. Since the x_{n_i} E-converge to x we can define an E-continuous curve $\alpha:[0,1] \to \mathcal{M}$ by

$$\alpha(t) = \begin{cases} \hat{\alpha}(t), & 0 < t \leq 1 \\ x, & t = 0; \end{cases}$$

α is also future-timelike by Lemma A.2.3.

Now, suppose $\alpha \cap V = \alpha \cap U$ for some E-open set U. Since the x_{n_i} are not in V, $x_{n_i} \notin \alpha \cap V$ for each i so $x_{n_i} \notin \alpha \cap U$ for each i. Thus, $\{x_{n_i}\} \subseteq \mathcal{M} - (\alpha \cap U) = (\mathcal{M} - \alpha) \cup (\mathcal{M} - U)$. But $x_{n_i} \in \alpha$ so we must have $x_{n_i} \in \mathcal{M} - U$. But $\mathcal{M} - U$ is E-closed and $\{x_{n_i}\}$ E-converges to x so $x \in \mathcal{M} - U$, i.e., $x \notin U$. Thus, $x \notin \alpha \cap U = \alpha \cap V$ and this is a contradiction since x is in both α and V. ∎

A number of basic topological properties of $\mathcal{M}^P$ follow immediately from Theorem A.3.2. Since the $N_\varepsilon^P(x)$ with ε rational form a local base at x, $\mathcal{M}^P$ is first countable (4.4(b) of [**Wi**]). Since P-open sets have nonempty E-interior, $\mathcal{M}^P$ is separable (5F of [**Wi**]). If R is a light ray in $\mathcal{M}$ and $x \in R$, then any $N_\varepsilon^P(x)$ intersects R only at x so, as a subspace of $\mathcal{M}^P$, R is discrete (4G of [**Wi**]). R is also P-closed since it is, in fact, E-closed and the P-topology is finer than the E-topology. Being separable and containing such large closed discrete subspaces prevents $\mathcal{M}^P$ from being normal (15.1 of [**Wi**]) since the Tietze Extension Theorem (15.8 of [**Wi**]) would require that all the continuous real-valued functions on any closed subspace extend to $\mathcal{M}^P$, but an uncountable closed discrete subspace has too many. In fact, it follows easily from our next lemma that $\mathcal{M}^P$ is not even regular (14.1 of [**Wi**]) and therefore certainly not normal (although it is Hausdorff since any two distinct points are contained in disjoint basic open sets).

Lemma A.3.3. *The closure in $\mathcal{M}^P$ of $N_\varepsilon^P(x)$ is $\mathrm{Cl}_E(N_\varepsilon^P(x)) - (\mathrm{bdy}_E(N_\varepsilon^E(x)) \cap \mathrm{bdy}_E(\mathcal{C}(x)))$ (see Figure A.3.3).*

Proof: Since P is finer than E, $\mathrm{Cl}_P(A) \subseteq \mathrm{Cl}_E(A)$ for any subset A of $\mathcal{M}$. Moreover, the points in $\mathrm{bdy}_E(N_\varepsilon^E(x)) \cap \mathrm{bdy}_E(\mathcal{C}(x))$ are not in $\mathrm{Cl}_P(N_\varepsilon^P(x))$ since, if y is such a point, $N_{\varepsilon/2}^P(y)$ does not intersect $N_\varepsilon^P(x)$ [the null cone at y is tangent to the surface of the Euclidean ball $N_\varepsilon^E(x)$ at such a y] (see Figure A.3.4.). Thus, $\mathrm{Cl}_P(N_\varepsilon^P(x)) \subseteq \mathrm{Cl}_E(N_\varepsilon^P(x)) - (\mathrm{bdy}_E(N_\varepsilon^E(x)) \cap \mathrm{bdy}_E(\mathcal{C}(x)))$. But the reverse containment is also clear since, if y is in the set on the right-hand side, every $N_\delta^P(y)$ intersects $N_\varepsilon^P(x)$. ∎

From Lemma A.3.3 it is clear that $\mathcal{M}^P$ is not regular since no $N_\varepsilon^P(x)$ contains a $\mathrm{Cl}_P(N_\delta^P(x))$. Moreover, since any P-compact set is necessarily E-compact and no $\mathrm{Cl}_P(N_\varepsilon^P(x))$ is E-compact (or even E-closed) we find

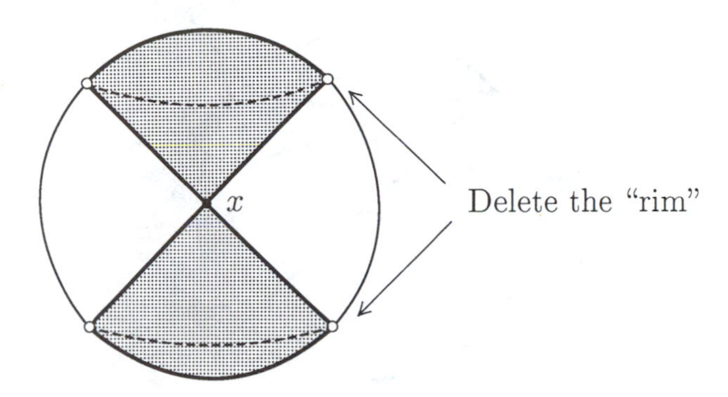

Figure A.3.3

that no point in $\mathcal{M}^P$ has a compact neighborhood. In particular, $\mathcal{M}^P$ is not locally compact (18.1 of [**Wi**]).

Exercise A.3.2. Show that $\mathcal{M}^P$ is not countably compact (17.1 of [**Wi**]), Lindelöf (16.5 of [**Wi**]), or second countable (16.1 of [**Wi**]).

In order to investigate the connectivity properties of $\mathcal{M}^P$ and for other purposes as well we will need to determine the P-continuous curves in $\mathcal{M}$.

Lemma A.3.4. *Let I be a nondegenerate interval in $\mathbb{R}$ and $\alpha : I \to \mathcal{M}$ a curve. Then:*

1. *If α is P-continuous, then it is E-continuous*
2. *If α is timelike, then it is P-continuous.*

Proof: (1) Let U be an E-open set in $\mathcal{M}$. Then U is P-open. Since α is P-continuous, $\alpha^{-1}(U)$ is open in I so α is E-continuous.

(2) Assume α is timelike (and therefore E-continuous by definition). Let V be a P-open set in $\mathcal{M}$. We show that $\alpha^{-1}(V)$ is open in I. By definition of the P-topology there exists an E-open set U in $\mathcal{M}$ such that $\alpha \cap V = \alpha \cap U$. Thus, $\alpha^{-1}(V) = \alpha^{-1}(\alpha \cap V) = \alpha^{-1}(\alpha \cap U) = \alpha^{-1}(U)$

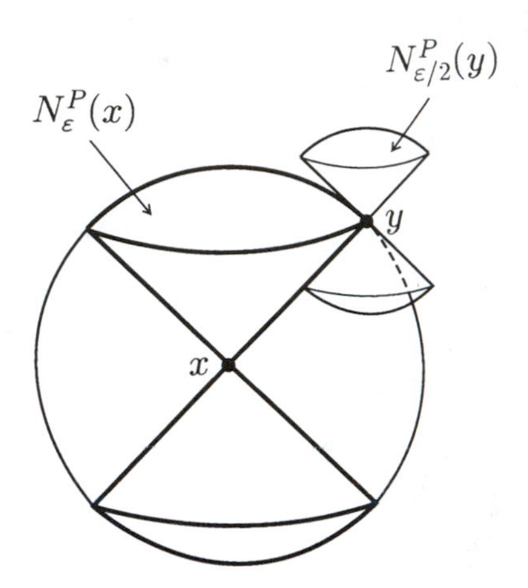

Figure A.3.4

which is open in I since α is E-continuous. ■

It is not quite true that a P-continuous curve must be timelike, but almost. We define a *Feynman path*[2] in $\mathcal{M}$ to be an E-continuous curve $\alpha : I \to \mathcal{M}$ with the property that for each t_0 in I there exists a connected relatively open subset U of I containing t_0 such that

$$\alpha(U) \subseteq \mathcal{C}(\alpha(t_0)) .$$

Observe that, since $\mathcal{C}(\alpha(t_0))$ is a P-open subset of $\mathcal{M}$, any P-continuous curve in $\mathcal{M}$ is necessarily a Feynman path. We show that the converse is also true.

Theorem A.3.5. *A curve $\alpha : I \to \mathcal{M}$ is P-continuous if and only if it is a Feynman path.*

Proof: All that remains is to prove that a Feynman path $\alpha : I \to \mathcal{M}$ is P-continuous. Fix a $t_0 \in I$. We show that α is P-continuous at t_0. For this let $N^P_\varepsilon(\alpha(t_0))$ be a basic P-neighborhood of $\alpha(t_0)$. Now, $\alpha^{-1}(N^P_\varepsilon(\alpha(t_0))) = \alpha^{-1}(N^E_\varepsilon(\alpha(t_0)) \cap \mathcal{C}(\alpha(t_0))) = \alpha^{-1}(N^E_\varepsilon(\alpha(t_0))) \cap$

[2]Being essentially timelike, but zigzaging with respect to time orientation, they resemble the Feynman track of an electron.

$\alpha^{-1}(\mathcal{C}(\alpha(t_0)))$. Since α is a Feynman path there exists a connected, relatively open subset U_1 of I containing t_0 such that U_1 is contained in $\alpha^{-1}(\mathcal{C}(\alpha(t_0)))$. Since α is E-continuous by definition, there exists a connected, relatively open subset U_2 of I containing t_0 such that $U_2 \subseteq \alpha^{-1}(N_\varepsilon^E(\alpha(t_0)))$. Thus, if $U = U_1 \cap U_2$ we have $t_0 \in U \subseteq \alpha^{-1}(N_\varepsilon^P(\alpha(t_0)))$ so $\alpha(U) \subseteq N_\varepsilon^P(\alpha(t_0))$ and α is P-continuous at t_0. ■

Since any two points in $N_\varepsilon^P(x)$ can be joined by a Feynman path (in fact, by a timelike segment or two such segments "joined" at x), $\mathcal{M}^P$ is locally pathwise connected (27.4 of [**Wi**]). Moreover, since any straight line in $\mathcal{M}$ can be approximated by a Feynman path, $\mathcal{M}^P$ is also pathwise connected (27.1 of [**Wi**]) and therefore connected (27.2 of [**Wi**]).

Our next objective is to show that a P-homeomorphism $h: \mathcal{M}^P \to \mathcal{M}^P$ of $\mathcal{M}^P$ onto itself carries timelike curves onto timelike curves, i.e., that $\alpha: I \to \mathcal{M}$ is timelike if and only if $h \circ \alpha: I \to \mathcal{M}$ is timelike. We prove this by characterizing timelike curves entirely in terms of set-theoretic and P-topological notions that are obviously preserved by P-homeomorphisms.

Theorem A.3.6. *A curve $\alpha : I \to \mathcal{M}$ is timelike if and only if the following two conditions are satisfied:*

1. *α is P-continuous and one-to-one*
2. *For every t_0 in I there exists a connected, relatively open subset U of I containing t_0 and a P-open neighborhood V of $\alpha(t_0)$ in $\mathcal{M}$ such that:*
   - *(a) $\alpha(U) \subseteq V$*
   - *(b) Whenever t_0 is in the interior of I and a and b are in U and satisfy $a < t_0 < b$, then every P-continuous curve in V joining $\alpha(a)$ and $\alpha(b)$ passes through $\alpha(t_0)$.*

Proof: First assume α is timelike. Since the proofs are the same in the two cases we will assume that α is future-timelike. Then α is P-continuous by Lemma A.3.4(2) and one-to-one by Lemma A.2.1 so (1) is satisfied. Now fix a t_0 in I and select $U \subseteq I$ as in the definition of future-timelike at t_0. Let $V = \mathcal{C}(\alpha(t_0))$. Then V is a P-open neighborhood of $\alpha(t_0)$ with $\alpha(U) \subseteq V$ so part (a) of (2) is satisfied. Next suppose t_0 is in the interior of U and let a and b be in U with $a < t_0 < b$. Then $\alpha(a) \in \mathcal{C}_T^-(\alpha(t_0))$ and $\alpha(b) \in \mathcal{C}_T^+(\alpha(t_0))$. Suppose $\gamma : [c,d] \to \mathcal{M}$ is a P-continuous curve in V with $\gamma(c) = \alpha(a)$ and $\gamma(d) = \alpha(b)$. By P-continuity, $\gamma[c,d]$ is a connected subspace of $\mathcal{M}^P$ (26.3 of [**Wi**]). But if $\alpha(t_0)$ were not in the image of γ, then $\gamma[c,d] = [\gamma[c,d] \cap \mathcal{C}_T^-(\alpha(t_0))] \cup [\gamma[c,d] \cap \mathcal{C}_T^+(\alpha(t_0))]$ would be a disconnection (26.1 of [**Wi**]) of $\gamma[c,d]$. Thus, (b) of (2) is also satisfied.

Conversely, suppose $\alpha : I \to \mathcal{M}$ satisfies (1) and (2). Then α is E-continuous by Lemma A.3.4. We show that α is timelike at each t_0 in the interior of I and appeal to Lemma A.2.3. Let U and V be as in (2). Assume without loss of generality that V is a basic open neighborhood $N_\varepsilon^P(\alpha(t_0))$. Let $U^- = \{t \in U : t < t_0\}$ and $U^+ = \{t \in U : t > t_0\}$. Select $a \in U^-$ and $b \in U^+$. Since α is one-to-one, $\alpha(a) \neq \alpha(t_0)$ and $\alpha(b) \neq \alpha(t_0)$ so $\alpha(a)$ and $\alpha(b)$ both lie in $\mathcal{C}_T^-(\alpha(t_0)) \cup \mathcal{C}_T^+(\alpha(t_0))$. Assuming that $\alpha(a)$ is in $\mathcal{C}_T^-(\alpha(t_0))$ we show that α is future-timelike at t_0 [if $\alpha(a) \in \mathcal{C}_T^+(\alpha(t_0))$ the same proof shows that α is past-timelike at t_0]. If $\alpha(b)$ were also in $\mathcal{C}_T^-(\alpha(t_0))$ we could construct a Feynman path from $\alpha(a)$ to $\alpha(b)$ that is contained entirely in $N_\varepsilon^P(\alpha(t_0)) \cap \mathcal{C}_T^-(\alpha(t_0))$. But such a Feynman path would be a P-continuous curve in V joining $\alpha(a)$ and $\alpha(b)$ which could not go through $\alpha(t_0)$, thus contradicting part (b) of (2). Thus, $\alpha(b) \in \mathcal{C}_T^+(\alpha(t_0))$. We conclude that $\alpha(U^-) \cap \mathcal{C}_T^-(\alpha(t_0)) \neq \emptyset$ and $\alpha(U^+) \cap \mathcal{C}_T^+(\alpha(t_0)) \neq \emptyset$. Since α is one-to-one, $\alpha(t_0) \notin \alpha(U^-)$ and $\alpha(t_0) \notin \alpha(U^+)$. But α is P-continuous so $\alpha(U^-)$ and $\alpha(U^+)$ are both connected subspaces of $\mathcal{M}^P$ and so we must have $\alpha(U^-) \subseteq \mathcal{C}_T^-(\alpha(t_0))$ and $\alpha(U^+) \subseteq \mathcal{C}_T^+(\alpha(t_0))$, i.e., α is future-timelike at t_0. ■

Corollary A.3.7. *If $h : \mathcal{M}^P \to \mathcal{M}^P$ is a P-homeomorphism of $\mathcal{M}^P$ onto itself, then a curve $\alpha : I \to \mathcal{M}$ is timelike if and only if $h \circ \alpha : I \to \mathcal{M}$ is timelike.*

Proof: Conditions (1) and (2) of Theorem A.3.6 are both obviously preserved by P-homeomorphisms. ■

Corollary A.3.8. *If $h : \mathcal{M}^P \to \mathcal{M}^P$ is a P-homeomorphism of $\mathcal{M}^P$ onto itself, then h carries $\mathcal{C}_T(x)$ bijectively onto $\mathcal{C}_T(h(x))$ for every x in $\mathcal{M}$.*

Exercise A.3.3. Prove Corollary A.3.8. ■

We wish to show that a P-homeomorphism either preserves or reverses the order $\ll$. First, the local version.

Lemma A.3.9. *Let $h : \mathcal{M}^P \to \mathcal{M}^P$ be a P-homeomorphism and x a fixed point in $\mathcal{M}$. Then either*

1. $h(\mathcal{C}_T^-(x)) = \mathcal{C}_T^-(h(x))$ *and* $h(\mathcal{C}_T^+(x)) = \mathcal{C}_T^+(h(x))$ *or*
2. $h(\mathcal{C}_T^-(x)) = \mathcal{C}_T^+(h(x))$ *and* $h(\mathcal{C}_T^+(x)) = \mathcal{C}_T^-(h(x))$.

Proof: Suppose there exists a p in $\mathcal{C}_T^+(x)$ with $h(p) \in \mathcal{C}_T^+(h(x))$ [the argument is analogous if there exists a p in $\mathcal{C}_T^+(x)$ with $h(p) \in \mathcal{C}_T^-(h(x))$].

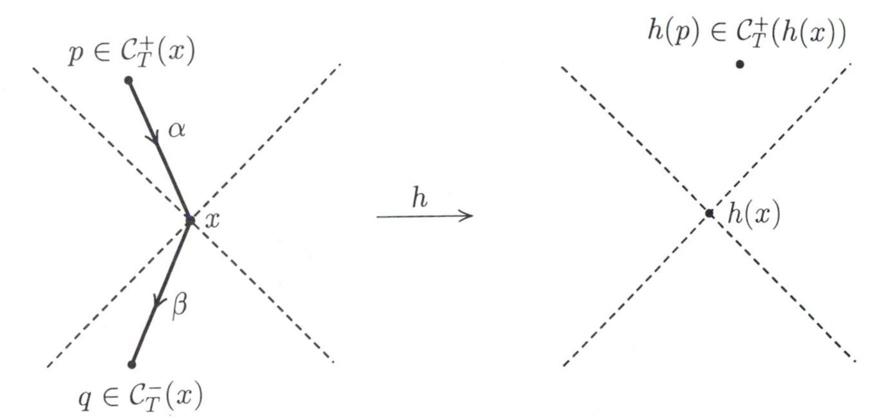

Figure A.3.5

Exercise A.3.4. Show that $h(\mathcal{C}_T^+(x)) \subseteq \mathcal{C}_T^+(h(x))$.

Now let q be in $\mathcal{C}_T^-(x)$. We claim that $h(q)$ is in $\mathcal{C}_T^-(h(x))$. Let α and β be past-timelike curves from p to x and x to q respectively. Let γ be the past-timelike curve from p to q consisting of α followed by β. Then, by Corollary A.3.7, $h \circ \alpha$, $h \circ \beta$ and $h \circ \gamma$ are all timelike. By Lemma A.2.3, $h \circ \gamma$ is either everywhere past-timelike or everywhere future-timelike. But $h \circ \alpha$ is past-timelike since $h(x) \ll h(p)$ and $h \circ \gamma$ initially coincides with $h \circ \alpha$ so it too must be past-timelike. By Theorem A.2.2, $h(q) \ll h(x)$, i.e., $h(q) \in \mathcal{C}_T^-(h(x))$. As in Exercise A.3.4 it follows that $h(\mathcal{C}_T^-(x)) \subseteq \mathcal{C}_T^-(h(x))$. But Corollary A.3.8 then gives $h(\mathcal{C}_T^+(x)) = \mathcal{C}_T^+(h(x))$ and $h(\mathcal{C}_T^-(x)) = \mathcal{C}_T^-(h(x))$. ■

With this we can now prove our major result.

Theorem A.3.10. *If $h : \mathcal{M}^P \to \mathcal{M}^P$ is a P-homeomorphism of $\mathcal{M}^P$ onto itself, then h either preserves or reverses the order $\ll$, i.e., either*

1. *$x \ll y$ if and only if $h(x) \ll h(y)$ or*
2. *$x \ll y$ if and only if $h(y) \ll h(x)$.*

Proof: Let $S = \{x \in \mathcal{M} : h$ preserves $\ll$ at $x\}$. We will show that S is open in $\mathcal{M}^P$. The proof that $\mathcal{M}^P - S$ is open in $\mathcal{M}^P$ is the same

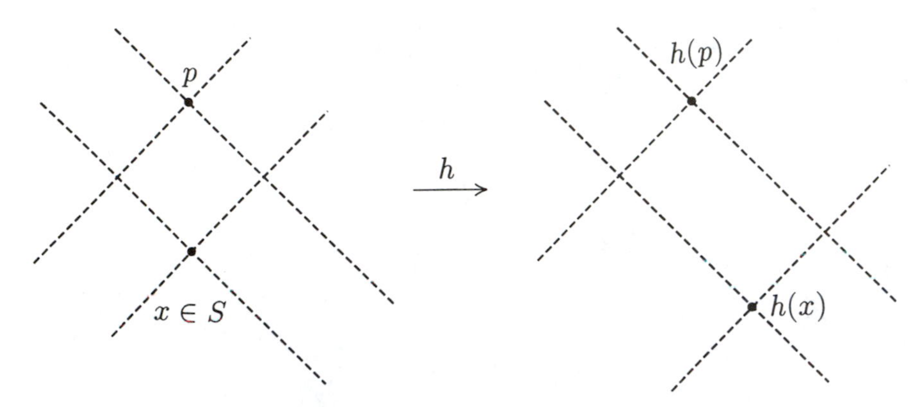

Figure A.3.6

so connectivity of $\mathcal{M}^P$ implies that either $S = \emptyset$ or $S = \mathcal{M}$. Suppose then that $S \neq \emptyset$ and select an arbitrary $x \in S$. Then $\mathcal{C}(x)$ is a P-open set containing x. We show that $\mathcal{C}(x) \subseteq S$ and conclude that S is open. To see this suppose $p \in \mathcal{C}_T^+(x) \subseteq \mathcal{C}(x)$ [the proof for $p \in \mathcal{C}_T^-(x)$ is similar]. Now, $x \in S$ implies $h(p) \in \mathcal{C}_T^+(h(x))$ (see Figure A.3.6.). By Lemma A.3.9, $h(\mathcal{C}_T^+(p))$ equals either $\mathcal{C}_T^+(h(p))$ or $\mathcal{C}_T^-(h(p))$. But the latter is impossible since $\mathcal{C}_T^+(p) \subseteq \mathcal{C}_T^+(x)$ implies $h(\mathcal{C}_T^+(p)) \subseteq h(\mathcal{C}_T^+(x)) = \mathcal{C}_T^+(h(x))$. Thus, $h(\mathcal{C}_T^+(p)) = \mathcal{C}_T^+(h(p))$ so p is in S as required. ∎

From Theorem A.3.10 and Exercise 1.6.3 we conclude that if $h: \mathcal{M}^P \to \mathcal{M}^P$ is a P-homeomorphism, then either h or $-h$ is a causal automorphism.

Exercise A.3.5. Show that if $h: \mathcal{M} \to \mathcal{M}$ is a causal automorphism, then h and $-h$ are both P-homeomorphisms. *Hint:* Zeeman's Theorem 1.6.2.

Now, if X is an arbitrary topological space the set $H(X)$ of all homeomorphisms of X onto itself is called the *homeomorphism group of* X (it is closed under the formation of compositions and inverses and so is indeed a group under the operation of composition). If G is a subset of $H(X)$ we will say that G *generates* $H(X)$ if every homeomorphism of X onto itself can be written as a composition of elements of G. We now know that $H(\mathcal{M}^P)$ consists precisely of the maps $\pm h$ where h is a causal automorphism and Zeeman's Theorem 1.6.2 describes all of these.

Theorem A.3.11. *The homeomorphism group* $H(\mathcal{M}^P)$ *of* $\mathcal{M}^P$ *is generated by translations, dilations and (not necessarily orthochronous) orthogonal transformations.*

Modulo translations and nonzero scalar multiplications, $H(\mathcal{M}^P)$ is essentially just the Lorentz group $\mathcal{L}$.

Appendix B

Spinorial Objects

B.1 Introduction

Here we wish to examine in some detail the mathematical origin and physical significance of the "essential 2-valuedness" of spinors, to which we alluded in Section 3.5. A genuine understanding of this phenomenon depends on topological considerations of a somewhat less elementary nature than those involved in Appendix A. Thus, in Section B.3, we must assume a familiarity with point-set topology through the construction of the fundamental group and its calculation for the circle (see Sections 32-34 of [**Wi**] or Sections 1-4 of [**G**]). The few additional homotopy-theoretic results to which we must appeal can all be found in Sections 5-6 of [**G**].

As we left it in Chapter 3, Section 5, the situation was as follows: Each nonzero spin vector ξ^A uniquely determines a future-directed null vector v and a 2-dimensional plane $\mathcal{F}$ spanned by v and a spacelike vector w orthogonal to v. The pair $(v, \mathcal{F})$ is called the null flag of ξ^A, with v the flagpole and $\mathcal{F}$ the flag. A phase change (rotation) $\xi^A \to e^{i\theta}\xi^A$ $(\theta \in \mathbb{R})$ of the spin vector ξ^A yields another spin vector with the same flagpole v as ξ^A, but whose flag is rotated around this flag pole by 2θ relative to the flag of ξ^A. The crucial observation is that if ξ^A undergoes a continuous rotation $\xi^A \to e^{i\theta}\xi^A$, $0 \leq \theta \leq \pi$, through π, then the end result of the rotation is a *new* spin vector $e^{i\pi}\xi^A = -\xi^A$, but the *same* null flag. Let us reverse our point of view. Regard the null flag $(v, \mathcal{F})$ as a concrete geometrical representation of the spin vector ξ^A in much the same way that a "directed line segment" represents a vector in classical physics and Euclidean geometry. One then finds oneself in the awkward position of having to concede that rotating this geometrical object by 2π about some axis yields an object apparently indistinguishable from the first, but representing, not ξ^A, but $-\xi^A$. One might seek additional geometrical data to append to the null flag (as we added the flag when we found that the flagpole itself did not uniquely determine ξ^A) in order to distinguish the object representing ξ^A from that representing $-\xi^A$. It is clear, however, that if "geometrical data" is to be understood in the usual sense, then any such data would also be returned to its original value after a rotation of 2π. The sign ambiguity in our geometrical representation of spin vectors seems unavoidable, i.e., "essential". Perhaps even more curious is the fact that a further rotation of the flag by 2π (i.e., a total rotation of 4π) corresponds to $\theta = 2\pi$ and so returns to us the original spin vector $\xi^A = e^{i(2\pi)}\xi^A$ *and* the original null flag.

This state of affairs is quite unlike anything encountered in classical physics or geometry. By analogy, one would have to imagine a "vector" and its geometrical representation as a directed line segment with the property that, by rotating the arrow through 2π about some axis one obtained the geometrical representation of some other "vector". But, of course, classical Euclidean vector (and, more generally, tensor) analysis is built on the premise that this cannot be the case. Indeed, a vector (tensor) is just a carrier of some representation of the rotation group and the element of the rotation group corresponding to rotation by 2π about any axis is the identity. This is, of course, just a mathematical reflection of the conventional wisdom that rotating an isolated physical system through 2π yields a system that is indistinguishable from the first.

B.2 The Spinning Electron and Dirac's Demonstration

"Conventional wisdom" has not fared well in modern physics so it may come as no surprise to learn that there are, in fact, physical systems at the subatomic level whose state *is* altered by a rotation of the system through 2π about some axis, but is returned to its orginal value by a rotation through 4π. Indeed, any of the elementary particles in nature classified as a Fermions (electrons, protons, neutrons, neutrinos, etc.) possess what the physicists call "half-integer spin" and, as a consequence, their quantum mechanical descriptions ("wave functions") behave in precisely this way (a beautifully lucid and elementary account of the physics involved here is available in Volume III of the Feynman Lectures on Physics [**Fe**]). That the spin state of an electron behaves in this rather bizarre way has been known for many years, but, because of the way in which quantum mechanics decrees that physical information be extracted from an object's wave function, was generally thought to have no observable consequences. More recently it has been argued that it is possible, in principle, to construct devices in which this behavior under rotation is exhibited on a macroscopic scale (see [**AS**], [**KO**] and [**M**]). These constructions, however, depend on a rather detailed understanding of how electrons are described in quantum mechanics. Fortunately, Paul Dirac has devised a remarkably ingenious demonstration involving a perfectly mundane macroscopic physical system in which "something" in the system's state is altered by rotation through 2π, but returned to its original value by a 4π rotation. Next we describe the so-called "Dirac Scissors Problem" and, in the next section, investigate the mathematics behind the phenomenon.

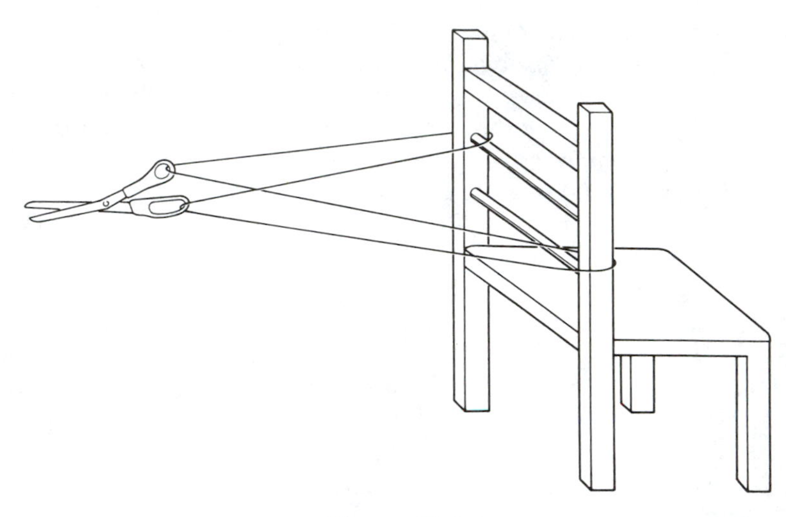

Figure B.2.1

The demonstration involves a pair of scissors, a piece of (elastic) string and a chair. Pass the string through one finger hole of the scissors, then around one arm of the chair, then through the other fingerhole and around the other arm of the chair and then tie the two ends of the string together (see Figure B.2.1). The scissors is now rotated about its axis of symmetry through 2π (one complete revolution). The strings become entangled and the problem is to disentangle them by moving only the string, holding the scissors and chair fixed (the string needs to be elastic so it can be moved around these objects, if desired). Try it! No amount of manuvering, simple or intricate, will return the strings to their original, disentangled state. This, in itself, is not particularly surprising perhaps, but now repeat the exercise, this time rotating the scissors about its axis through *two* complete revolutions (4π). The strings now appear even more hopelessly tangled, but looping the string just once over the pointed end of the scissors (counterclockwise if that is the way you turned the scissors) will return them to their original condition.

One is hard-pressed not to be taken aback by the result of this little game, but, in fact, there are even more dramatic demonstrations of the same phenomenon. Imagine a cube (with its faces numbered, or painted different colors, so that one can keep track of the rotations it experiences). Connect each corner of the cube to the corresponding corner of a room with elastic string (see Figure B.2.2). Rotate the cube by 2π about any axis. The strings become tangled and no manipulation of the strings that leaves

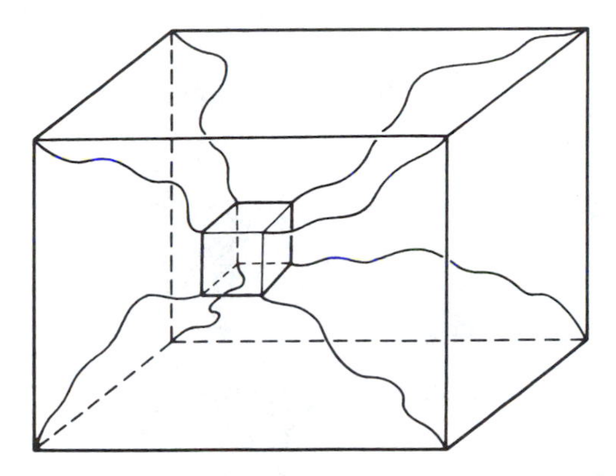

Figure B.2.2

the cube (and the room) fixed will untangle them. Rotate by another 2π about the same axis for a total rotation of 4π and the tangles apparently get worse, but a carefully chosen motion of the strings (alone) will return them to their original state (the appropriate sequence of manuvers is shown in Figure 41.6 of [**MTW**]).

In each of these situations there is clearly "something different" about the state of the system when it has undergone a rotation of 2π and when it has been rotated by 4π. Observe also that, in each case, the "system" is more than just an isolated pair of scissors or a cube, but includes, in some sense, the way in which that object is "connected" to its surroundings. In the next section we return to mathematics to show how all of this can be said precisely and, indeed, how the mathematics itself might have suggested the possibility of such phenomena and the relevance of spinors to their description.

B.3 Homotopy in the Rotation and Lorentz Groups

We begin by establishing some notation and terminology and briefly reviewing some basic results related to the notion of "homotopy" in topology (a good, concise source for all of the material we will need is [**G**], Sections

1-6). Much of what we have to say will be true in an arbitrary topological space, but this much generality is not required and tends to obscure fundamental issues with tiresome technicalities. For this reason we shall restrict our attention to the category of "connected topological manifolds". A Hausdorff topological space X is called an (*n-dimensional*) *topological manifold* if each $x \in X$ has an open neighborhood in X that is homeomorphic to an open set in $\mathbb{R}^n$ (18.3 of [**Wi**] or (6.8) of [**G**]). A *path* in X is a continuous map $\alpha : [0,1] \to X$. If $\alpha(0) = x_0$ and $\alpha(1) = x_1$, then α is a *path from* x_0 *to* x_1 in X and X is *path connected* if such a path exists for every pair of points $x_0, x_1 \in X$ (27.1 of [**Wi**]).

Exercise B.3.1. Show that a topological manifold X that is connected (26.1 of [**Wi**]) is necessarily path connected. *Hint:* Fix an arbitrary $x_0 \in X$ and show that the set of all $x_1 \in X$ for which there is a path in X from x_0 to x_1 is both open and closed.

Henceforth, "space" will mean "connected topological manifold".

Let α_0 and α_1 be two paths in X from x_0 to x_1. We say that α_0 and α_1 are (*path*) *homotopic* (*with endpoints fixed*) if there exists a continuous map $H : [0,1] \times [0,1] \to X$, called a *homotopy from* α_0 *to* α_1, which satisfies

$$H(s,0) = \alpha_0(s),$$

$$H(s,1) = \alpha_1(s),$$

$$H(0,t) = x_0,$$

$$H(1,t) = x_1$$

for all s and t in $[0,1]$. In this case we write $\alpha_0 \simeq \alpha_1$. For each t in $[0,1]$, $\alpha_t(s) = H(s,t)$ defines a path in X from x_0 to x_1 and, intuitively, one regards H as providing a "continuous deformation" of α_0 into α_1 through the family $\{\alpha_t : t \in [0,1]\}$ of paths. $\simeq$ is an equivalence relation on the set of all paths from x_0 to x_1 and we denote the equivalence class of a path α by $[\alpha]$. The *inverse* of a path α from x_0 to x_1 is the path α^{-1} from x_1 to x_0 defined by $\alpha^{-1}(s) = \alpha(1-s)$. One verifies that $\alpha_0 \simeq \alpha_1$ implies $\alpha_0^{-1} \simeq \alpha_1^{-1}$ so one may define the inverse of a homotopy equivalence class by $[\alpha]^{-1} = [\alpha^{-1}]$. If α is a path from x_0 to x_1 in X and β is a path from x_1 to x_2 in X, then the *product path* $\beta\alpha$ from x_0 to x_2 is defined by

$$(\beta\alpha)(s) = \begin{cases} \alpha(2s), & 0 \leq s \leq \frac{1}{2} \\ \beta(2s-1), & \frac{1}{2} \leq s \leq 1\,. \end{cases}$$

Again, $\alpha_0 \simeq \alpha_1$ and $\beta_0 \simeq \beta_1$ imply $\beta_0\alpha_0 \simeq \beta_1\alpha_1$ so one may define the product of the homotopy equivalence classes $[\alpha]$ and $[\beta]$ by $[\beta][\alpha] = [\beta\alpha]$, provided the initial point of all the paths in $[\beta]$ coincides with the terminal point of all the paths in $[\alpha]$. A *loop at* x_0 is a path from $\alpha(0) = x_0$ to

$\alpha(1) = x_0$. Then α^{-1} is also a loop at x_0. Moreover, if β is another loop at x_0, then $\beta\alpha$ is defined and is also a loop at x_0. Letting

$$\pi_1(X, x_0) = \{[\alpha] : \alpha \text{ is a loop at } x_0\},$$

one finds that the operations $[\alpha]^{-1} = [\alpha^{-1}]$ and $[\beta][\alpha] = [\beta\alpha]$ give $\pi_1(X, x_0)$ the structure of a group with identity element $[x_0]$, where we are here using x_0 to designate also the *constant* (or *trivial*) *loop* at x_0 defined by $x_0(s) = x_0$ for all s in $[0,1]$. $\pi_1(X, x_0)$ is called the *fundamental group of X at x_0*. If x_0 and x_1 are any two points in X and γ is a path in X from x_1 to x_0 (guaranteed to exist by Exercise B.3.1), then $[\alpha] \to [\gamma^{-1}\alpha\gamma]$ is an isomorphism of $\pi_1(X, x_0)$ onto $\pi_1(X, x_1)$. For this reason one generally writes $\pi_1(X)$ for any one of the isomorphic groups $\pi_1(X, x)$, $x \in X$, and calls $\pi_1(X)$ the *fundamental group of X*. Obviously, homeomorphic spaces have the same (that is, isomorphic) fundamental groups. More generally, any two homotopically equivalent ((3.6) of [**G**]) spaces have the same fundamental groups.

A space is said to be *simply connected* if its fundamental group is isomorphic to the trivial group, i.e., if every loop is homotopic to the trivial loop (somewhat loosely one says that "every closed curve can be shrunk to a point"). Any Euclidean space $\mathbb{R}^n$ is simply connected ((3.2) of [**G**]), as is the n-sphere $S^n = \{(x^1, \ldots, x^{n+1}) \in \mathbb{R}^{n+1} : (x^1)^2 + \cdots + (x^{n+1})^2 = 1\}$ for any $n \geq 2$ (see Exercise B.3.5 and (4.13) of [**G**]). For $n = 1$, however, the situation is different. Indeed, the fundamental group of the circle, $\pi_1(S^1)$, is isomorphic to the additive group $\mathbb{Z}$ of integers ((4.4) of [**G**]). Essentially, a loop in S^1 is characterized homotopically by the (integer) number of times it wraps around the circle (positive in one direction and negative in the other).

Exercise B.3.2. Let X and Y be two topological manifolds of dimensions n and m respectively. Show that $X \times Y$, provided with the product topology, is a topological manifold of dimension $n + m$.

It is not difficult to show ((4.8) of [**G**]) that the fundamental group of a product $X \times Y$ is isomorphic to the direct product of the fundamental groups of X and Y, i.e., $\pi_1(X \times Y) \cong \pi_1(X) \times \pi_1(Y)$. In particular, the fundamental group of the torus $S^1 \times S^1$ is $\mathbb{Z} \times \mathbb{Z}$.

In order to calculate several less elementary examples and, in the process, get to the heart of the connection between homotopy and spinorial objects, we require the notion of a "universal covering manifold". As motivation let us consider again the circle S^1. This time it is convenient to describe S^1 as the set of all complex numbers of modulus one, i.e., $S^1 = \{z \in \mathbb{C} : z\bar{z} = 1\}$. Define a map $p : \mathbb{R} \to S^1$ by $p(\theta) = e^{2\pi\theta i} = \cos(2\pi\theta) + i\sin(2\pi\theta)$. Observe that p is continuous, carries $0 \in \mathbb{R}$ onto $1 \in S^1$ and, in effect, "wraps" the real line around the circle. Notice also that each $z \in S^1$ has a neighborhood U is S^1 with the property that $p^{-1}(U)$ is a disjoint

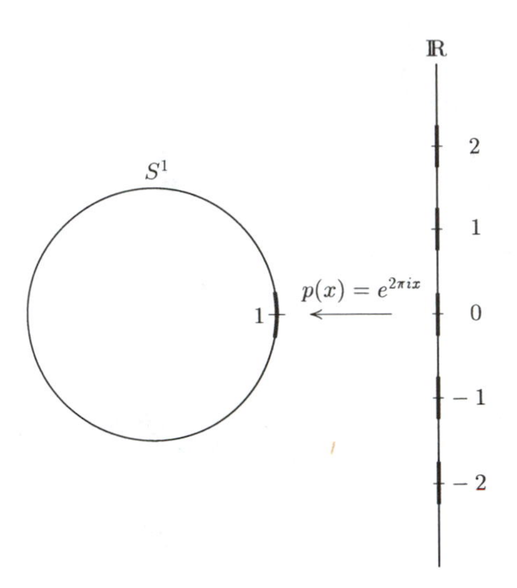

Figure B.3.1

union of open sets in $\mathbb{R}$, each of which is mapped homeomorphically by p onto U (this is illustrated for $z = 1$ in Figure B.3.1). In particular, the "fiber" $p^{-1}(z)$ above each $z \in S^1$ is discrete. Now let us consider a homeomorphism ϕ of $\mathbb{R}$ onto itself which "preserves the fibers of p", i.e., satisfies $p \circ \phi = p$, so that $r \in p^{-1}(z)$ implies $\phi(r) \in p^{-1}(z)$. We claim that such a homeomorphism is uniquely determined by its value at $0 \in \mathbb{R}$ (or at any other single point in $\mathbb{R}$), i.e., that if ϕ_1 and ϕ_2 are two p-fiber preserving homeomorphisms of $\mathbb{R}$ onto $\mathbb{R}$ and $\phi_1(0) = \phi_2(0)$, then $\phi_1 = \phi_2$. To see this let $E = \{e \in \mathbb{R} : \phi_1(e) = \phi_2(e)\}$. Then $E \neq \emptyset$ since $0 \in E$ and, by continuity of ϕ_1 and ϕ_2, E is closed in $\mathbb{R}$. Since $\mathbb{R}$ is connected the proof will be complete if we can show that E is open. Thus, let e be a point in E so that $\phi_1(e) = \phi_2(e) = r$ for some $r \in \mathbb{R}$. Notice that $p(\phi_1(e)) = p(r)$ and $p(\phi_1(e)) = p(e)$ imply that $p(e) = p(r)$. Now select open neighborhoods V_e and V_r of e and r which p maps homeomorphically onto a neighborhood U of $p(e) = p(r)$. Let $V = V_e \cap \phi_1^{-1}(V_r) \cap \phi_2^{-1}(V_r)$. Then V is an open neighborhood of e contained in V_e and with $\phi_1(V) \subseteq V_r$ and $\phi_2(V) \subseteq V_r$. For each $v \in V$, $p(\phi_1(v)) = p(\phi_2(v))$ since both equal $p(v)$. But $\phi_1(v), \phi_2(v) \in V_r$ and, on V_r, p is a homeomorphism so $\phi_1(v) = \phi_2(v)$. Thus, $V \subseteq E$ so E is open. We find then that the homeomorphisms of $\mathbb{R}$ that preserve the fibers of p are completely determined by their values at 0. Since the elements in a single fiber $p^{-1}(z)$ clearly differ by integers, the value of a p-fiber preserving homeomorphism of $\mathbb{R}$ at 0 is an integer.

Exercise B.3.3. For each integer n let $\phi_n: \mathbb{R} \to \mathbb{R}$ be the translation of $\mathbb{R}$ by n, i.e., $\phi_n(y) = y + n$ for each $y \in \mathbb{R}$. Show that the set $\mathcal{C}$ of p-fiber preserving homeomorphisms of $\mathbb{R}$ is precisely $\{\phi_n : n \in \mathbb{Z}\}$. Observe, moreover, that $\phi_n \circ \phi_m = \phi_{n+m}$ so, as a group under the operation of composition, $\mathcal{C}$ is isomorphic to the additive group $\mathbb{Z}$ of integers, i.e., to $\pi_1(S^1)$.

Distilling the essential features out of this last example leads to the following definitions and results. Let X be a connected topological manifold. A *universal covering manifold* for X consists of a pair $(\tilde{X}, p)$, where $\tilde{X}$ is a simply connected topological manifold and $p: \tilde{X} \to X$ is a continuous surjection (called the *covering map*) with the property that every $x \in X$ has an open neighborhood U such that $p^{-1}(U)$ is a disjoint union of open sets in $\tilde{X}$, each of which is mapped homeomorphically onto U by p. Every connected topological manifold has a universal covering manifold $(\tilde{X}, p)$ ((6.8) of [**G**]) that is essentially unique in the sense that if $(\tilde{X}', p')$ is another, then there exists a homeomorphism ψ of $\tilde{X}'$ onto $\tilde{X}$ such that $p \circ \psi = p'$ ((6.4) of [**G**]). A homeomorphism ϕ of $\tilde{X}$ onto itself that preserves the fibers of p, i.e., satisfies $p \circ \phi = p$, is called a *covering transformation* and the collection $\mathcal{C}$ of all such is a group under composition. Moreover, $\mathcal{C}$ is isomorphic to $\pi_1(X)$ ((5.8) of [**G**]). $\mathcal{C}$ is often easier to contend with than $\pi_1(X)$ and we will now use it to compute the examples of real interest to us.

We shall construct these examples "backwards", beginning with a space $\tilde{X}$ that will eventually be the universal covering manifold of the desired example X, which is defined as a quotient (9.1 of [**Wi**]) of $\tilde{X}$. First take $\tilde{X}$ to be the 2-sphere $S^2 = \{(x^1, x^2, x^3) \in \mathbb{R}^3 : (x^1)^2 + (x^2)^2 + (x^3)^2 = 1\}$ with the topology it inherits as a subspace of $\mathbb{R}^3$.

Exercise B.3.4. Show that S^2 is a (Hausdorff, 2-dimensional, connected, compact) topological manifold. *Hint:* Show, for example, that, on the upper hemisphere $\{(x^1, x^2, x^3) \in S^2 : x^3 > 0\}$, the projection map $(x^1, x^2, x^3) \to (x^1, x^2)$ is a homeomorphism onto the unit disc $(x^1)^2 + (x^2)^2 < 1$.

Exercise B.3.5. Show that S^n is a (Hausdorff, n-dimensional, connected, compact) topological manifold for any $n \geq 1$.

Now define an equivalence relation $\sim$ on S^2 by identifying antipodal points, i.e., if $y, z \in S^2$, then $y \sim z$ if and only if $z = -y$. Let $[y]$ denote the equivalence class of y, i.e., $[y] = \{y, -y\}$, and denote by $\mathbb{R}P^2$ the set of all equivalence classes. Define $p: S^2 \to \mathbb{R}P^2$ by $p(y) = [y]$ for every $y \in S^2$ and provide $\mathbb{R}P^2$ with the quotient topology determined by p [i.e., $U \subseteq \mathbb{R}P^2$ is open if and only if $p^{-1}(U)$ is open in S^2]. $\mathbb{R}P^2$ is then called the *real projective plane*.

Exercise B.3.6. Show that $\mathbb{R}P^2$ is a (Hausdorff, 2-dimensional, connected, compact) topological manifold.

Now, since S^2 is simply connected and $p\colon S^2 \to \mathbb{R}P^2$ clearly satisfies the defining condition for a covering map and since universal covering manifolds are unique we conclude that

$$\widetilde{\mathbb{R}P^2} \cong S^2.$$

But then $\pi_1(\mathbb{R}P^2)$ is isomorphic to the group of p-fiber preserving homeomorphisms $\phi\colon S^2 \to S^2$ of S^2. We claim that this group contains precisely two elements, namely, the identity map $[\phi_0(y) = y$ for every $y \in S^2]$ and the antipodal map $[\phi_1(y) = -y$ for every $y \in S^2]$. To see this observe that the fibers of p are just pairs of antipodal points $\{y, -y\}$ so such a ϕ must, for each $y \in S^2$, satisfy either $\phi(y) = y$ or $\phi(y) = -y$ and so $S^2 = \{y \in S^2 : \phi(y) = y\} \cup \{y \in S^2 : \phi(y) = -y\}$. Since both of these sets are obviously closed, connectivity of S^2 implies that one is $\emptyset$ and the other is S^2 as required. Thus, $\pi_1(\mathbb{R}P^2)$ has precisely two elements and so must be isomorphic to the group of integers mod 2, i.e.,

$$\pi_1\left(\mathbb{R}P^2\right) \cong \mathbb{Z}_2.$$

It will be important to us momentarily to observe that there is another way to construct $\mathbb{R}P^2$. For this we carry out the identification of antipodal points on S^2 in two stages. First identify points on the lower hemisphere $(x^3 < 0)$ with their antipodes on the upper hemisphere $(x^3 > 0)$, leaving the equator $(x^3 = 0)$ fixed. At this point we have a copy of the closed upper hemisphere $(x^3 \geq 0)$ which, by projecting into the x^1x^2-plane, is homeomorphic to the closed disc $(x^1)^2 + (x^2)^2 \leq 1$. To obtain $\mathbb{R}P^2$ we now need only identify antipodal points on the boundary circle $(x^1)^2 + (x^2)^2 = 1$. This particular construction can be refined to yield a "visualization" of $\mathbb{R}P^2$ (see Chapter 1, Volume 1, of [**Sp**]). Visualization here is not easy, however. Indeed, given a little thought, $\pi_1(\mathbb{R}P^2) = \mathbb{Z}_2$ is rather disconcerting. Think about some loop in $\mathbb{R}P^2$ that is *not* homotopically trivial, i.e., cannot be "shrunk to a point in $\mathbb{R}P^2$" (presumably because it "surrounds a hole" in $\mathbb{R}P^2$). Traverse the loop twice and (because $1 + 1 = 0$ in $\mathbb{Z}_2$) the resulting loop *must* be homotopically trivial. What happened to "the hole"? Think about it (especially in light of our second construction of $\mathbb{R}P^2$).

Exercise B.3.7. Define *real projective 3-space* $\mathbb{R}P^3$ by beginning with the 3-sphere $S^3 = \{(x^1, x^2, x^3, x^4) \in \mathbb{R}^4 : (x^1)^2 + (x^2)^2 + (x^3)^2 + (x^4)^2 = 1\}$ in $\mathbb{R}^4$ and identifying antipodal points $(y \sim -y)$. Note that $\widetilde{\mathbb{R}P^3} = S^3$ and conclude that $\pi_1(\mathbb{R}P^3) \cong \mathbb{Z}_2$. Also observe that $\mathbb{R}P^3$ can be obtained by identifying antipodal points on the boundary $(x^1)^2 + (x^2)^2 +$

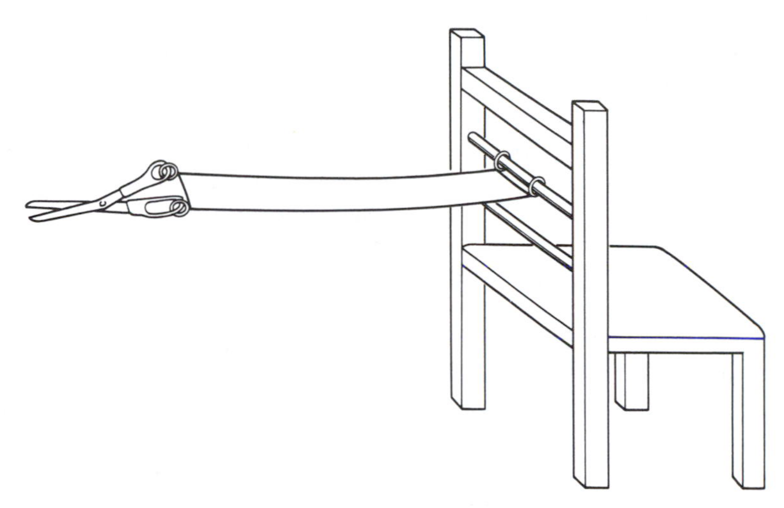

Figure B.3.2

$(x^3)^2 = 1$ of the closed 3-dimensional ball $(x^1)^2 + (x^2)^2 + (x^3)^2 \leq 1$.

In an entirely analogous manner one defines $\mathbb{R}P^n$ for any $n \geq 2$ and shows that $\pi_1(\mathbb{R}P^n) \cong \mathbb{Z}_2$.

Exercise B.3.8. What happens when $n = 1$?

Now let us return to the Dirac experiment. As with any good magic trick, some of the paraphenalia is present only to divert the attention of the audience. Notice that none of the essential features of the apparatus are altered if we imagine the strings glued (in an arbitrary manner) to the surface of an elastic belt so that we may discard the strings altogether in favor of such a belt connecting the scissors and the chair (see Figure B.3.2). Rotate the scissors through 2π and the belt acquires one twist which cannot be untwisted by moving the belt alone. Rotate through 4π and the belt has two twists that can be removed by looping the belt once around the scissors.

Regarding the scissors as a rigid, solid body in 3-space we now introduce what the physicists would call its "configuration space". Fix some position of the scissors in space as its "original" configuration. Any continuous motion of the scissors in space will terminate with the scissors in some new configuration which can be completely described by giving a point in $\mathbb{R}^3$ (e.g., the location of the scissors' center of mass) and a rotation that would carry the original orientation of the scissors onto its new orientation. This second element of the description we specify by giving an element of the *rotation group* $SO(3)$, i.e., the set of all 3×3 unimodular orthogonal matrices [when viewed as a subgroup of the Lorentz group we denoted $SO(3)$ by $\mathcal{R}$; see Section 1.3]. Thus, the configuration space of our scissors is taken to be $\mathbb{R}^3 \times SO(3)$.

In configuration space $\mathbb{R}^3 \times SO(3)$ a continuous motion of the scissors in space is represented by a continuous curve. In particular, if the initial and final configurations are the same, by a loop. Consider, for example, some point x_0 in $\mathbb{R}^3 \times SO(3)$, i.e., some initial configuration of the scissors. A continuous rotation of the scissors through 2π about some axis is represented by a loop at x_0 in $\mathbb{R}^3 \times SO(3)$. Dirac's ingenious demonstration permits us to actually "see" this loop. Indeed, let us visualize Dirac's apparatus with the belt having one "twist". Now imagine the scissors free to slide along the belt toward the chair. As it does so it completes a rotation through 2π. When it reaches the chair, translate it (without rotation) back to its original location and one has traversed a loop in configuration space. Similarly, for a rotation through 4π. Indeed, it should now be clear that any position of the belt can be viewed as representing a loop in $\mathbb{R}^3 \times SO(3)$ (slide the scissors along the belt then translate it back). Now imagine yourself manipulating the belt (without moving scissors or chair) in an attempt to untwist it. At each instant the position of the belt represents a loop in $\mathbb{R}^3 \times SO(3)$ so the process itself may be thought of as a continuous sequence of loops (parametrized, say, by time t). If you succeed with such a sequence of loops to untwist the belt you have "created" a homotopy from the loop corresponding to the belt's initial configuration to the trivial loop (no rotation, i.e., no twists, at all). What Dirac seems to be telling us then is that the loop in $\mathbb{R}^3 \times SO(3)$ corresponding to a 2π rotation is *not* homotopically trivial, but that corresponding to a rotation through 4π *is* homotopic to the trivial loop.

It is clearly of some interest then to understand the "loop structure", i.e., the fundamental group, of $\mathbb{R}^3 \times SO(3)$. Notice that $SO(3)$ does indeed have a natural topology. The entries in a 3×3 matrix can be strung out into a column matrix which can be viewed as a point in $\mathbb{R}^9$. Thus, $SO(3)$ can be viewed as a subset of $\mathbb{R}^9$ and therefore inherits a topology as a subspace of $\mathbb{R}^9$. A considerably more informative "picture" of $SO(3)$ can be obtained as follows: Every rotation of $\mathbb{R}^3$ can be uniquely specified by an axis of rotation, an angle and a sense of rotation about the axis. We claim that all of this information can be codified in a single object, namely, a vector $\vec{n}$ in $\mathbb{R}^3$ of magnitude at most π. Then the axis of rotation is the line along $\vec{n}$, the angle of rotation is $|\vec{n}|$ and the sense is determined by the "right-hand rule". Notice that a rotation along $\vec{n}$ through an angle θ with $\pi \leq \theta \leq 2\pi$ is equivalent to a rotation along $-\vec{n}$ through $2\pi - \theta$ so the restriction on $|\vec{n}|$ is necessary (although not quite sufficient) to ensure that the correspondence between rotations and vectors be one-to-one. The set of vectors $\vec{n}$ in $\mathbb{R}^3$ with $|\vec{n}| \leq \pi$ is just the closed ball of radius π about the origin. However, a rotation about $\vec{n}$ through π is the same as a rotation about $-\vec{n}$ through π so antipodal points on the boundary of this ball represent the same rotation and therefore must be identified in order that this correspondence with rotations be bijective. Carrying out this identification yields, according to Exercise B.3.7, real projective 3-space (topologically, the radius of the ball is irrelevant, of course). One can write out analytically the one-to-one correspondence we have just described geometrically to show that it is, in fact, continuous as a map from $\mathbb{R}P^3$ to $SO(3) \subseteq \mathbb{R}^9$. Since $\mathbb{R}P^3$ is compact (being a continuous image of S^3), we find that $SO(3)$ is homeomorphic to $\mathbb{R}P^3$. In particular, $\pi_1(SO(3)) \cong \pi_1(\mathbb{R}P^3) \cong \mathbb{Z}_2$. Thus, $\pi_1(\mathbb{R}^3 \times SO(3)) \cong \pi_1(\mathbb{R}^3) \times \pi_1(SO(3)) \cong \{0\} \times \mathbb{Z}_2$ so

$$\pi_1(\mathbb{R}^3 \times SO(3)) \cong \mathbb{Z}_2$$

and our suspicions are fully confirmed. In quite a remarkable way, the topology of the rotation group is reflected in the physical situation described by Dirac.

Exercise B.3.9. In $\mathbb{Z}_2$, $1+1+1=1$ and $1+1+1+1=0$. More generally, $2n+1=1$ and $2n=0$. What does this have to say about the scissors experiment?

But what has all of this to do with spinors? The connection is perhaps best appreciated by way of a brief digression into semantics. We have called $\mathbb{R}^3 \times SO(3)$ the "configuration space" of the object we have under consideration (the scissors). In the classical study of rigid body dynamics, however, it might equally well have been called its "state space" since, neglecting the object's (quite complicated) internal structure, it was (tacitly) assumed that the physical state of the object was entirely determined by its configuration in space. Suppressing the (topologically trivial and physically uninteresting) translational part of the configuration (i.e., $\mathbb{R}^3$), the body's "state" was completely specified by a point in $SO(3)$. Based on our observations in Section 3.1, we would phrase this somewhat more precisely by saying that all of the physically significant aspects of the object's condition (as a rigid body) should be describable as carriers of some representation of $SO(3)$ (keep in mind that, from our point of view, a rotated object is just the same object viewed from a rotated frame of reference). We shall refer to such quantities (which depend only on the object's configuration and not on "how it got there") as *tensorial objects*.

But the conclusion we draw from the Dirac experiment is that there may well be more to a system's "state" than merely its "configuration". This additional element has been called (see [**MTW**]) the *version* or *orientation-entanglement relation* of the system and its surroundings and at times it must be taken into account, e.g., when describing the quantum mechanical state of an electron with spin. Where is one to look for a mathematical model for such a system's "state" if now there are two, where we thought there was one? The mathematics itself suggests an answer. Indeed, the universal covering manifold of $SO(3)$ (i.e., of $\mathbb{R}P^3$) is S^3 and is, in fact, a *double cover,* i.e., the covering map is precisely two-to-one, taking the same value at y and $-y$ for each $y \in S^3$. Will S^3 do as the "state space"? That this idea is not the shot-in-the-dark it may at first appear will become apparent once it has been pointed out that we have actually seen all of this before.

Recall that, in Section 1.7, we constructed a homomorphism, called the spinor map, from $SL(2,\mathbb{C})$ onto $\mathcal{L}$ that was also precisely two-to-one and carried the unitary subgroup SU_2 of $SL(2,\mathbb{C})$ onto the rotation subgroup $\mathcal{R}$ [i.e., $SO(3)$] of $\mathcal{L}$.

Exercise B.3.10. Let $\underline{I} = \left[\begin{smallmatrix} 1 & 0 \\ 0 & 1 \end{smallmatrix}\right]$, $\underline{i} = \left[\begin{smallmatrix} i & 0 \\ 0 & -i \end{smallmatrix}\right]$, $\underline{j} = \left[\begin{smallmatrix} 0 & 1 \\ -1 & 0 \end{smallmatrix}\right]$ and $\underline{k} = \left[\begin{smallmatrix} 0 & i \\ i & 0 \end{smallmatrix}\right]$. Show that $\underline{I}$, $\underline{i}$, $\underline{j}$ and $\underline{k}$ are all in SU_2 and that, moreover, any $A \in SU_2$ is uniquely expressible in the form

$$A = a\underline{I} + b\underline{i} + c\underline{j} + d\underline{k},$$

where $a, b, c, d \in \mathbb{R}$ and $a^2 + b^2 + c^2 + d^2 = 1$. Regard SU_2 as a subset of $\mathbb{R}^8$ by identifying

$$A = \begin{bmatrix} a + bi & c + di \\ -c + di & a - bi \end{bmatrix}$$

with the column matrix $\mathrm{col}\,[a\ b\ c\ d\ \ -c\ d\ a\ \ -b] \in \mathbb{R}^8$ and define a map from SU_2 into $\mathbb{R}^4$ that carries this column matrix onto $\mathrm{col}\,[a\ b\ c\ d] \in \mathbb{R}^4$. Show that this map is a homeomorphism of SU_2 onto $S^3 \subseteq \mathbb{R}^4$. Finally, observe that the restriction of the spinor map to SU_2 is a continuous map onto $\mathcal{R}$ [i.e., $SO(3)$] which satisfies the defining property of a covering map for $SO(3)$.

Thus we find that SU_2 and the restriction of the spinor map to it constitute a concrete realization of the universal covering manifold for $SO(3)$ and its covering map. Old friends, in new attire. And now, how natural it all appears. Identify a "state" of the system with some $\tilde{y} \in SU_2$. This corresponds to some "configuration" $y \in SO(3)$ (the image of $\tilde{y}$ under the spinor map). Rotating the system through 2π corresponds to a *loop* in $SO(3)$ which, in turn, lifts ((5.2) of [**G**]) to a *path* in SU_2 from $\tilde{y}$ to $-\tilde{y}$ (a different "state"). Further rotation of the system through 2π traverses the loop in $SO(3)$ again, but, in SU_2, corresponds to a path from $-\tilde{y}$ to $\tilde{y}$ and so a rotation through 4π returns the original "state".

Exercise B.3.11. For each $t \in \mathbb{R}$ define a matrix $A(t)$ by

$$A(t) = \begin{bmatrix} e^{\frac{t}{2}i} & 0 \\ 0 & e^{-\frac{t}{2}i} \end{bmatrix}.$$

Show that $A(t) \in SU_2$ and that its image under the spinor map is the rotation

$$R(t) = \begin{bmatrix} \cos t & -\sin t & 0 & 0 \\ \sin t & \cos t & 0 & 0 \\ 0 & 0 & 1 & 0 \\ 0 & 0 & 0 & 1 \end{bmatrix}.$$

Hint: Take $\theta = \phi_2 = 0$ and $\phi_1 = t$ in Exercise 1.7.7.

Exercise B.3.12. Show that $\alpha\colon [0, 2\pi] \to SU_2$ defined by $\alpha(t) = A(t)$ for $0 \leq t \leq 2\pi$ is a path in SU_2 from the identity $I_{2\times 2}$ to $-I_{2\times 2}$ whose image under the spinor map is a loop $\text{Spin} \circ \alpha$ at $I_{4\times 4}$ (which is not nullhomotopic in $\mathcal{R}$). On the other hand, $\beta: [0, 4\pi] \to SU_2$ defined by $\beta(t) = A(t)$ for $0 \leq t \leq 4\pi$ is a loop at $I_{2\times 2}$ in SU_2 and its image $\text{Spin} \circ \beta$ is also a loop at $I_{4\times 4}$ (which is nullhomotopic in $\mathcal{R}$).

Mathematical quantities used to describe various aspects of the system's condition are still determined by the state of the system, but now we take this to mean that they should be expressible as carriers of some representation of SU_2. (Incidentally, any discomfort one might feel about the apparently miraculous appearance at this point of a group structure for the covering space should be assuaged by a theorem to the effect that this too is "essentially unique"; see (6.11) of [**G**].) Although we shall not go into the details here, it should come as no surprise to learn that the universal cover of the entire Lorentz group $\mathcal{L}$ consists of $SL(2, \mathbb{C})$ and the spinor map so that to obtain a relativistically invariant description of, say, the state of an electron, one looks to the representations of $SL(2, \mathbb{C})$, that is, to the 2-valued representations of $\mathcal{L}$ (see Section 1.7). Quantities such as the wave function of an electron (which depend not only on the object's configuration, but also on "how it got there") we call *spinorial objects* and are described mathematically by carriers of the representations of $SL(2, \mathbb{C})$, i.e., by spinors.

References

[**AS**] Aharonov, Y. and L. Susskind, "Observability of the sign change of spinors under 2π rotations", Phys. Rev., 158(1967), 1237-1238.

[**A**] Alphors, L., *Complex Analysis,* McGraw-Hill, New York, 1979.

[**BJ**] Bade, W. L. and H. Jehle, "An introduction to spinors", Rev. Mod. Phys., 25(1953), 714-728.

[**B**] Bolker, E. D., "The spinor spanner", Amer. Math. Monthly, 80(1973), 977-984.

[**C**] Cartan, E., *The Theory of Spinors,* M.I.T. Press, Cambridge, MA, 1966.

[**E**] Einstein, A., et al., *The Principle of Relativity,* Dover, New York, 1958.

[**Fa**] Fadell, E., "Homotopy groups of configuration spaces and the string problem of Dirac", Duke Math. J., 29(1962), 231-242.

[**Fe**] Feynman, R. P., R. B. Leighton and M. Sands, *The Feynman Lectures on Physics,* Vol. III, *Quantum Mechanics,* Addison-Wesley, Reading, MA, 1966.

[**FL**] Freedman, M. H. and Feng Luo, *Selected Applications of Geometry to Low-Dimensional Topology,* A.M.S. University Lecture Series, American Mathematical Society, Providence, RI, 1989.

[**GMS**] Gelfand, I. M., R. A. Minlos and Z. Ya. Shapiro, *Representations of the Rotation and Lorentz Groups and their Applications,* Pergamon, New York, 1963.

[**G**] Greenberg, M., *Lectures on Algebraic Topology,* W.A. Benjamin, New York, 1967.

[**HKM**] Hawking, S. W., A. R. King and P. J. McCarthy, "A new topology for curved spacetime which incorporates the causal, differential and conformal structures", J. Math. Phys., 17(1976), 174-181.

[**H**] Herstein, I. N., *Topics in Algebra,* Blaisdell, Waltham, MA, 1964.

[IS] Ives, H. E. and G. R. Stilwell, "Experimental study of the rate of a moving atomic clock", J. Opt. Soc. Am., 28(1938), 215; 31(1941), 369.

[KO] Klein, A. G. and G. I. Opat, "Observability of 2π rotations: A proposed experiment", Phys. Rev. D, 11(1975), 523-528.

[K] Kuiper, N. H., *Linear Algebra and Geometry,* North Holland, Amsterdam, 1965.

[La] Lang, S., *Linear Algebra,* Springer-Verlag, New York, 1987.

[LU] Laporte, O. and G. E. Uhlenbeck, "Application of spinor analysis to the Maxwell and Dirac equations", Phys. Rev., 37(1931), 1380-1397.

[LY] Lee, T. D. and C. N. Yang, "Parity nonconservation and a two-component theory of the neutrino", Phys. Rev., 105(1957), 1671-1675.

[Le] Lenard, A., "A characterization of Lorentz transformations", Amer. J. Phys., 19(1978), 157.

[M] Magnon, A. M. R., "Existence and observability of spinor structure", J. Math. Phys., 28(1987), 1364-1369.

[MTW] Misner, C. W., K. S. Thorne and J. A. Wheeler, *Gravitation,* W. H. Freeman, San Francisco, 1973.

[N_1] Naber, G. L., *Topological Methods in Euclidean Spaces,* Cambridge University Press, Cambridge, England, 1980.

[N_2] Naber, G. L., *Spacetime and Singularities,* Cambridge University Press, Cambridge, England, 1988.

[Ne] Newman, M. H. A., "On the string problem of Dirac", J. London Math. Soc., 17(1942), 173-177.

[Par] Parrott, S., *Relativistic Electrodynamics and Differential Geometry,* Springer-Verlag, New York, 1987.

[Pay] Payne, W. T., "Elementary spinor theory", Amer. J. Phys., 20(1952), 253-262.

[Pen] Penrose, R., "The apparent shape of a relativistically moving sphere", Proc. Camb. Phil. Soc., 55(1959), 137-139.

[PR] Penrose, R. and W. Rindler, *Spinors and Spacetime,* Vols. I-II, Cambridge University Press, Cambridge, England, 1984, 1986.

[R] Robb, A. A., *Geometry of Space and Time,* Cambridge University Press, Cambridge, England, 1936.

[Sa] Salmon, G., *A Treatise on the Analytic Geometry of Three Dimensions,* Vol. 1, Chelsea, New York.

[Sp] Spivak, M., *A Comprehensive Introduction to Differential Geometry,* Vols. I-V, Publish or Perish, Houston, TX, 1975.

[Sy] Synge, J. L., *Relativity: The Special Theory,* North Holland, Amsterdam, 1972.

[TW] Taylor, E. F. and J. A. Wheeler, *Spacetime Physics,* W. H. Freeman, San Francisco, 1963.

[V] Veblen, O., "Spinors", Science, 80(1934), 415-419.

[We] Weyl, H., *Space-Time-Matter,* Dover, New York, 1952.

[Wi] Willard, S., *General Topology,* Addison-Wesley, Reading, MA, 1970.

[Z_1] Zeeman, E. C., "Causality implies the Lorentz group", J. Math. Phys., 5(1964), 490-493.

[Z_2] Zeeman, E. C., "The topology of Minkowski space", Topology, 6(1967), 161-170.

Symbols

$\mathcal{M}$	Minkowski spacetime, 1, 9
$\mathcal{O}, \hat{\mathcal{O}}, \ldots$	observers, 2
$\Sigma, \hat{\Sigma}, \ldots$	spatial coordinate systems, 2
c	speed of light, 3
$\mathcal{S}, \hat{\mathcal{S}}, \ldots$	frames of reference, 3
$x^a, \hat{x}^a, \ldots$	spacetime coordinates, 3, 10
Λ^T	transpose of Λ, 5,14
η	5,10
$g(v, w) = v \cdot w$	value of the inner product g on (v, w), 7
$\mathcal{W}^{\perp}$	orthogonal complement of $\mathcal{W}$, 7
$\mathcal{Q}$	quadratic form determined by g, 7
$v^2 = \mathcal{Q}(v) = v \cdot v$	7
$\{e_a\}, \{\hat{e}_a\}, \ldots$	orthonormal bases, 8, 9, 12
$\delta_{ab} = \delta^{ab} = \delta^a{}_b = \delta_a{}^b$	4×4 Kronecker delta
$\eta_{ab} = \eta^{ab}$	entries of η, 10
$C_N(x_0)$	null cone at x_0, 11
$R_{x_0,x}$	null worldline through x_0 and x, 11
$\Lambda = [\Lambda^a{}_b]$	matrix of an orthogonal transformation, 13

$[\Lambda_a{}^b]$	inverse of $[\Lambda^a{}_b]$, 15
$\mathcal{L}_{GH}$	general homogeneous Lorentz group, 15
$\mathcal{C}_T(x_0)$	time cone at x_0, 17
$\mathcal{C}_T^{\pm}(x_0)$	future and past time cones at x_0, 17
$\mathcal{C}_N^{\pm}(x_0)$	future and past null cones at x_0, 17
$\mathcal{L}$	Lorentz group, 21
$\mathcal{R}$	rotation subgroup of $\mathcal{L}$, 21
$\vec{u}, \vec{\hat{u}}, \ldots$	velocity 3-vectors, 22, 23
β	relative speed of $\mathcal{S}$ and $\hat{\mathcal{S}}$, 23, 28
γ	$(1-\beta^2)^{-\frac{1}{2}}$, 23, 28
$\vec{d}, \vec{\hat{d}}, \ldots$	direction 3-vectors, 23
$\Lambda(\beta)$	boost, 28
θ	velocity parameter, 29
$L(\theta)$	hyperbolic form of $\Lambda(\beta)$, 30
$\tau(v)$	duration of v, 47
$\Delta\tau = \tau(x - x_0)$	48
$\alpha'(t)$	velocity vector of the curve α, 51
$L(\alpha)$	proper time length of α, 52
$\tau = \tau(t)$	proper time parameter, 55
$U = \alpha'(\tau)$	world velocity of α, 56
$A = \alpha''(\tau)$	world acceleration of α, 56
$\gamma(\vec{u}, 1) = U$	57

$S(x - x_0)$	proper spatial separation, 61
$\ll$	chronological precedence, 64
$<$	causal precedence, 64
A^{CT}	conjugate transpose of A, 75
$\mathbb{C}^{2\times 2}$	set of complex 2×2 matrices, 75
$\mathcal{H}_2$	Hermitian elements of $\mathbb{C}^{2\times 2}$, 75
σ_a	Pauli spin matrices, 75, 178
$SL(2, \mathbb{C})$	special linear group, 75
Λ_A	image of A under spinor map, 77
$A(\theta)$	maps onto $L(\theta)$ under spinor map, 78
SU_2	special unitary group, 78
R_x^-	past null direction through x, 80
S^-	celestial sphere, 80
(α, m)	material particle, 87
m	proper mass of (α, m), 88
$P = mU$	world momentum of (α, m), 88
$\vec{p}$	relative 3-momentum, 88
$(\vec{p}, m\gamma) = P$	88
E	total relativistic energy, 89
(α, N)	photon, 90
$\vec{e}, \vec{\hat{e}}, \ldots$	direction 3-vectors of (α, N), 89
N	world momentum of (α, N), 90

$\epsilon, \hat{\epsilon}, \ldots$	energies of (α, N), 90
$\nu, \hat{\nu}, \ldots$	frequencies of (α, N), 90
$\lambda, \hat{\lambda}, \ldots$	wavelengths of (α, N), 91
h	Planck's constant, 91
$(\mathcal{A}, x, \tilde{\mathcal{A}})$	contact interaction, 93
m_e	mass of the electron, 95
$\approx$	is approximately equal to
(α, m, e)	charged particle, 100
e	charge of (α, m, e), 100
$\vec{E}$	electric field 3-vector, 102
$\vec{B}$	magnetic field 3-vector, 102
$\operatorname{rng} T$	range of T, 104
$\ker T$	kernel of T, 104
$\operatorname{tr} T$	trace of T, 118
$N_\varepsilon^E(x_0)$	open Euclidean ε-ball about x_0, 126, 212
$f_{,\alpha} = \frac{\partial f}{\partial x^\alpha}$	126
$p \xrightarrow{F} F(p)$	assignment of a linear transformation to p, 126
$\operatorname{div} F$	divergence of $p \xrightarrow{F} F(p)$, 127
$\tilde{F}$	bilinear form associated with F, 128
$d\tilde{F}$	exterior derivative of $\tilde{F}$, 130
ϵ_{abcd}	Levi-Civita symbol, 130

${}^*\tilde{F}$	dual of $\tilde{F}$, 131
$L_{ab} = L(e_a, e_b)$	components of the bilinear form L, 144
$GL(n, \mathbb{R})$	real general linear group of order n, 146
$GL(n, \mathbb{C})$	complex general linear group of order n, 146
D	a group representation, 146
$D_\Lambda = D(\Lambda)$	image of Λ under D, 146
$\mathcal{M}^*$	dual of the vector space $\mathcal{M}$, 147
$\{e^a\}$	basis for $\mathcal{M}^*$ dual to $\{e_a\}$, 147
v^*	element $u \to v \cdot u$ of $\mathcal{M}^*$ for $v \in \mathcal{M}$, 147
$v_a = \eta_{a\alpha} v^\alpha$	components of v^* in $\{e^a\}$, 147
$\otimes$	tensor (or outer) product, 148, 149, 175
$\mathcal{T}^r_s$	vector space of world tensors on $\mathcal{M}$, 149
$L^{a_1 \cdots a_r}{}_{b_1 \cdots b_s}$	components of $L \in \mathcal{T}^r_s$, 149
$\mathcal{M}^{**} = (\mathcal{M}^*)^*$	second dual of $\mathcal{M}$, 150
x^{**}	element $f \to f(x)$ of $\mathcal{M}^{**}$ for $x \in \mathcal{M}$, 150
Spin	the spinor map, 151
P_{mn}	space of polynomials in z and $\bar{z}$, 152, 153, 158
$D^{(\frac{m}{2}, \frac{n}{2})}$	spinor representation of type (m, n), 153
$A, B, C, \ldots$	spinor indices taking the values 1, 0, 155-160
$\dot{X}, \dot{Y}, \dot{Z}, \ldots$	conjugated spinor indices taking the values $\dot{1}$, $\dot{0}$, 156-160

$G = [G_A{}^B]$	element of $SL(2, \mathbb{C})$, 156
$\bar{G} = [\bar{G}_{\dot{X}}{}^{\dot{Y}}]$	conjugate of G, 156
ß	spin space, 161
$<, >$	skew-symmetric "inner product" on ß, 161
$\{s^A\}, \{\hat{s}^A\}, \ldots$	spin frames, 161-162
$\phi_A, \hat{\phi}_A, \ldots$	components of $\phi \in$ ß, 162
ß*	dual of ß, 164
$\{s_A\}, \{\hat{s}_A\}, \ldots$	dual spin frames, 164
δ^A_B	2×2 Kronecker delta, 164
ϕ^*	element $\psi \rightarrow < \phi, \psi >$ of ß* for $\phi \in$ ß, 164
$\phi^A, \hat{\phi}^A, \ldots$	components of $\phi^* \in$ ß*, 164
$[\mathcal{G}^A{}_B]$	transposed inverse of $[G_A{}^B]$, 164
ϕ^{**}	element $f \rightarrow f(\phi)$ of ß** for $\phi \in$ ß, 165
$\bar{\text{ß}} = \text{ß} \times \{1\}$	"conjugate" of ß, 166
$\bar{\phi}$	$(\phi, 1) \in \bar{\text{ß}}$ for $\phi \in$ ß, 166
$\{\bar{s}^{\dot{X}}\}, \{\bar{\hat{s}}^{\dot{X}}\}, \ldots$	conjugate spin frames, 167
$\bar{\phi}_{\dot{X}}, \bar{\hat{\phi}}_{\dot{X}}, \ldots$	components of $\bar{\phi}$, 167
$\bar{\mathcal{G}} = [\bar{\mathcal{G}}^{\dot{X}}{}_{\dot{Y}}]$	conjugate of $\mathcal{G} = [\mathcal{G}^A{}_B]$, 167
$\bar{\text{ß}}^*$	dual of $\bar{\text{ß}}$, 167
$\{\bar{s}_{\dot{X}}\}, \{\bar{\hat{s}}_{\dot{X}}\}, \ldots$	dual conjugate spin frames, 167

d_E	E-distance, 210
$N_\varepsilon^E(x_0)$	open Euclidean ε-ball about x_0, 211
$\mathcal{M}^E$	$\mathcal{M}$ with the Euclidean topology, 211
$\mathrm{Cl}_E\, A$, $\mathrm{bdy}_E\, A, \ldots$	Euclidean closure, boundary,... of $A \subseteq \mathcal{M}$, 211
$\mathcal{C}(x)$	$\mathcal{C}_T^-(x) \cup \mathcal{C}_T^+(x) \cup \{x\}$, 216
$N_\varepsilon^P(x_0)$	P-open ε-ball about x_0, 216
$\mathcal{M}^P$	$\mathcal{M}$ with the path topology, 216
$\mathrm{Cl}_P\, A$, $\mathrm{bdy}_P\, A, \ldots$	path closure, boundary,... of $A \subseteq \mathcal{M}$, 218
$H(X)$	homeomorphism group of X, 224, 225
$[\alpha]$	homotopy class of the path α, 230
α^{-1}	inverse of the path α, 230
$\beta\alpha$	product of the paths α and β, 230
$\pi_1(X, x_0)$	fundamental group of X at x_0, 231
$\pi_1(X)$	fundamental group of X, 231
$(\tilde{X}, p)$	universal covering manifold of X, 233
$\mathbb{R}P^2$	real projective plane, 233
$\mathbb{Z}_2$	group of integers mod 2, 234
$\mathbb{R}P^3$	real projective 3-space, 234
$\mathbb{R}P^n$	real projective n-space, 235
$SO(3)$	rotation group, 236

Index

Applied Mathematical Sciences

(continued from page ii.)

52. *Chipot:* Variational Inequalities and Flow in Porous Media.
53. *Majda:* Compressible Fluid Flow and System of Conservation Laws in Several Space Variables.
54. *Wasow:* Linear Turning Point Theory.
55. *Yosida:* Operational Calculus: A Theory of Hyperfunctions.
56. *Chang/Howes:* Nonlinear Singular Perturbation Phenomena: Theory and Applications.
57. *Reinhardt:* Analysis of Approximation Methods for Differential and Integral Equations.
58. *Dwoyer/Hussaini/Voigt (eds):* Theoretical Approaches to Turbulence.
59. *Sanders/Verhulst:* Averaging Methods in Nonlinear Dynamical Systems.
60. *Ghil/Childress:* Topics in Geophysical Dynamics: Atmospheric Dynamics, Dynamo Theory and Climate Dynamics.
61. *Sattinger/Weaver:* Lie Groups and Algebras with Applications to Physics, Geometry, and Mechanics.
62. *LaSalle:* The Stability and Control of Discrete Processes.
63. *Grasman:* Asymptotic Methods of Relaxation Oscillations and Applications.
64. *Hsu:* Cell-to-Cell Mapping: A Method of Global Analysis for Nonlinear Systems.
65. *Rand/Armbruster:* Perturbation Methods, Bifurcation Theory and Computer Algebra.
66. *Hlavácek/Haslinger/Necasl/Lovísek:* Solution of Variational Inequalities in Mechanics.
67. *Cercignani:* The Boltzmann Equation and Its Applications.
68. *Temam:* Infinite Dimensional Dynamical Systems in Mechanics and Physics.
69. *Golubitsky/Stewart/Schaeffer:* Singularities and Groups in Bifurcation Theory, Vol. II.
70. *Constantin/Foias/Nicolaenko/Temam:* Integral Manifolds and Inertial Manifolds for Dissipative Partial Differential Equations.
71. *Catlin:* Estimation, Control, and the Discrete Kalman Filter.
72. *Lochak/Meunier:* Multiphase Averaging for Classical Systems.
73. *Wiggins:* Global Bifurcations and Chaos.
74. *Mawhin/Willem:* Critical Point Theory and Hamiltonian Systems.
75. *Abraham/Marsden/Ratiu:* Manifolds, Tensor Analysis, and Applications, 2nd ed.
76. *Lagerstrom:* Matched Asymptotic Expansions: Ideas and Techniques.
77. *Aldous:* Probability Approximations via the Poisson Clumping Heuristic.
78. *Dacorogna:* Direct Methods in the Calculus of Variations.
79. *Hernández-Lerma:* Adaptive Markov Processes.
80. *Lawden:* Elliptic Functions and Applications.
81. *Bluman/Kumei:* Symmetries and Differential Equations.
82. *Kress:* Linear Integral Equations.
83. *Bebernes/Eberly:* Mathematical Problems from Combustion Theory.
84. *Joseph:* Fluid Dynamics of Viscoelastic Fluids.
85. *Yang:* Wave Packets and Their Bifurcations in Geophysical Fluid Dynamics.
86. *Dendrinos/Sonis:* Chaos and Socio-Spatial Dynamics.
87. *Weder:* Spectral and Scattering Theory for Wave Propagation in Perturbed Stratified Media.
88. *Bogaevski/Povzner:* Algebraic Methods in Nonlinear Perturbation Theory.
89. *O'Malley:* Singular Perturbation Methods for Ordinary Differential Equations.
90. *Meyer/Hall:* Introduction to Hamiltonian Dynamical Systems and the N-body Problem.
91. *Straughan:* The Energy Method, Stability, and Nonlinear Convection.
92. *Naber:* The Geometry of Minkowski Spacetime.